RESOURCE ASSESSMENT IN FORESTED LANDSCAPES

RESOURCE ASSESSMENT IN FORESTED LANDSCAPES

David D. Reed
School of Forestry and Wood Products
Michigan Technological University
Houghton, Michigan

Glenn D. Mroz
School of Forestry and Wood Products
Michigan Technological University
Houghton, Michigan

John Wiley & Sons, Inc.
New York • Chichester • Weinheim • Brisbane • Singapore • Toronto

This text is printed on acid-free paper.

Library of Congress Cataloging in Publication Data:

Reed, David D.

Resource assessment in forested landscapes / David D. Reed and Glenn D. Mroz.

p. cm.

Includes bibliographical references and index.

ISBN 0-471-15582-9 (alk. paper)

1. Forest and forestry—Mensuration. 2. Forest and forestry—Valuation. 3. Forest landscape management. 4. Forest and forestry—Great Lakes Region—Mensuration. 5. Forest and forestry—Great Lakes Region—Valuation. 6. Forest landscape management—Great Lakes Region. I. Mroz, Glenn D. II. Title.

SD551.R44 1997

634.9′28–dc21 97-1735

Printed in the United States of America

10 9 8 7 6 5 4 3 2 1

To our parents

Jim and Doss Reed;
and
Jim and Lorraine Mroz;

and families

Beth, Caroline, Kristen, Michelle;
and
Gail, Heather, Jennifer

Contents

Preface xi

PART I: Context 1

CHAPTER 1: Introduction 3

Basic Definitions 4
Spatial and Temporal Scale 5
Summary 10

CHAPTER 2: Landscape Components 11

Abiotic Environment 15
Biotic Resources 17
Social Context 19
Synthesis and Integration 20
Summary 22

PART II: Abiotic Environment 23

CHAPTER 3: Geology 25

Landforms and Deposits 26
Information Sources 30

Interpretations 31
Legal Considerations 35
Summary 36
Applications 36

CHAPTER 4: Soil 38

Soil Physical Properties 39
Soil Water 43
Soil Chemical Properties 48
Soil Surveys 49
Interpretation of Soil Survey Information 51
Coarse Woody Debris 54
Summary 60
Applications 61

CHAPTER 5: Climate 62

Temperature 63
Moisture 67
Light 73
Wind 75
Climate ClassiÆcation 78
Climate Change 81
Summary 82
Applications 83

CHAPTER 6: Hydrology 84

Watershed Characteristics 85
Stream Descriptors 98
Wetland Descriptors 101
Water Quality Measures 106
Best Management Practices 109
Summary 111
Applications 111

PART III: Biotic Resources 113

CHAPTER 7: Vegetative Community 115

Historic Information 116
Qualitative Descriptions of Vegetation 117
Quantitative Measures of Community Composition and Structure 123
Ecosystem Similarity 136
Landscape Structure 137

Summary 141
Applications 141

CHAPTER 8: Forest Overstory 143

Measuring Harvested Products 144
Individual Tree Measurements 159
Inventory Planning 172
Forest Overstory Inventory Methods 174
Summarizing Forest Overstory Information 192
Summary 194
Applications 194

CHAPTER 9: Wildlife Resources 196

Habitat Assessment 197
Estimation of Population Size 206
Threatened and Endangered Species 219
Community Structure 220
Summary 221
Applications 222

PART IV: Social Context 225

CHAPTER 10: Legal and Regulatory Environment 227

Federal Laws and Regulations 228
State Laws and Regulations 236
Local Laws and Regulations 240
Summary 242
Applications 242

CHAPTER 11: Recreation and Aesthetics 244

ClassiÆcation of Outdoor Recreational Activities 245
Concentrated Recreation 249
Dispersed Recreation 253
Aesthetics and the Visual Impact of Management Activity 255
Summary 259
Applications 259

CHAPTER 12: Prehistoric and Historic Land Use 261

Federal and State Historic Preservation Programs 262
Prehistoric Land Use 264
Historic Land Use 266
Summary 269
Applications 270

PART V: Integration and Synthesis 271

CHAPTER 13: Geographic Information Systems 273

Terminology 274
Structure 276
Errors and Accuracy 281
Analysis 291
Projections 295
Summary 297
Applications 297

CHAPTER 14: Integrated Environmental Assessment 299

Description of Resource Status and Condition 301
Description of Management Alternatives 301
Analysis of Management Alternatives 304
Selection of a Preferred Alternative 309
Summary 310
Applications 311

PART VI: Supplemental Material 313

CHAPTER 15: Basic Statistical Sampling Methods 315

Sampling Versus Complete Enumeration 316
Simple Random Sampling 317
Systematic Sampling 327
StratiÆed Random Sampling 329
Ratio and Regression Estimation 339
Cluster Sampling 347
Summary 350
Applications 351

APPENDIX A: Units of Measure 355

APPENDIX B: Common and Scientific Species Names 357

References 361

Index 379

Preface

This book is the outgrowth of our experiences in formal classroom instruction of undergraduate and graduate students and in continuing education experiences with practicing professionals. Over the last ten years, it has become apparent to us that (1) the scope of information that land managers must consider in the decision-making process has broadened considerably, with more detail needed about a greater variety of resources, (2) the need to consider management impacts at larger spatial and temporal scales than single stands or small management units is increasing, and (3) the disciplinary background of practicing land managers has changed to include a much greater proportion of professionals with formal training in botany, archaeology, social science, hydrology, law, and many other fields of study. These factors led us to develop this text, which we believe is appropriate for advanced undergraduate or graduate courses designed to integrate knowledge gained in specialized courses in different disciplines. In fact, we believe the best use of this text will be in classes where students are participating in field exercises, developing an integrated resource assessment for a specific area, or are working through case studies of actual examples of environmental assessment exercises. We also hope this book will provide a reference tool for practicing land managers and others responsible for designing and conducting resource assessment efforts.

This text is not designed to provide encyclopedic coverage of field procedures or detailed development of the theory supporting assessment methods for different landscape components. We do not believe that a single volume can provide comprehensive coverage of such a wide body of knowledge. Nor is this text designed to substitute for detailed texts covering forest measurements, soils, hydrology, and other disciplinary fields in specialized undergraduate courses. Instead, we designed this text to provide a review of the basic procedures for assessing a wide range of natural resources with the goal of providing information to support management decision making. We think that this is appropriate for students attempting to integrate knowledge from past courses in single disciplinary fields, as well as for practicing professionals needing to quickly find basic information about assessment procedures that may be outside their formal disciplinary training. For these reasons, readers with backgrounds in a field such as forest measurements may find the information on forest overstory inventory to be basic and lacking in comprehensive coverage of advanced topics. The same readers, however, may find the chapters on hydrology and soils to contain much information that is difficult for them to locate in a single source.

We elected to use examples from our experience in the Upper Great Lakes region in most chapters. While examples of almost every application could be found from any region in North America, much of the information useful in resource assessments is regionally specific, and it seemed appropriate to provide a set of examples that have at least some semblance of an internal thread of consistency. Instructors, students, and professionals in other regions should be able to locate similar sources of information in almost every case. In the same vein, the choice of measurement units is also somewhat arbitrary; SI units are, of course, the international and scientific standard. Common use in the United States is still based on English units, some of which, like the board foot, have no SI equivalent. We elected to concentrate on SI units except when there was no equivalent to the English unit, but to give English equivalents for many units in the text and to include applications using English units at the end of most chapters.

The social context for land management is rapidly changing, and information presented in this text, particularly information concerning the legal and regulatory environment in which land managers operate, may become dated quite quickly. Readers are warned to familiarize themselves with legal and regulatory requirements applicable to their situations and not to rely on this text as a primary source for such information.

Resource assessment has become a routine exercise for many public land management agencies. It is also a growing component of private land management. There is clearly a mandate from the public for land managers to consider the "big picture" when designing and evaluating management alternatives. We hope that this text will help students and professionals appreciate that the complexities of the landscape can be considered in an orderly fashion in crafting and evaluating management alternatives.

We would like to thank our families for their patience and understanding during the development of this manuscript. Steve Boyce provided encouragement and advice, without which this text probably would not have been written. Beth Jones Reed provided a tremendous amount of editorial assistance and many valuable comments and suggestions. Mike Hyslop developed the information illustrating the use of geographic information systems. Our friends and colleagues in the School of Forestry and Wood Products at Michigan Technological University and elsewhere tolerated numerous questions and participated in many discussions concerning the appropriate content of this book. Finally, the undergraduate and graduate students at Michigan Tech and the participants in continuing education courses with whom we have worked forced us to define the scope and sharpen our presentation of much of the material contained here.

David D. Reed
Glenn D. Mroz

PART I

Context

Chapter

1

Introduction

Earth's population is nearing six billion people and increasing rapidly (Population Reference Bureau, 1991). By the year 2040, the world's population is projected to be between 8 and 11 billion people. Utilization of resources to support this population is placing growing pressure on ecological systems. While conservation and recycling efforts will meet some of the demand for goods, the global challenge facing land managers is to develop the means to produce commodities needed by humans while maintaining global ecosystem integrity. At the same time, increasing affluence in the developed countries is building demand for tourism and leisure activity.

Not only is resource demand increasing because of population increases, but per capita consumption has also increased in recent years. The global demand for timber is increasing by 86 million cubic meters per year and is expected to reach 5.1 billion cubic meters per year by 2010 (FAO, 1995). While this is a staggering figure, the volume of wood utilized for fuelwood exceeds the total volume of wood harvested for all other reasons (FAO, 1992). In developing countries, 60 to 95% of total energy needs are met by fuelwood (Leach and Mearns, 1988). At least partially due to this increase in demand, global forest area declined by 4% from 1980 to 1990 (FAO, 1992).

Demand for forest-related commodities will continue to grow as long as population increases and per capita consumption does not decline. Not only are human populations dependent on the commodities available from forest ecosystems, but they also depend on the many other processes and functions of those ecosystems. Physical and biological processes produce noncommodity values critical to the maintenance of human populations; examples include the filtering of fresh water and the conservation of biodiversity. However, our understanding of these processes and the implementation of management activities that protect them have often lagged behind the knowledge and effort dedicated to extracting resources. Over the past two decades, much effort has been committed to narrowing this gap through the development of new knowledge required for sustainable forest management where resource extraction continues while ecological function is maintained.

To paraphrase W. E. Deming, one cannot truly understand a system without understanding the variation within the system. Just as manufacturing has been revolutionized in the last 20 years by a shift in focus from outputs to processes, so natural resource management is undergoing a similar shift. Whether under the name of *ecosystem management* or some other term, sustainable management of natural resources requires consideration of ecological processes at many levels of implementation. Commodity production must take place within a context requiring the maintenance of ecological function, the continued integrity of noncommodity biological populations, and the satisfaction of economic constraints on management. This implies that management must not focus on single operational units, but on the landscape-level production and the maintenance of a mix of values (Boyce, 1995).

Heightened public awareness of environmental disturbance associated with commodity production has led to calls for the preservation of an increasing area of forest land. But this has created other problems, because developed countries have the ability to acquire resources from global sources. In North America, in particular, it might be possible to preserve natural ecosystems by shifting commodity production to other countries. But this also shifts the environmental and ecological consequences of the high rate of commodity consumption to other regions of the globe. The net global effect may be to increase the overexploitation of already stressed ecosystems in other parts of the world (Sedjo et al., 1994).

There is definitely a need to preserve rare and ecologically valuable habitats. For most locations, particularly those with a long history of disturbance, the responsible course of action is to sustainably manage ecosystems, which entails some level of commodity production, while maintaining ecosystem function and integrity. A requirement for such management is a broad base of knowledge of both the physical and biological characteristics of specific areas and their role within the surrounding landscape. In short, a resource assessment is a necessary and key part of the planning process.

BASIC DEFINITIONS

Many definitions have been offered for terms and concepts related to biological systems and their management. To some extent, these definitions evolve over time as scientific understanding and the practical art of land management advance. The definitions of basic terms offered here provide a common basis of understanding for the discussion of resource assessment in forested landscapes.

An organism's *habitat* is the sum total of biotic and abiotic factors that affect the organism and upon which it depends for its survival. For trees, the term *site* is equivalent to *habitat* (Spurr and Barnes, 1980). A *population* is a group of potentially interbreeding members of a species. A *community* is a collection of populations of different species that exist in physical proximity and directly interact through the flow of energy or material, or indirectly interact through modification of the abiotic environment.

An *ecosystem* is a community coupled with the abiotic factors affecting the individual organisms (Odum, 1971). An ecosystem is not a closed system, in that material and energy are transferred to and from neighboring ecosystems. A collection of ecosystems that exist in physical proximity and that mutually transfer energy and material is a *landscape*. *Ecological processes* are the functions of abiotic or biotic ecosystem components which govern the flow of energy or material within the system and between neighboring systems.

Resource assessment is the process of obtaining knowledge of the abiotic, biotic, and social components of a landscape. It is far more than a simple descriptive process that is soon forgotten. The gathered information is used throughout the management decision-making process and the many years of resource management that follow. Abiotic characteristics place limitations on management options. For example, slope steepness and soil erosion potential may control the type of vegetation management that may be implemented at a given location. Existing biological systems also place constraints on management options, but there are many situations in which the existing biological system may be altered, either deliberately or incidentally, to improve habitat for favored species or to increase economic returns. The patchwork of farmland and forests throughout the Lake States, for example, has created ideal habitat for white-tailed deer. Another important consideration is that all resource management takes place within the context of human land use and the legal and regulatory framework developed by society (Figure 1.1). A complete assessment goes beyond describing current conditions; it includes analyses of the limitations placed on management options by abiotic, biotic, and social factors. Burkhart, Barrett, and Lund (1984) give general elements of a typical timber inventory plan; these are modified in Table 1.1 and placed within the context of the resource assessment process. *Resource planning* is the process of utilizing information developed during a resource assessment. The purpose is to develop and evaluate management alternatives to achieve the desired goals and objectives. *Resource management* is the process of developing and executing resource plans.

SPATIAL AND TEMPORAL SCALE

The spatial and temporal extent of an ecological process is referred to as its *scale*. In any ecosystem there are many processes operating simultaneously

TABLE 1.1 General Elements of a Resource Assessment Plan

	Element	Content
I.	Purpose or Goal	Problems and issues that will be addressed Landscape characteristics to be addressed Required accuracy and precision of all variables
II.	Background	General description of the operational area General description of the larger landscape Types of information already available Information that must be collected
III.	Information Required	Expected format of the assessment report All tables (complete with headings) to be compiled Titles of all maps to be produced Outline of the final document
IV.	Inventory Design	Sampling design for all variables Types of sample units that will be used Estimated numbers of sample units required
V.	Measurement Procedures	Documentation of field measurement protocols and standards Methods for assuring data quality, including data quality objectives Methods for establishing sample units Procedures for implementing the field measurement standards Methods of recording information in the field Listing and description of the equipment needed
VI.	Constraints	Resources available Time and effort estimates for each phase Due dates or dates for the completion of intermediate steps Labor available
VII.	Compilation Procedures	Methods of data analysis Description of the formulae used Methods of editing raw data Methods of linking existing data with new data Procedures for entering new data collected into digital format Description of computational programs
VIII.	Completion	Personnel responsible for reporting results Personal responsible for archiving information Linking of new data with existing information Methods of permanent data storage and archiving

FIGURE 1.1 Landscapes consist of abiotic, biotic, and social components whose interactions are mediated by ecological processes governing the flow of energy and materials within the system and with neighboring systems.

at many different scales (Table 1.2). Processes operating with rapid time steps and over short distances are *small-scale processes*. Examples include processes such as those that directly affect soil microorganism populations in decaying leaf litter. At the same time, *large-scale processes* are operating with much longer time steps and over much greater areas. Examples of large-scale processes include those affecting the community of overstory species, resulting in community succession to a different mix of species over time.

At first glance, the conventional biological hierarchy appears to be defined by spatial and temporal scales (Figure 1.2). It is true that, in general, the spatial and temporal scales of the relevant processes increase as the level of organization increases. But the relevant scale to address an issue is not simply a function of hierarchical level. Successional processes in grasslands operate at different time scales than those in forests; the spatial scale relevant to small mammal habitat is considerably different from that for large carnivores or ungulates. The use of a single spatial scale sometimes makes it easier to observe some processes, while obscuring others. In the end, scaling is largely a matter of human perception and values; the use of multiple scales is necessary to fully address resource management issues (Allen and Hoekstra, 1992).

The levels of a hierarchy are defined by the interactions among the components (Webster, 1979). At each new level of organization, new feedback mechanisms governing the interactions of the components are encountered. The behavior at one level of organization cannot be completely deduced or inferred from knowledge of processes operating at "higher" or "lower" levels of organization. As Webster (1979) stated:

TABLE 1.2 General Spatial and Temporal Time Scales of Some Abiotic and Biotic Processes Affecting Forest Ecosystems

Process	*Spatial Scale (m)*	*Temporal Scale (seconds)*
Leaf Physiology	10^{-5}–10^{-1}	10^{0}–10^{5}
Whole-Plant Physiology	10^{-2}–10^{1}	10^{3}–10^{6}
Plant Growth	10^{-2}–10^{1}	10^{6}–10^{8}
Regeneration	10^{-2}–10^{2}	10^{5}–10^{10}
Gap Dynamics	10^{0}–10^{2}	10^{9}–10^{10}
Natural Soil Development	10^{1}–10^{3}	10^{10}–10^{11}
Stand Dynamics	10^{2}–10^{3}	10^{9}–10^{10}
Thunderstorm	10^{2}–10^{3}	10^{3}–10^{4}
Land Use Changes	10^{2}–10^{4}	10^{6}–10^{8}
Cold Front	10^{5}–10^{7}	10^{4}–10^{5}
Natural Climate Fluctuations	10^{5}–10^{8}	10^{7}–10^{8}
Global Climate Change	10^{6}–10^{8}	10^{8}–10^{11}

Adapted from Turner, Dale, and Gardner (1989) and King, Emanuel, and O'Neill (1990).

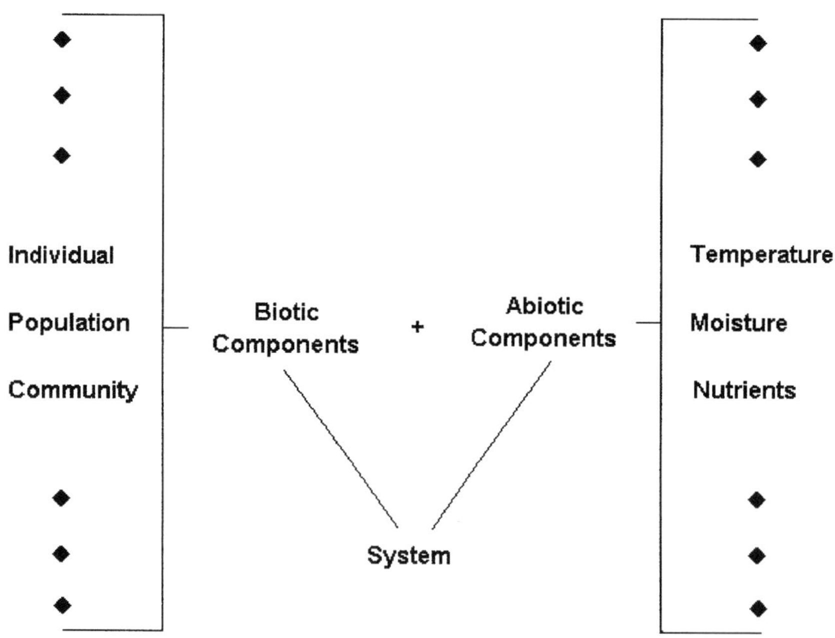

FIGURE 1.2 Biological units at different levels of the general biological hierachy interact with the abiotic environment to form complex systems.

> At each level, there are behaviors and laws relative to those behaviors which are specific to that level and not defined for other levels. These behaviors cannot be predicted from the laws of lower levels, but they are entirely consistent with lower level laws.

This implies, for example, that it is impossible to infer ecosystem response to a stress agent from controlled studies on individual organisms; trees in a forest are affected by different processes than those in growth chambers in a laboratory. For some purposes, though, there is little conceptual difference between a deciduous forest ecosystem and the ecosystem that exists within a moose's digestive tract. Clearly, then, the biological and social issues and the problems being addressed are the deciding factors concerning the appropriate scales of observation for an assessment.

Science is just beginning to develop insights into the processes operating at the landscape level of interacting ecosystems (Wiens, 1992). On the other hand, descriptive efforts have been under way for many years, and these have resulted in a great number of descriptions of the patterns exhibited among ecosystems at the landscape level. Still, there is a very poor understanding of the processes governing the formation of these patterns. Just as a functional problem of a single organ may be of great consequence to an organism, the situation in a single ecosystem within a landscape may affect or be affected by conditions in other ecosystems that may be great distances away. Erosion and sedimentation resulting from a poorly designed stream crossing in the headwaters of a watershed, for example, may affect fish populations, aquatic habitat, and hydrological relationships throughout the entire watershed. Similarly, silvicultural methods designed to promote the regeneration and recruitment of shade-tolerant deciduous tree species may be futile if the surrounding landscape supports large populations of herbivores. As evidence of the complexity of landscape-level interactions, McLaren and Peterson (1994) demonstrated that wolf populations and balsam fir regeneration and recruitment are linked on Isle Royale, Michigan; the linkage is governed by the herbivory of an intermediate trophic level (moose), and, at times, when the moose population is high, there is very little recruitment of balsam fir into the overstory because of the pressures of heavy winter feeding. This example reinforces the concept that maintenance of landscape-level ecological function is necessary for the continued production of forest-derived commodities.

Even low levels of human population are capable of significant landscape-level changes, as is evidenced by the apparent role of humans in the extinction of large Pleistocene mammals (Martin, 1984). Disturbances caused by increasing human utilization of forest commodities or by changes in land use may singly affect relatively small spatial areas, but they can also very quickly accumulate to result in severe disruption of landscape-level processes. The understanding of landscape processes and the impacts of their disruption on the sustainability of particular ecosystems is in its infancy. Sustainable resource utilization requires balancing the production of commodities necessary to support large human populations with the maintenance of ecological function at all levels of biological organization. This is becoming increasingly necessary as pressures on forested landscapes increase.

SUMMARY

Increasing pressures on natural resources resulting from increasing global population and consumption are forcing resource managers to move beyond just the consideration of commodity production and to develop methods of utilization that sustain the integrity and function of biological systems. This increased management complexity occurs within a context of physical, chemical, biological, and social constraints. Concentration on internal functioning, along with outputs, forces consideration of the continued viability of processes operating at many spatial and temporal scales. Resource assessment, which entails the description and analysis of the abiotic, biotic, and social factors operating in a landscape, provides the information necessary to develop management plans and to support management activities.

Chapter

2

Landscape Components

Landscapes consist of interacting ecosystems, and ecosystems consist of communities in combination with their abiotic environments. The boundaries of ecosystems are, however, often difficult to delineate. The relevant spatial extent, in fact, may vary, depending on the ecological process being considered. In many cases, boundaries are better thought of in terms of the strengths of interactions that cross them rather than as physical limits. Watershed boundaries are defined by physical features that limit the exchange of water; by definition, there is a negligible exchange of water across watershed boundaries until the areas converge in a higher-order watershed. Unlike the movement of water, the movement of wildlife or insect populations is not constrained by physical watershed boundaries. The deposition of materials in rain, snow, or dust is influenced by topography, but there is still transport of material across watershed boundaries and movement from one watershed to another. The topographical boundary of two watersheds, therefore, is a region where there is minimum hydrologic exchange, but there still continues to be a flow of energy and material between neighboring watersheds.

The definition of ecosystem boundaries is complicated by human impacts. In what was once a contiguous deciduous temperate forest containing a community of interacting flora and fauna, human-induced land use change may impose a pattern of agricultural fields and other features mixed with forests (Figure 2.1). Human impacts may also include favoring one or more species over others; an extreme example in forests is the establishment of plantations with associated vegetation control. Do these activities create new ecosystems with new boundaries? Or are the results merely different manifes-

FIGURE 2.1 A landscape that was originally contiguous forest now contains forests, agricultural land, and other features such as roads and utility corridors.

tations of a single ecosystem? The answer is scale-dependent, related to the particular processes being considered.

Human activities impose biologically artificial economic ownership patterns across a region. The metes and bounds surveys of the states along the Atlantic coast often utilized natural features, such as waterways, to define property lines. In most of the United States, however, the township and range surveys established as part of the U.S. Public Land Survey were conducted with little regard for natural landscape features. Today's survey lines and associated ownership patterns impose cultural boundaries on top of the already difficult-to-delineate natural ecosystems. Much of the disturbance associated with land use or road construction is framed within the cultural means of defining land ownership.

Resource managers often deal with relatively small spatial areas defined by economic ownership patterns. But in some places in North America there are large, contiguous areas of common ownership where the spatial extent of management concern may be based on the extent of relevant ecological processes. In most of North America, especially in the central and eastern parts of the continent, resource managers are forced by ownership patterns to consider relatively small operational areas that are located without regard to any natural ecological processes. Resource management in forested landscapes must often occur within a mosaic of ownership and land use patterns (Figure 2.2). In such a landscape, ecological processes of material and energy flow extend beyond the culturally defined boundaries of an operational area. Resource managers may have no input into management decisions regarding

FIGURE 2.2 The ownership pattern in one six-mile by six-mile township illustrates the mixture of public and private ownership within a forested landscape; the black areas are public land, the shaded areas are owned by various industrial foresty concerns, and the white areas are owned by private individuals.

neighboring areas, but all landowners must manage their lands within the framework of federal, state, and local laws and regulations that govern land use and resource management. This situation makes resource assessment a critical process in management planning. Without control, or even knowledge, of planned management for neighboring locations within a landscape, there are cases in which a comprehensive resource assessment provides managers with what may be the only available knowledge on which to base plans and management activities.

This may seem like an impossible situation; management of natural ecosystems is difficult to plan and administer even in ideal settings. How can a manager make intelligent decisions in the face of such uncertainty? Forest management has traditionally focused on site-specific management options with minimal consideration of the surrounding landscape. Multiple-use management, or, as sometimes phrased, managing for the greatest good for the greatest number of people in the long run, considers many human uses of forest resources in addition to commodity production. The consideration of ecological integrity and function in a landscape setting is, in many ways, an evolutionary shift in management emphasis and not as revolutionary as sometimes claimed. After all, land managers have worked within a legal and regulatory framework for centuries (Manwood, 1598), and Evelyn (1664) noted the need to consider the longer-term impacts of management activity on continued productivity.

Resource assessments describe the abiotic and biotic characteristics of a landscape as well as the social context in which management occurs. These factors all place limits on land-use and management options. Abiotic characteristics frame the environment upon which all land use is based. Whether an area contains a biodiversity reserve, an agricultural field, a commodity-driven intensive forestry operation, or vacation condominiums, the basic abiotic characteristics of geology, soil, climate, and hydrology form the context within which plant and animal life survive and reproduce.

The biotic resources of a region consist of the flora and fauna that inhabit the area. Forests are, by definition, communities in which the dominant overstory species are long-lived trees. Commodity production from forests is derived primarily from the vegetation, mainly in the form of wood products: veneer, lumber, cordwood, chips, fuelwood, and other forms of wood fiber. In addition, specialty products such as boughs and greenery, Christmas trees, foods, medicines, and other products are commercially harvested from forests. Forests do not exist in a vacuum. Instead, they are components of the surrounding landscape. Neighboring areas may also be mostly forested, but the surrounding dominant communities might be quite different agricultural, prairie, or savannah systems. Most human classification of ecosystems and biomes is based on the characteristics of the dominant vegetation, not on the associated fauna. Wildlife populations, defined in the broadest sense and not just as game species, utilize plant communities for food and shelter while playing critical ecological roles in the transfer of energy and materials. Wildlife can also have significant impacts on the structure and composition of the plant community itself. The sustainable utilization of forest resources over the long term is dependent on maintaining the function and integrity of the full range of biotic components of a landscape.

The social context within which land ownership and management occurs includes the legal and regulatory environment, which places constraints on management options. In addition, it is also important to consider aesthetic values and human use of the landscape for recreational purposes. Current human interaction may, in effect, place further limits on management options beyond formal legal or regulatory constraints. The landscape also contains a record of past historic and prehistoric human use and provides a reservoir of cultural history and values. In the United States, land managers must comply with federal and state historic preservation laws, which may mandate consideration of these landscape values in the planning and execution of management activity in many areas.

This mixture of abiotic, biotic, and cultural characteristics and features, then, forms the context for resource management in forested landscapes. Assessing these attributes, organizing the information, and evaluating the resulting impacts and limitations on management pose major challenges for managers. Following brief descriptions of these characteristics and features, each will be examined in greater detail in succeeding chapters.

ABIOTIC ENVIRONMENT

Geology

The underlying geology of an area forms the physical features of the landscape. The topography of a landscape is shaped by forces of uplift and degradation, which have acted over hundreds of millions of years. Geologically recent glacial activity in much of North America has formed many of the physiographic features on the landscape today. Geologic formations provide the parent material for soil development, and this can greatly impact soil fertility. Topography and the underlying geology govern water flow within the landscape and may provide unique habitat for plant communities. In much of central North America, for instance, eastern red cedar communities dominate areas where limestone occurs near the surface; jack pine is an edaphic climax species on dry, sandy glacial outwash plains in the Great Lakes region. Slope and aspect affect the temperature and amount of sunlight available to the plant community at a given location, play a large role in the transport and availability of water, and at the same time may place limitations on management activities, affecting the types of vehicles and equipment that can be used in an area. The underlying bedrock and surface geology define the physical shape and appearance of a landscape and form the basis for understanding many other landscape components.

Soils

The soil is the basis for plant productivity, providing the mechanical support, nutrients, and water that plants depend on for their needs. Mechanical support is important for tree stability, and the soil's mechanical properties are also a factor in management concerns such as road construction and equipment operation. Soil nutrient pools are dependent on the parent material from which the soil is derived and on the nutrients introduced by atmospheric deposition. A soil's water holding capacity is a function of its texture and porosity and, combined with precipitation, governs the amount of water available to support a plant community on a site. The soil is the conduit for lateral water flow through a landscape; discontinuities between soil horizons or between soils and bedrock often form the regions of lateral water flow. Soil horizons that restrict the vertical flow of water may cause seasonally high water table levels, which can then restrict the use of an area for recreation or harvesting for large parts of the year and place limits on the biotic community. The depth of the aerated zone of the soil above the water table impacts effective rooting depth and may greatly influence the species composition of a forest community. The soil, therefore, forms the basis for plant growth and development and plays a significant role in defining the hydrologic properties of a landscape.

Climate

Temperature and the amount of available water are two critical abiotic factors that affect physiological activity. *Climate* is the long-term characterization of seasonal temperature and precipitation patterns. *Weather* is the short-term realization of climate; temporal deviations from long term climate, as in the case of severe local drought, may be governed by global processes. Climatic factors govern the geographic range limits of a number of species and are strongly coupled with the spatial extent of biomes (e.g., tundra, boreal forest, or temperate deciduous forest). Regional and national monitoring networks have operated for more than 100 years in many locations, and climate patterns are well characterized over most of North America. Regional patterns in climate are locally impacted by topography, such as when rain shadows are created by mountain ranges in western North America. Large bodies of water can also greatly influence regional climate; areas in the lee of the Great Lakes receive large amounts of snowfall, but may have longer growing seasons and greater average annual temperatures than areas to the south that are not impacted by the temperature-mediating effect of the Great Lakes. Climate and weather conditions define the ambient environment in which flora and fauna grow and reproduce.

Hydrology

Water is necessary to support life. It is the scarcity of water that causes jack pine to be the dominant plant species on glacial outwash plains in the Great Lakes region. It is the abundance and distribution of water that supports wetland plant communities. Individual species are adapted to certain limits of water availability, and the range of species with the potential to occupy a site is largely governed by the hydrologic properties of the site. The movement of water across the landscape is a major factor in the transport of energy and nutrients. Water cannot move uphill and its directional flow plays a large role in the directional flows of energy and material. Water may also serve as an agent for transporting vast quantities of soil material through the landscape. Geologic erosion is a natural process and is always occurring to some extent. Disturbance of soils and alteration in water flow may result in dramatic changes in erosion and sedimentation across a landscape. Physical site alterations, such as the establishment of drainage ditches, may also dramatically alter the local water regime, which impacts the plant communities and plant productivity in the surrounding area. The seasonal presence of large quantities of water may place limitations on the equipment suitable for use on a site; severe rutting from the inappropriate use of equipment when soil is saturated may alter the lateral flow of water through the soil. This may cause increased water retention upslope of a rut, with associated plant mortality resulting from saturated conditions, and reduced water availability downslope, possibly resulting in decreased productivity and increased moisture stress. Water is itself critical to sustaining life, but it is also a major agent of material transport within landscapes.

BIOTIC RESOURCES

Vegetative Community

The presence or absence of individual plant species at a given location is dependent on many factors: seed availability, past land use, soils, climate, and hydrology, among others. Species individually respond to these various factors, but certain groups of species may often be found together in regions with similar abiotic environments. It is these plant associations that occur repeatedly across a region that form the basis of classification systems based on vegetative association. Such systems are often based to a greater degree on herbaceous indicator species rather than overstory species. Braun-Blanquet (1928) in Europe and Daubenmire (1952) in the United States were instrumental in developing methods of studying vegetative associations. The utility of delineating vegetative associations to resource management is based on the underlying concept that plant communities integrate abiotic site factors, and the presence or absence of species can be strong indicators of the abiotic environment. In North America, vegetative associations have been used as indicators of the abiotic environment to a great extent in the Intermountain area of the western United States and in the Great Lakes region. There is a very long tradition of phytosociological investigation in Europe, particularly in Scandinavia and in the former communist countries of Central and Eastern Europe. Systems for classifying vegetative associations are artificial constructs based on human perceptions, but they can be of great value in providing information regarding general abiotic characteristics, the response of plant communities to the abiotic environment, and the degree of past disturbance. Vegetation associations can also be good indicators of possible habitats for threatened and endangered plant species. More quantitative descriptions of the vegetative community include indices of species diversity and methods to quantify the similarity of two different plant communities. Systems of classifying vegetative association and quantitative measures of community structure have been related to abiotic factors and to overstory productivity to provide interpretive information for land managers.

Forest Overstory

The dominant vegetation in forests consists of the species comprising the forest overstory. The forest overstory is the vegetation component most often utilized for commodity production. The major commodities produced in forested systems in North America are usually timber-related items, such as veneer, sawlogs, cordwood, and wood chips. Globally, fuelwood is the most important commodity derived from forests (FAO, 1992). Economic returns may also be achieved from specialty wood products and nonfiber products such as boughs and greenery, Christmas trees, and food products, to name a few. Yet in predominantly forested landscapes, the economic value of agricultural crop production may exceed the value of forest-related commodity production in most years. Commodity production is important because it often provides the only economic return to the owner of forested land. This factor is obviously critical on private lands where taxes must be paid, but it is often important on public lands as well. Some economic benefits may accrue to landowners from recreational use, but in most cases they are relatively small as compared with the possible returns from commodity production. Short-term economic considerations may encourage the overutilization of wood products, when a longer view might postpone immediate returns in favor of future benefits. Sustainable management implies that economic utilization and commodity production will not sacrifice ecological integrity or the maintenance of ecological function in the landscape.

Wildlife Resources

In its most general sense, the term *wildlife* includes all faunal species occupying a landscape, not just game species. Many wildlife species have little economic value and little potential for providing economic returns to landowners. At the same time, the ecological and environmental value of these species may be enormous. Their role in the transport of energy and material within a landscape or in impacting the dynamics of vegetative populations can be quite significant. Disruption of these functions may have severe consequences, impacting the economic feasibility of commodity production. There are also strong linkages between terrestrial and aquatic populations. One of the most well-recognized cases is the role of forest vegetation in providing suitable stream habitats for salmon reproduction in the Pacific Northwest. In this case, there are serious trade-offs between commercial forest product utilization, commercial fisheries, and electrical power generation. Managing such a complex and spatially extensive system in such a way as to maintain the economic and ecological viability of all species is extremely difficult and is as much an exercise in managing social issues as an effort to address biological concerns (Lee, 1993).

SOCIAL CONTEXT

Legal and Regulatory Environment

Management of all lands in North America, regardless of ownership, is constrained to some extent by federal, state or province, and, possibly, local laws and regulations concerned with land use and resource management. For example, the U.S. federal Clean Water Act of 1977 and related state water quality acts apply to all ownerships. Management activities in federally designated wilderness areas and National Parks are narrowly focused, but landscape-level ecological issues must still be considered, as witnessed by the Yellowstone fires of 1988 that burned inside and outside the park. There are numerous federal laws that define the goals and activities of federal land management agencies (e.g., USDA Forest Service, 1983). On the other hand, private landowners almost always pay taxes, which is an incentive to manage land in such a way as to achieve at least a minimal level of economic return at some point. Tax policy may have great impact in the selection and economic viability of management options. In many states, forest practices are regulated through a permitting process or through voluntary guidelines for landowners to follow. In some states, legislation concerning commercial forest land may grant special tax status in return for concessions from landowners to allow public access or to place constraints on management practices. In addition, local areas may have zoning regulations that constrain management options. The entire body of federal, state, and local land management legislation is designed to protect the general societal interest in environmental quality and land use activity. Managers must be aware of these limitations and design activities to meet management goals and yet ensure compliance with legal and regulatory requirements.

Recreation

Recreational activity takes many forms, which vary in their level of impact on ecological processes within a landscape. Areas that tend to concentrate recreational use, such as campgrounds or scenic areas, may in effect place constraints on the management of surrounding areas because of aesthetic considerations. Tourism is an important component of many rural economies, and attractions such as hiking and snowmobiling trails can affect management decisions to a far greater extent than their relative area within the landscape would indicate. Management decisions may be specifically designed to enhance recreational opportunities, promoting, for instance, habitat management activities that favor game species. In some parts of the United States, landowners have great discretion in regulating the use of their property by others and may be able to generate additional income through the leasing of hunting rights. In other states, commercial forest acts declare that all registered commercial forest land must allow open public access, though limits may be placed on the types of permitted activities. In most areas, this is also true for public lands. Aesthetic considerations, economic opportunities for the landowner, and legal rights to access all impact the range of feasible management options within a forested landscape.

Prehistoric and Historic Land Use

Most of North America has been inhabited for 10,000 years or more, and even longer habitation periods are possible for nonglaciated areas (Denevan, 1992). For most of this time, population levels were much lower than today's, but humans still had significant impacts on the landscape. Examples include the intensively farmed areas of the Midwest and the Southwest and the areas that were repeatedly burned to aid hunting or to encourage berry production (Whitney, 1994). Traces of prehistoric human occupation and utilization of the landscape may provide important and rare clues into the life and times of prehistoric peoples, especially when viewed within the context of their locations. The same is true for historic times, where clues to human use of resources can provide valuable insights that are missing from sometimes biased written records. It is not necessary to preserve all traces of past land use, but it is also not appropriate to obliterate all such traces in the interest of short-term utilization. In many cases, relatively minor and inexpensive modification of planned activities can preserve evidence of past land use while placing very little constraint on current practice. There are state and federal laws regulating the preservation of culturally significant areas and artifacts with which resource managers must comply (National Park Service, 1993). Past human use can also have modified the landscape, impacting vegetation and other factors such as drainage patterns. Current and future management options may be limited by past physical alteration of the landscape. Vegetative community structure and composition are directly impacted by many management activities; the development of current management alternatives is facilitated by the understanding of past treatments and their resulting impacts on the landscape.

SYNTHESIS AND INTEGRATION

Geographic Information Systems

The information required to assess the abiotic, biotic, and social factors affecting the management options for a relatively small parcel within a forested landscape can be substantial. The same is true for assessing the environmental effects of a particular construction project or other site-specific manipulation. It is very difficult to coherently assemble and interpret this vast quantity of information. Yet legislation such as the National Environmental Policy Act of 1969 requires that these assessments be conducted for all federal projects in the United States. In the not-too-distant past, the ability to assemble and integrate massive amounts of spatially explicit information was restricted by the available technology. In fact, such efforts were often limited to numerous meticulously hand-drawn maps, which would be laid side by side in order to facilitate visual assessment. In the last few years the development of geographic information systems has dramatically changed the situation. Now it is relatively easy to assemble and digitally manipulate spatially explicit information for large areas. Computational routines that allow overlays of different types of information and rapid changes in scaling have had tremendous impacts on the ability to synthesize and interpret environmental information. Today the available technology does not limit the process; in fact, the technology for combining and assembling spatial information has far outpaced the development of analytical methods for synthesizing and integrating multiple types of information.

Geographic information systems are relatively complex computer software programs designed to facilitate the manipulation and analysis of spatially explicit information. Various landscape characteristics are represented as different layers of information, which can be combined to identify the spatial extent of certain conditions or combinations of conditions on the landscape. Ironically, one inherent danger to the use of geographic information systems is due to their utility. It is very easy to assemble and use digital information, but it is relatively difficult to reasonably consider the reliability of the information. This is especially true when the system is capable of producing full-color, publication-quality maps. For classification-type data, such as maps of soil types, a common mapping standard is to require 90% or more of the area be correctly classified. Even with this level of accuracy, combinations of different types of data (e.g., soil maps, vegetation cover types, and ownership patterns) can lead to the swift compounding of cumulative error. Such interpretive issues cannot be ignored by resource managers or others conducting integrated environmental assessments. Yet in spite of this drawback, geographic information systems are powerful tools that have revolutionized resource managers' abilities to develop plans for the sustainable management of ecosystems and landscapes.

Integrated Environmental Assessment

Integrated environmental assessments include a resource assessment, as discussed here, along with a description of management alternatives, identification of the economic and environmental consequences of each alternative, and the selection of a preferred alternative. Integrated assessments are required for federally supported projects by the National Environmental Policy Act of 1969. There are additional state laws that may require integrated assessments for certain types of management activities. Private resource-management organizations may be required to develop integrated environmental assessments in certain areas or for certain categories of activity. Increasingly, however, it may be advantageous, from a marketing standpoint, for an industrial concern to be able to demonstrate that its resource acquisition and land management activities are sustainable over the long term by developing integrated assessments for commodity-related activities. Growing efforts to establish widely accepted means of certification to indicate that wood products have been produced in a sustainable fashion are also leading to increased applications of integrated environmental assessment techniques on private lands.

SUMMARY

A landscape is made up of interacting ecosystems. Every landscape is a mixture of abiotic, biotic, and social components that serve to define options available to resource managers. Abiotic components include the underlying geology, soils, climate, and hydrological conditions in a landscape. Biotic components include the flora and fauna occupying the landscape for all or part of the year. And all land management and human landscape use takes place within a social context. Integrating these multiple factors is a difficult process, which has become easier with the development of geographic information systems. These systems are not cure-alls, and care must be taken in the interpretation of such information. They have, however, greatly increased the ability to assemble and analyze spatially explicit information. Information obtained in a resource assessment forms a basis for the development of possible management alternatives. The ability to evaluate the economic and environmental consequences of each alternative leads to the development of a preferred alternative. The entire process should balance the interacting forces in a landscape, allowing managers to forge a plan for sustainable resource management.

PART II

Abiotic Environment

Chapter

3

Geology

Surface geologic features have a direct effect on vegetation and ecosystem response to management activity or natural disturbance. A description of the surface geology of an area is often helpful in understanding ecological processes within a landscape and anticipating ecological response to treatment. It is also helpful in the development and evaluation of management alternatives. *Physiography* is a description of the morphology of the earth's surface, and *geomorphology* is the study of the processes by which different *landforms* are created. It is the surface geological features that support ecosystems, and these surficial features are formed through the actions of many geological processes. Such processes may be grouped under the broad headings of uplift processes, or those involved in the creation of new landforms, and degradation processes, or those involved in the modification of existing landforms. These processes function within the earth's crust, interacting through mechanisms such as climate and gravity to shape the features on the earth's surface.

Geology is the field of science that deals with the study of the earth, particularly the outer layers of the earth's crust including rocks, soil, and other physical features. Within the context of resource assessment, the geology of an area usually refers to a wide array of physical characteristics, including the underlying bedrock features, surface deposits of mineral materials from which soils have developed, and outstanding surface features such as rock cliffs and sinkholes. Because biological productivity and ecosystem management is largely a surface phenomenaon, geological information included in resource assessments usually deals with features closest to the surface.

For the most part, geological information is useful in a resource assessment to understand landscape pattern and structure, including such things as the soil patterns on the landscape and the patterns of water movement through the landscape. Many ecological classification systems, such as the national system currently under development by the U.S. Forest Service (McNab and Avers, 1994), include landform as a component of the classification system (see Chapter 7). Surface features may be used to locate deposits of sand and gravel materials that may be useful for road construction. Surface landforms also define the character of the landscape and have a major influence on the design of visually pleasing land management treatments (Lucas, 1991). It is rarely necessary for resource assessments to describe deeper geologic structures or mineral rights in great detail. Such information may be needed, for example, in the consideration of the installation of deep water wells, and in those cases in which management may be impacted by subsurface exploration or the extraction of minerals or oil. This chapter concentrates on the description of some common landform types and sources of geological information, together with their use, importance, and relevance in resource assessments. For a more detailed discussion of geology, landform processes, and geomorphic systems, refer to a basic geological text such as presented by Ritter (1986), Graf (1987), or Duff (1993).

LANDFORMS AND DEPOSITS

Geologic deposits have a significant effect on soil, hydrology, vegetation, biological productivity, and management. In describing a landscape in a resource assessment, classification of the types of geologic deposits that are present may aid in understanding the physical processes acting within the landscape. Many landforms are familiar, such as plains, mountains, and valleys, but there are literally hundreds of different landforms that are characteristic of only limited geographic areas. The types of plant communities that develop on landforms depend, to a large extent, on the type of parent material available for soil development. Brady (1990) developed a system of classifying parent material based on whether it is sedentary, meaning it formed in place, or whether it was transported to its current location. Transported material is further stratified by the mode of deposition (Table 3.1).

As mentioned, sedentary parent materials are those that have not been transported. The underlying rock weathers in place, and the soil develops in and from the weathered parent rock. These types of parent materials may be found on many landforms. The residual soil that develops on sedentary landforms has many characteristics that are directly related to the properties of the underlying geologic material. North American examples of regions with sedentary parent material include the Appalachian Mountains, the Piedmont Plateau, and the Great Plains. A wide variety of landforms in these areas have soils derived from residual parent materials.

TABLE 3.1 Classification of Soil Parent Material and Some Common Related Landforms (see Brady (1990) for Detailed Discussion)

Parent Material Group	*Transport Mechanism*	*Type of Deposit*	*Common Landforms*
Sedentary	None	Residual	Any landform without evidence of transported parent material
Transported	Gravity	Colluvial	Mountain slopes, benches, saddles, talus slopes
	Water	Alluvial	Floodplains, deltas, fans
		Marine	Coastal plains
		Lacustrine	Lake plains
	Ice	Glacial	Unstratified—Moraines, drumlins Stratified—Outwash, eskers
	Wind	Eolian	Loess, dunes

Transported geologic materials are divided according to the transport agent. *Colluvial* deposits consist of mineral materials that have essentially been dislodged at some height in the landscape and carried to a lower area by gravity. These deposits are usually found on slopes in mountainous areas. They consist of a wide variety of particle sizes, ranging from clay to sand, depending on the characteristics of the parent material; they also have a large component of coarse fragments (see Chapter 4). Because of their rocky nature, areas with colluvium are often forested because they may not be suitable for other uses.

Water forms a variety of different landforms. Water erodes mineral material, directly forming erosional features such as gullies, canyons, and ravines. But the eroded material is then carried and deposited in another location, forming depositional features such as floodplains or lake plains. *Alluvial* or *fluvial* deposits are mineral materials that have been transported, primarily by streams and rivers, forming landforms such as floodplains, deltas, and alluvial fans.

Floodplains are formed along the lengths of streams and rivers as a result of the deposition of sediment on the valley floor. Floodplains consist of fine material, eroded upstream or from surrounding higher ground, that has been deposited during periods of flooding as the water level recedes. For assessment purposes, a floodplain may be considered to be part of the river or stream; it is there because of flooding and will continue to be periodically flooded in the future, affecting many land uses and the feasibility of many management alternatives. Floodplains are generally very productive ecosystems and are frequently managed as wetlands (see Chapter 6).

Yet not all sediment carried by streams and rivers is deposited on floodplains. Deltas are formed when sediment carried by a stream or river settles to the bottom at the point where the stream or river joins a larger body of slower-moving water. In these areas, the water velocity diminishes and there less energy to carry sediment, resulting in the deposition of fine material and the formation of a delta. These deposits usually consist of fine-textured material and often form areas that may be classified as wetlands (see Chapter 6).

Alluvial fans are somewhat similar to deltas, except that they form in terrestrial settings. In mountainous areas or hilly terrain, steep, narrow channels that open up to a broader channel result in lower water velocity. Alluvial fans often form when material carried by the water is deposited in these settings. The channels forming alluvial fans may not be permanent, but may sometimes carry water for only part of the year. These deposits also consist of fine-textured material and are sometimes very productive.

Marine deposits are found in coastal areas and usually form relatively flat plains. They develop from sediment that has been carried to the ocean by rivers and streams and deposited in the rather shallow water along the ocean margin. Various uplift processes raise the marine sediments higher than sea level so that they are now dry. The soils are variable, depending on the source of the sediment. They are often quite productive and may support a wide variety of forested ecosystems.

Much of North America is covered in deposits called *glacial drift.* Drift consists of mineral material that was deposited as a result of the glacial activity that dominated the continent during the Pleistocene, which ended about 10,000 years ago. Drift may be classified as either nonstratified or stratified, depending on whether the depositional processes resulted in a sorting of the mineral material by particle size.

Nonstratified drift is material that was transported by a glacier and deposited by the ice with a minimum of water action. Often called *till,* it contains a mixture of particle sizes ranging from sand-, silt-, and clay-sized particles to cobbles, rocks, and large boulders. Till is deposited in an unsorted state, with all sizes of particles and fragments mixed together. Till may be further divided according to whether it was deposited under the weight of the overlying ice (basal or lodgement till) or whether it was lowered onto the bedrock as the ice melted from below it (ablation till).

Moraines are landforms that are composed of deposits of nonstratified glacial drift. Moraines usually consist of till, but their morphology varies greatly. For example, lateral and end moraines, which develop at the margins of active glaciers, form large hills or ridges that sometimes rise several hundreds of meters above the surrounding area. They are easily distinguishable from the surrounding terrain on soil and topographic maps. Ground moraines, which develop from debris released from the bottom of a glacier as it melts, form level to gently sloping *till plains.* In glaciated regions, moraines are often the more productive landforms; the soils that develop on them tend to be finer textured because of the mixture of sand-, silt-, and clay-sized particles in the till (see Chapter 4).

Drumlins are landforms that are sometimes associated with ground moraines. They are relatively small hills in the otherwise flat landscape of a till plain and are usually composed of clayey till material. Drumlins are longer than they are wide, and the long axis lies in the direction in which the ice moved. Drumlins are of interest in assessment and landscape management because they may support much different vegetation than the surrounding area.

Stratified drift differs from till in that it is deposited by meltwater from a glacier, rather than by the ice itself. Stratified drift has been sorted into layers of similar-sized sand and gravel particles by the action of the moving water. This stratified material is often called *glaciofluvial sediment,* or *outwash.* It is usually coarse textured, because it is deposited by fast-moving meltwater that does not allow clay and silt particles to settle out; finer material is carried a greater distance by the streamwater. As a result, outwash consists of various sand- and gravel-sized materials that are arranged in layers, depending on the speed of the depositing water, with coarser material settling out faster, near the source of the meltwater. When great rivers flowed from the glaciers, they formed *outwash plains,* which are level to gently sloping landforms that developed at the front of the ice margin. These plains are made up of layers of coarse-textured stratified drift. They tend to have low water-holding capacity and are subject to drought, depending on their relative elevation in the landscape. If outwash plains occur at relatively high topographic positions, they may be very dry and receive water only through precipitation. If they are at a low position in the landscape, they may receive water flowing from adjacent hills and drought may be much less severe. In either case, because sand particles in eastern North America are commonly quartz, the soils in glacial outwash plains tend to have low nutrient levels and are not particularly fertile.

An *esker* is a less common glaciofluvial landform that consists of stratified drift. Eskers are usually recognizable on the landscape as long ridges that may appear similar to some types of moraines, but eskers consist of sorted sands and gravel with occasional large, coarse fragments. Eskers develop from streams flowing on or below the bases of glaciers; the ridge marks the location of the streambed after the ice melted around it. Eskers have variable productivity, but they are often sought out as sources of road construction materials.

Other glacial features include *lacustrine* deposits, which are formed when fine-textured mineral material carried by glacial meltwater is deposited in lake water; the deposits are later exposed when the lake level recedes. Such areas are common in the Great Lakes region because lake levels fluctuated greatly during the last ice advance (Wisconsin Period). Numerous proglacial lakes developed and flooded large areas of the landscape, then drained as the ice that blocked flow channels melted. Lacustrine deposits usually form level to gently sloping plains with fine-textured clay soils. They have often eroded subsequent to their deposition and have variable productivity.

Wind is another transport agent that is responsible for the formation of some landforms. Although not associated with a single type of landform, *loess* is a windblown or *eolian* deposit of primarily silt-sized material. Loess is common in proglacial areas. Large areas of the Midwest and the Great Plains states are covered with loess that was carried by wind blowing off the top of the continental ice sheets. Loess deposits may also be found in glaciated areas as a cap deposited on top of other landforms. Because of their fine texture, soils that develop in loess are generally productive owing to their good water- and nutrient-holding properties.

In many coastal areas, including along the shorelines of the Great Lakes, windblown sand has formed dunes. These hilly features are usually in fairly close proximity to present coasts and shorelines, but may also be found along the shorelines of old glacial lakes. Dunes are dominated by coarse sand. Vegetation covering dunes may range from forest to grasslands. Because they are mostly sand, dunes are often subject to drought and are not very fertile. Once disturbed, they are easily eroded and may be very difficult to stabilize. Many states have laws that control management activities and regulate land use on dunes.

INFORMATION SOURCES

There are a variety of sources of information that can be used in describing the geology and landform features of a resource assessment area. In the United States, information concerning the topography of an area may be obtained from U.S. Geological Survey (USGS) maps. These maps, often referred to as quadrangle sheets or quad sheets, are commonly produced at scales of 1 : 24,000 (7.5-minute series) and 1 : 62,500 (15-minute series). At a scale of 1 : 24,000, one inch on a map represents 24,000 inches on the ground, or 2,000 feet; at a scale of 1 : 62,500, one inch on a map represents 5,208 feet, or approximately one mile. Topographic information is also available in digital form, as in digital elevation models, which may be incorporated into geographic information systems (see Chapter 13). USGS maps are useful in determining elevations, slope, aspect, and the drainageways for an area and are a primary source for determining watershed boundaries and divides (see Chapter 6).

There are also national geological classifications. At scales of 1 : 7,500,000, these maps identify the broad context of the geology of an assessment area (U.S. Geological Survey, 1966). In addition, there are national classifications and descriptions of physiography (Hunt, 1967), as well as geomorphology and landforms (Graf, 1987; Thelin and Pike, 1991). These are useful in establishing the broader context of geological features on the landscape of interest.

When an assessment area lies within one state, there are other sources of information and maps at the state level that may be used in the assessment process. Each state has a Geological Survey office charged with the responsibility of mapping geologic resources. These maps often represent both surface and subsurface geology of an area at a much smaller scale (1 : 500,000), have higher resolution, and contain more detail than the national maps (e.g., Seagall and Wilson, 1987; Mudrey et al., 1982). There are also many other publications describing geology at the state level, some of which are written in the form of nontechnical guides and are easily used by nonspecialists (e.g., Wright, 1987; 1990). For even more detailed information on the geology of an assessment area, there may be maps and scientific literature available at even smaller scales that contain a large body of site-specific information (e.g., Farrand and Bell, 1982; Clayton, 1984).

In many resource assessments, soil survey reports (see Chapter 4) produced by the Natural Resource Conservation Service (NRCS), formerly known as the Soil Conservation Service, are a good initial source of geological information. These reports, produced for individual counties, usually include a brief description of the surficial and bedrock features, including their origin, age, location, and economic significance. They also include a description of major drainageways, streams, and lakes. To describe the larger context of the landscape, the reports contain general soils maps, which have a strong link to surface geology and landforms. These general maps are usually made at a scale of 1 : 500,000 or less. The soil survey reports contain more specific soil maps, often printed on copies of aerial photographs, at scales of approximately 1 : 20,000 for an entire county. Since soil-mapping units are often defined according to slope and landform considerations, the soil descriptions included in the reports, as well as the maps, are helpful in understanding the surface geology. Each soil survey report also includes a section on soil formation, which links the soil in the county to the geologic material from which it is derived. In addition, the reports discuss the origin of the soils and their relationship to the physiography of the county. Because the soil survey reports contain such a volume of information on surface geology, they are often the only source of surficial geological information utilized in resource assessments. Greater detail about subsurface or surface features is usually not required unless there are specific concerns related to oil and gas or mineral exploration, mining, or associated surface land use within a landscape.

INTERPRETATIONS

The degree to which geologic and topographic information is used in resource assessments, and the amount of information presented, will vary, depending on the specific management goals within the landscape and the proposed management alternatives. In most forest management applications, the primary concern is to obtain a description of the relationships among geology, landform, and soils in order to understand both the inherent productivity of different landforms and the movement of water on the landscape. Clearly, a great deal of the interpretation will be based on the type of mineral material present within the landscape.

Other topographic information that may be determined from this type of data includes such variables as elevation, slope, and aspect. *Elevation* is the height above sea level; this information may be directly read from USGS topographic maps. Meteorology is tightly coupled with elevation; air temperature, for instance, decreases at an average rate of 1°C for every 100 meters of elevation (0.6°C per 100 m if condensation occurs). The more elevation varies across the landscape, the more it can influence growing conditions by affecting local weather (see Chapter 5). Cooler temperatures will affect the rate of plant growth as well as the length of the growing season. Temperature may also affect the amount and type of precipitation; cooler temperatures can correspond to a greater proportion of precipitation in the form of snow. This, in turn, may influence streamflow, recreational opportunities, wildlife habitat, and species composition. In mountainous areas, this is clearly a concern in resource management. It is also an important concern for areas on the lee side of the Great Lakes, where lake-effect snows are enhanced by the local topography, with some areas receiving as much as ten meters (400 inches) of snowfall in some years.

Elevation can be expressed by either the average or the range of the values present within a landscape. When elevation is of particular concern, the geographic area falling within different elevation ranges may be determined. In using a USGS quadrangle sheet, or other topographic map, this area can be estimated using a dot grid. This is simply a transparent sheet of plastic with a gridded series of dots printed on it (Figure 3.1). The plastic sheet is placed over the topographic map; the number of dots falling in the area

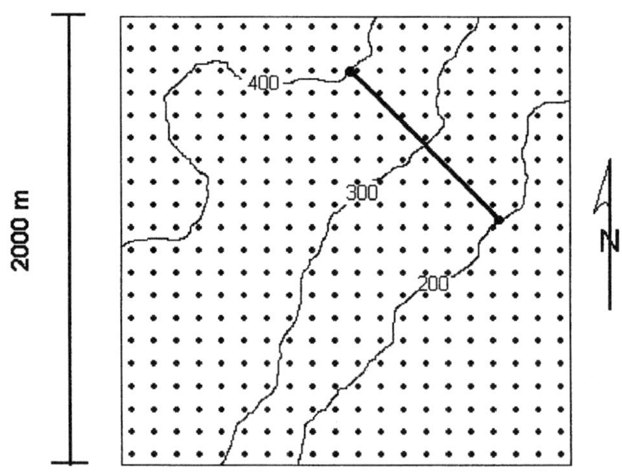

FIGURE 3.1 Illustration of the use of a dot grid to estimate area and the estimation of average slope and aspect from a topographic map.

TABLE 3.2 Topographic Information for the Area Illustrated in Figure 3.1

Elevation Range (m)	*Area Within the Elevation Range (hectares)*	*Percentage of Area Within the Elevation Range*	*Percentage of the Total Area Above the Lower Limit of the Elevation Range*
100–200	100	25.0	100.0
200–300	100	25.0	75.0
300–400	150	37.5	50.0
400–500	50	12.5	12.5
Total	400	100.0	—

within the elevation range of interest indicates the geographic area within that range (Table 3.2). Accuracy is improved when smaller grid sizes (implying that there are more dots on the sheet) are used. These results may be expressed using a hypsometric curve, which illustrates the relationship between elevation and the geographic area above that elevation (Figure 3.2). If there are important threshold elevations within a landscape, such as the maximum elevation at which a particular species can survive and reproduce, a hypsometric curve provides the means to evaluate the area of suitable habitat within the landscape. These relationships are also useful when threshold elevations influencing ecosystem productivity, timing or duration of planting seasons, or wildlife migrations exist within a landscape.

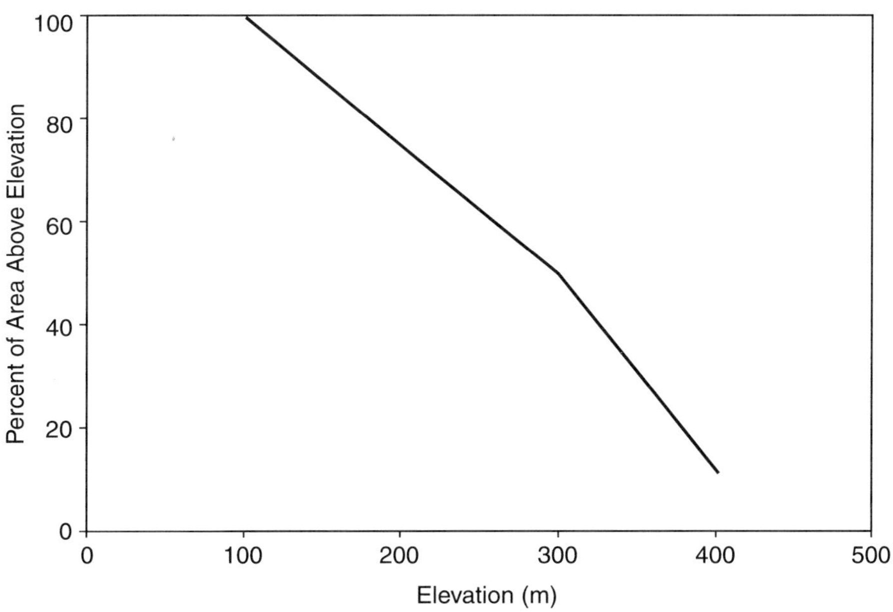

FIGURE 3.2 A hypsometric curve constructed using the information in Table 3.2 for the map illustrated in Figure 3.1.

Slope and aspect are two important measures of topographic position that are easily obtained from a topographic map. *Slope* is determined by dividing the difference in elevation by the horizontal distance between two points:

$$\text{Slope \%} = \frac{\Delta \text{ elevation}}{\text{distance}} \cdot (100) \qquad [3.1]$$

where the change in elevation and the distance are measured in the same units (meters or feet). Note that Equation [3.1] expresses slope as a percentage value. In Figure 3.1, the slope between the two points connected by the 1,000-meter solid line is calculated by dividing the change in elevation by the distance between the two points:

$$\begin{aligned} \text{Slope \%} &= \frac{(400 - 200)}{1000} \cdot (100) \qquad [3.2] \\ &= 20\% \end{aligned}$$

Equation [3.1] may be used to estimate the average slope between any two points in a landscape. Slope is also often reported as an average value for a particular landform or soil mapping unit (see Chapter 4). Slope is an important consideration in planning access roads, trails, and operability of equipment. It is also extremely important in considering how treatments may affect the way water moves across the landscape and in evaluating erosion potential (see Chapter 6).

Aspect describes the general direction in which a particular slope is facing; aspect is usually expressed as a cardinal direction or an azimuth. The aspect may be estimated in the field by measuring the compass bearing or azimuth from the upper to the lower slope; this measurement should be made as close to perpendicular across the topographic contours as possible. In Figure 3.1, the aspect of the slope between the two points connected by the solid line is approximately 135°, or S 45° E. Most ecological interpretations of aspect concern energy relationships within a landscape. South-facing slopes, for example, are usually warmer and drier in the Northern Hemisphere, because they face the sun and solar energy is concentrated there. The steeper the slope, the more solar energy is concentrated. North-facing slopes are usually cooler and moister, because they face away from the sun and receive less direct solar radiation. The warmest slopes have a southwestern aspect, facing the sun during the warmest part of the day. The coolest slopes have a northeastern aspect, receiving mostly indirect sunlight during the coolest part of the early morning. For a given soil and slope percentage, productivity, fire risk, and species composition may all be heavily dependent on aspect, owing to associated temperature and moisture conditions.

Topographic and geologic maps are commonly used to delineate watershed boundaries within an assessment area (see Chapter 6). Watersheds are areas of land that literally shed water so that all of it passes a central point, referred to as the mouth. Watersheds exist at many differ scales, whereby each watershed is part of an even larger network. Watershed delineation is accomplished by locating the streams within a landscape on a topographic map and finding the higher elevations that separate one watershed from another. Geologic maps may then be used to identify subsurface features that influence water movement. Together, these can produce a reasonably accurate assessment of surface and subsurface directional flow of water on a landscape without requiring costly monitoring efforts. However, where water amount or quality is a serious concern, or where contamination of water with pollutants may be a problem, as in the placement of waste disposal or sewage treatment facilities, on-site monitoring wells usually have to be installed to verify hydrologic characteristics and to ensure that possible contamination or other problems will be detected as soon as possible, allowing the least expensive mitigation measures to be implemented.

LEGAL CONSIDERATIONS

The rights to ownership of the minerals lying below the surface of the land, referred to as *mineral rights,* are not always conveyed to the purchaser in a sale of real estate. In such cases, the buyer holds title only to the *surface rights,* or the use of the surface for all the normal purposes associated with land ownership. The right to extract minerals from subsurface deposits or entitlement to royalties from mineral extraction may be held by other individuals, corporations, or state or federal governments. In many situations in the United States, prior owners have retained the mineral rights, and these may have been sold or leased to third parties; in some states, when land has reverted to state ownership because of nonpayment of taxes, the mineral rights have also reverted to the state. In any case, the owner of the mineral rights can grant leases to others for the purpose of mineral exploration and extraction, and the owner of the surface rights must allow such activities to take place.

The division of surface and mineral rights varies by state. In Michigan, for example, the right to conduct mining activities to extract sand and gravel is included in the surface rights, as is the right to install a water well. The right to explore for, or install, gas and oil wells, or to conduct other types of mining activity, belongs to the owner of the mineral rights.

A component of most mineral laws provides the holder with the right to access the minerals from the surface. From a land management standpoint, the title to mineral rights is important because development of mines and wells can have obvious consequences to the landscape surface. Exploitation of mineral deposits may preclude the use of land for other purposes. For example, oil and gas drilling in the eastern United States often requires clearings for roads, pipelines, and well sites. Management of the landscape for species needing interior forest conditions, such as pine marten, may not be possible in an area above an oil and gas field because of the disturbance and edge created by land clearing associated with oil and gas extraction. On the other hand, the area may be well suited for intensive forest cultivation or management for wildlife species that are habitat generalists, such as white-tailed deer.

If there is a question as to the ownership of mineral rights, it will be necessary to conduct a fairly intensive title search. All real estate transactions, including the particulars of rights sold in various transactions, are recorded in the Register of Deeds office in local county courthouses. Using the legal description of the land, it is possible to search all past records of ownership and deeds conveyed. This is a complex undertaking, because parcels of land often change hands and may be divided or combined into smaller or larger parcels over time. A standard title search used for real estate transactions does not normally report mineral rights, so the process takes extra effort to carefully trace each transaction. Professional abstracting services are available to do these searches and should be used when high-value areas are being developed or when there is concern that extraction of minerals may conflict with a proposed land use.

SUMMARY

Landform plays a critical role in many ecological processes. Physical characteristics such as slope and aspect are associated with many ecological properties and concerns, including soil erodibility, energy relationships, the risk of fire occurrence, biological productivity, and others. Soil characteristics are largely dependent on the larger landform, as are the hydrological characteristics of a landscape. For the most part, a resource assessment utilizes geological information for an understanding of landscape pattern and structure, including soil patterns and water movement through the landscape. Many ecological classification systems, for example, include landform in the classification criteria. Resource assessments associated with surface land management usually concentrate on surface geology. Usually, deeper geologic structures or mineral rights are described in great detail only in special cases, such as where deep water wells are considered or where surface land management may be impacted by oil and gas or mineral exploration or extraction.

APPLICATIONS

3.1 Obtain a copy of a national map of geologic features, a state map for your state, and a soil survey report for one county. At what scales are the maps presented? What types of information are available at the different scales?

3.2 Obtain a county soil survey report from your state. What types of geologic deposits are present in the county? What general soil properties tend to be associated with each one?

3.3 Using a USGS quadrangle sheet, outline a four-section area and delineate five elevation classes. Note that it may be necessary to combine several contours into each elevation class, depending on the characteristics of the area. Construct a hypsometric curve (as in Figure 3.2) for the area. What ecological questions in the area may be addressed by this information?

3.4 Obtain a USGS quadrangle sheet; delineate the boundaries of the watershed for a stream flowing within the area. What is the area of the watershed? What is the average slope from the highest point in the watershed to the mouth? Estimate the average aspect for two slopes within the watershed.

3.5 What is the relationship between surface rights and mineral rights in your state? What rights to water or surface deposits (sand and gravel) are associated with surface rights? What type of access must the owner of surface rights provide to the holder of mineral rights for the purposes of mineral exploration or extraction?

Chapter

4

Soil

Soil and climate are the two key factors that control the productive potential of landscapes. Climate dictates growing conditions through such components as temperature, precipitation, and solar radiation. But within a climatic region, the soil is the major determinant of local plant productivity. *Soil,* which is defined as the weathered outer portion of the earth's crust, sustains life by providing plants with mechanical support, water, and nutrients. It is a mixture of mineral particles, organic materials, and pores that are filled with air and water; interspersed within this matrix are a diversity of plant roots and micro- and macroorganisms. Soil varies greatly across the landscape, and it supports a wide diversity of above- and belowground flora and fauna.

Forest soils differ from those in other landscapes because they have developed under the influence of forest vegetation. Forest soils have developed with large amounts of wood and plant material that have been added as litter, usually over long periods of time. The decay of this material and the resulting cycling of carbon and nutrients depends on soil fauna, which are different from those that might be found in a grassland area or alpine meadow, for example. Many tree species are also more deeply rooted than herbaceous or grassland vegetation, so that forest soils tend to be affected or developed to a greater depth than soils developing in the presence of more shallowly rooted plants. Unlike agricultural soils, forest soils are also mixed or cultivated to a much lesser degree. As a result, the surface of forest soil is much more variable than that of agricultural soils. Although some forest soils are very productive, as a group they often are the soils in a landscape that are too poor to support the cultivation of crops. This may be due to factors such as high rock content, excessive drainage, low nutrient content, a high water table, or steep slopes. Because of these possible limiting factors, resource managers must pay special attention to the soil and the potential limitations it may impose on management in forested landscapes.

Knowledge of the characteristics of different soils is key to understanding the basic concepts of site productivity; some soils can support very productive communities with a large number of species, whereas others can support only very simple plant communities with slow growth rates. The soil also places limitations on management options across landscapes. Some soils, for example, can support access roads and heavy machinery year-round, while others may have seasonal restrictions owing to high water content. Some soils may be well suited to hiking or off-road-vehicle trails, while other soils can turn to quagmires or suffer severe erosion under such stresses. Regardless of the management objectives for an area, efforts to sustain the productivity of a forested landscape over long periods of time demand attention to proper soil management. This chapter reviews a few of the key soil properties that affect management and sources of information that are of value in the resource assessment process. More extensive discussions of soil properties and interpretations and management may be found in Pritchett and Fisher (1987) and Brady (1990).

SOIL PHYSICAL PROPERTIES

Soil texture, the relative proportions of sand-, silt-, and clay-sized mineral particles in the soil, is perhaps the most important single descriptor of soil properties used in the assessment of upland soils. *Sand* particles are the largest (0.05–2 mm), followed by *silt* (0.002–0.05 mm), and *clay* (<0.002 mm). Particles larger than 2 mm are termed *coarse fragments* and are not included in the determination of texture, but are usually reported separately for a given soil. The percentages of sand, silt, and clay in a soil may be used to determine the soil textural class (Figure 4.1). Trained individuals can often accurately identify soil texture class by feel in the field.

The class names for soils form a gradient from coarse to fine in the following sequence: sand, loamy sand, sandy loam, loam, silt loam, silty clay loam, clay loam, sandy clay, silty clay, and clay (Brady, 1990). Coarse soils, such as sands, are made up of relatively large particles and have low amounts of surface area per gram of soil, low total pore space, a large proportion of macropores (which are responsible for drainage and aeration), low nutrient- and water-holding capacity, rapid water infiltration rates, and high bearing strength to support loads such as vehicles. Fine soils have smaller soil particles with very high amounts of surface area per gram of soil, high total pore space, a large proportion of micropores (which are responsible for soil water storage over long periods), high nutrient- and water-holding capacity, and low bearing strength, especially when wet.

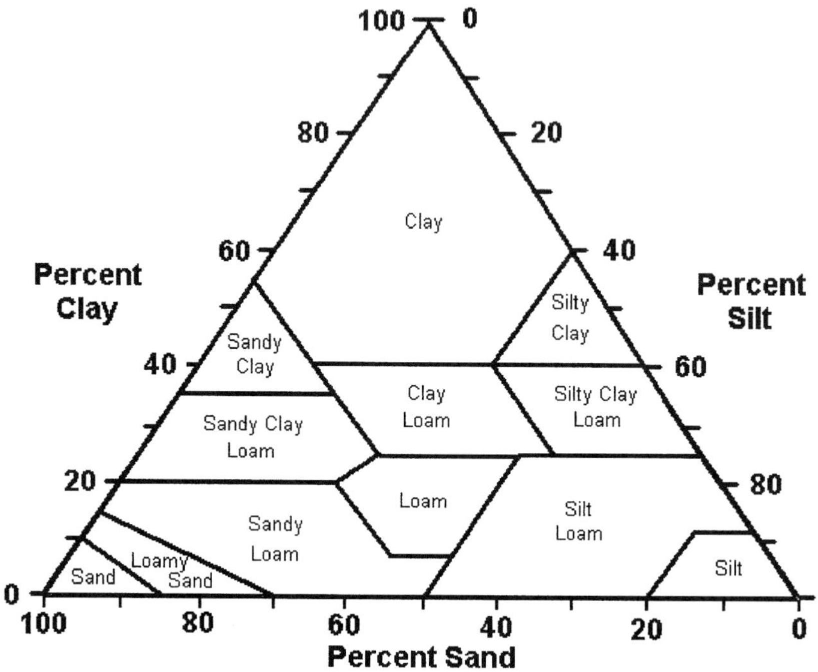

FIGURE 4.1 The relationship between soil texture (expressed in percentage of clay, silt, and sand in the soil) and soil textural class.

Whereas *texture* refers to the proportions of different sizes of mineral soil particles, *soil structure* describes the larger aggregates formed when soil particles are joined together. The particles may be held together by several different materials: (1) organic material, as is the case with *granular* structures; (2) clay, which is common with *platy* structures; or (3) deposits of iron and aluminum oxides (*sesquioxides*), as is common with the *subangular blocky* structures of some subsurface horizons in the forests of eastern North America. Soil may also be structureless; sands and silts with no structure are often referred to as *single grained,* while unstructured clays are referred to as *massive.*

Soil structure modifies the properties associated with texture. For example, a fine textured soil will have high moisture holding capacity because of the large amount of soil particle surface area and the large proportion of micropores associated with the small soil particles; if the same soil has a granular structure, it will also have macropores around each structural unit, which will improve the aeration and drainage of the soil. Conversely, a structureless sandy soil that has a large proportion of macropores will be well drained; if the same soil has a granular structure, it will have micropores within the structures that will retain water. In both cases, structure improves the soil, helping to temper soil water retention and drainage. Since granular structures are held together with organic matter, nutrient-holding capacity is also enhanced.

Soil texture is changed only by processes that operate over geological time scales; it is not altered by management activity. Soil structure, on the other hand, may be altered within short periods of time by management activity. A management practice that leads to increased soil organic matter will generally help to preserve or improve soil structure. Practices that lead to increased rates of soil organic matter decomposition, such as removal of the vegetation and subsequent soil warming or mixing of the surface soil horizons, have the potential to decrease soil structure and, hence, adversely affect many soil chemical properties and the physical properties associated with water movement and retention.

Soil physical properties are determined in the field by excavating a hole called a *soil pit*. After a soil pit is dug, the exposed vertical face in the hole is termed a *soil profile*. Soil profiles expose the various soil layers, which are called *horizons*. These horizons may be differentiated in the field based on color, texture, or structural differences and generally reflect soil-forming processes (Figure 4.2). Soil color is described using standard reference colors known as Munsell color charts. These colors are determined through the use of a numerical system based on the hue, value, and chroma (USDA Soil Conservation Service, 1975). The hue is the dominant spectral (rainbow)

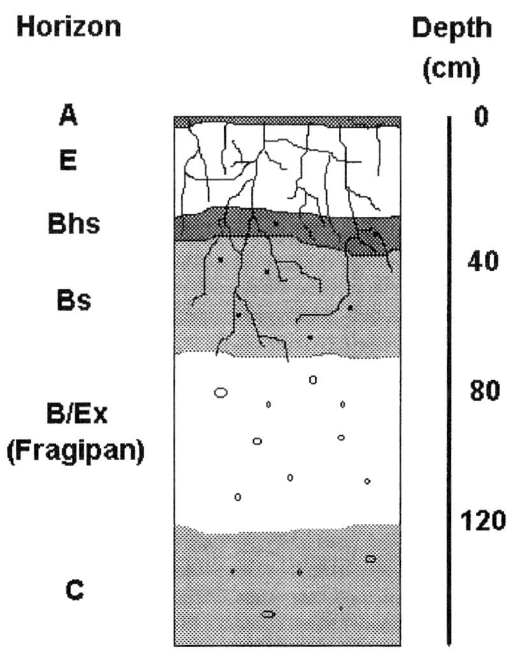

FIGURE 4.2 Stylized profile of the Munising soil series (Houghton County, Michigan), a typical forest upland soil from eastern North America (adapted from Schwenner, 1991).

color and is related to the wavelength of reflected light; value describes the amount of light reflected, or brightness; and chroma is a measure of the clarity or strength of the spectral color. Soil color is one of the primary indicators of soil horizons and is also used in wetland determination.

The surface of a forest soil, the *forest floor,* consists of organic materials deposited on the mineral soil surface and are designated as *O horizons*. These O horizons can be further subdivided based on the degree of decomposition of the organic material present in the horizon. The presence of identifiable organic fibers is the indicator of three different types of O horizons: (1) fibric, or O_i (2/3 fibers by mass); (2) hemic, or O_e (1/3 to 2/3 fibers by mass); and (3) sapric, or O_a (less than 1/3 fibers by mass). Fibric horizons consist of relatively fresh litter, whereas organic material in sapric horizons is greatly decomposed. Forest floor horizons become well developed, as determined by overall depth, when the rate of deposition of organic matter as litter exceeds the decomposition rate. The decomposition of litter by microorganisms is generally impeded by a litter substrate that is either acidic (such as with pine foliage), has a high carbon-to-nitrogen ratio, or is frequently cold or saturated with water. These conditions often lead to relatively thick forest floor horizons.

In some soils, the decomposition of litter is quite rapid and organic matter is often incorporated into the surface mineral soil by soil fauna. This results in a dark-colored mineral surface horizon described as an *A horizon.* This horizon is generally rich in organic matter, has granular structure, is adequately drained, and is relatively high in nutrients and available water for plants. A rather light-colored surface horizon, or *E horizon,* occurs in soils when the organic matter is not incorporated in the mineral soil but, in fact, has *eluviated* or moved to a lower horizon. This is common in sandy-textured soils supporting conifers and hardwoods in many parts of the world. An E horizon is generally indicative of soils with slow rates of organic matter decomposition and is often found in soils with well-developed forest floor horizons.

Underlying the O, A, and E surface horizons are *B horizons.* B horizons, or *illuviated* horizons, represent a zone of accumulation; organic and mineral materials that have been moved from overlying horizons through leaching collect in these B horizons. A subscript is used to describe the dominant material that has collected in a B horizon: B_s describes a horizon in which sesquioxides of iron and aluminum have collected; B_t describes a horizon with illuviated clay; B_h describes a horizon with elevated humus levels; and B_x describes a horizon that is dense or has a high bulk density, usually described as a *fragipan.* Fragipans usually have low permeability and may cause impeded drainage. They may also be difficult for roots to penetrate, limiting the soil rooting volume because of mechanical impedance or low aeration levels. These properties also limit the water and nutrients available to plants. At the bottom of the profile, beneath the B horizons, is the unweathered parent material from which the soil developed; this is referred to as the *C horizon.* It should be noted that some soils with limited depth may lack the lower horizons. A complete description and list of soil horizons may be found in Soil Taxonomy (USDA Soil Conservation Service, 1975).

SOIL WATER

Soil water is held in pores for varying time periods, depending on the sizes of the pores within the soil. Water held for a long time period is said to be retained by the soil, while water held for shorter time periods is said to be detained. The pore spaces that hold water against the force of gravity are called *micropores,* and they are responsible for longer-term water storage. *Macropores* are sufficiently large for gravitational pull to overcome the water's attraction to the soil surfaces, so the water drains from the soil, allowing air into the pore spaces. Hence, micropores retain water important for plant growth. Macropores merely detain water and are important for drainage and aeration of the soil.

The amount of water held in the soil may be estimated in a variety of ways. The simplest is a gravimetric determination. A soil sample is taken, weighed, dried in an oven at 105°C, and reweighed. Water content is calculated on a weight basis:

$$\theta_w = \frac{W_w}{W_s} \qquad [4.1]$$

where θ_w is the soil water content on a weight basis, W_s is the dry weight of the soil (g), and W_w is the weight of the water (g) in the original sample, which is simply the wet weight minus the dry weight. The result of Equation [4.1] may be multiplied by 100 to convert water content to a percentage. As an example, if a soil sample weighs 50 g when wet and 40 g when oven-dried, the weight of the water in the soil was 10 g and the water content was:

$$\begin{aligned} \theta_w &= \frac{10}{40} \\ &= 0.25 \text{ or } 25\% \end{aligned} \qquad [4.2]$$

To calculate water content on a volume basis, the weight of the soil water (from Equation [4.1]) is multiplied by the bulk density of the soil:

$$\theta_v = \theta_w \cdot D_b \qquad [4.3]$$

where θ_v is the soil water content on a volume basis and D_b is the soil bulk density in g/cm^3. *Bulk density* is a measure of the mass of solids per unit volume of soil. It is measured by taking an undisturbed soil sample of known volume, oven-drying the sample, and dividing the weight of the oven-dried sample by the known volume. For example, if the bulk density of a soil is

1.25 g/cm^3 and the water content of the soil is 25 %, or 0.25, the volumetric water content from Equation [4.3] is estimated to be:

$$\begin{aligned}\theta_v &= 0.25 \cdot 1.25 \\ &= 0.31 \text{ g/cm}^3\end{aligned} \quad [4.4]$$

This implies that 0.31 g of water are stored in each cubic centimeter of soil. The volumetric water content from Equation [4.3] may be multiplied by the depth of the soil profile represented by the sample to calculate total water content. If the sample had been taken from a 100 cm soil profile, then 31 cm of water would have been held in the profile.

One way to classify soil water is based on its energy status, or the tension with which it is held in the soil. This is a useful system for assessment purposes because it is a method of determining the amount of water in the soil that is available to plants. When the soil is saturated, meaning that all the soil pores are filled, water moves through the soil under the force of gravity. But as the soil water drains and the soil dries, water is still held in the soil in close association with the solid surfaces of the soil particles. This water is said to be under tension, because it is held above the saturated zone in the soil, adheres to the soil particle surfaces, and occupies the smaller pore spaces, or micropores. This tension is sometimes referred to as the *matric potential*, because it reflects the attraction of the soil water to the surfaces of the soil particle matrix. In an intuitive sense, this water is suspended in the soil, much like water is suspended on leaf surfaces at the beginning of a rain event. Equations [4.1] and [4.3] may be used to determine the amount of soil water storage or available water at different tension levels.

The amount of water stored in micropores and macropores can be estimated using a variety of techniques ranging from gravimetric measures to very sophisticated electronic instrumentation. However it is determined, it may then be described in terms useful for interpretive purposes. The soil is saturated, or at *maximum retentive capacity*, when all soil pores are filled with water. At the top of the saturated zone, the water tension is said to be at atmospheric pressure (0.101 MPa or 1 atmosphere); water below the surface of the saturated zone is at greater than atmospheric pressure because of the mass of the overlying water. Water in the macropores is termed *gravitational water* and is usually not considered to be available for plant use, because its residence time is short. *Field capacity* describes a soil water state that occurs a few days after a rainfall when rapid drainage through the macropores slows to negligible levels; at this time the micropores are filled with water and the macropores are essentially empty. The water held in the micropores at this point may be used by plants, and its energy status, or matric potential, ranges from −0.1 to −0.3 MPa, depending on the texture of the soil. If extensive drying takes place, the soil water reaches the *permanent wilting point* or *wilting coefficient*, a low water level at which plants cannot obtain enough water to maintain pressure in their cells, and the plants wilt.

The tension of the soil water at this point is estimated to be about −1.5 MPa (−15 atmospheres) for most mesophytic plants. If soil drying continues, soil water eventually will exist only in close association with the solid surfaces of clay particles; this water is termed hygroscopic and is held at tensions greater than −3.1 MPa (−31 atmospheres). Hygroscopic water moves only if the soil particle moves.

Realizing these limits is important, because they have many implications for management and they define the potential uses of soil. In a biological sense, field capacity and permanent wilting point define the limits of water availability for plants. The difference in the amount of water held at field capacity and the amount of water held at the permanent wilting point is the maximum amount of water the soil can hold that is available for plant uptake.

The difference between maximum retentive capacity and field capacity is used to estimate drainage class. The *drainage class* of a soil describes the natural frequency and duration of water retention in the soil. It is strongly related to soil texture, which governs the water-holding capacity of the soil, slope position, which is associated with water movement within the landscape, and climate, which controls evapotranspiration as well as precipitation frequency and duration. The Natural Resource Conservation Service recognizes seven soil drainage classes (USDA Soil Conservation Service, 1992), which range from dry to wet as follows:

- *Excessively well drained.* Water flows very rapidly through the soil; typical of coarse textured, rocky, or shallow soils; there are no mottles present.
- *Somewhat excessively well drained.* Water flows rapidly through the soil; typical of sandy soils and steep slopes; there are no mottles present.
- *Well drained.* Water flows readily, but not rapidly, from the soil; water is available to plants for most of the growing season, but wetness does not inhibit root growth; typical of medium-textured soils; these soils are mainly free of mottles.
- *Moderately well drained.* Water flows through the soil slowly during parts of the year; the soil is wet for a short time during the growing season, which may affect the growth of mesophytic plants. These soils often have a slowly pervious subsoil layer that impedes drainage, or they periodically receive high amounts of rainfall.
- *Somewhat poorly drained.* Water is removed slowly from the soil, so that the soil is wet for significant parts of the growing season; this wetness is usually reflected in a shift from mesophytic to hydrophytic plants. These soils are common in areas with a high water table, impermeable or slowly permeable subsoil layers, or constant rainfall.
- *Very poorly drained.* Water drains so slowly from these soils that it remains at or near the surface of the soil for most of the growing season; hydrophytic plants are common.

A key feature in this classification is the presence or absence of mottles in the soil. *Mottles* are irregular spots of a different color than the surrounding horizon. In the system used to describe soil color, mottles display low chroma colors, having chroma values of 2 or less. They are the result of a change in the valence of iron, which is abundant in soil; depending on soil aeration, iron changes from an oxidized state, in which it is rust colored (ferric, or Fe^{+++}), to a reduced state, in which it is a greenish to grayish blue color (ferrous, or Fe^{++}). The oxidized form is prominent in well-aerated, or *aerobic*, conditions. The ferrous form is more prominent in *anaerobic* conditions, where oxygen is lacking because of the high water levels in the soil. If the water table fluctuates with alternating anaerobic and aerobic conditions, there are portions of the soil where both forms of iron will be present. This condition is indicated by colored mottles in the soil profile. If the soil is flooded most of the time, a more uniformly colored *gley* horizon results, which indicates the more permanent anaerobic conditions found with high water levels. Gley horizons tend to have even lower chroma values (1 or less) than mottles. Mottles and gley horizons are often referred to as redoximorphic soil features because the processes of oxidation and reduction are controlling the morphology. These are important on-site indicators of water relationships in the soil and are used in identifying the hydric soils associated with wetlands (Chapter 6).

Each soil type described by the NRCS in soil surveys is categorized by drainage class. Drainage classification is a key to interpreting soil capabilities for any number of uses, ranging from estimating forest productivity to locating a septic field for a recreation area or constructing a wetland for mitigation purposes. Forests occur across the entire spectrum of drainage classes, and suitable management practices differ greatly among soils that are well drained as compared with those that are very poorly drained. For example, forests on well-drained soils may usually be harvested at any time of the year, because the soil tends to be resistant to disturbance owing to its low water-holding capacity and coarse texture. Conversely, most best management practices for poorly drained soils in the northern forest regions specify winter harvesting, when the soil is frozen, or harvesting when the soil is extremely dry and may better resist compaction and disturbance. Soil water, therefore, not only affects plant productivity, but may also affect the type and timing of prudent management practices.

Permeability, the rate at which water can move downward through a soil, is another important indicator of soil-water relationships (Table 4.1). Permeability provides an estimate of the speed at which material may be carried by water movement through the soil and into groundwater or adjacent surface water in streams and lakes. Estimation of the movement of fertilizers or herbicides, or the determination of the suitability of a site for a septic field, depends on soil permeability.

A process that can directly affect soil structure, water-holding capacity, drainage, and aeration is *compaction.* Soil compaction is a result of the breakdown of soil structure and is indicated by an increase in bulk density. Compaction can be caused by the weight of heavy equipment, such as used

TABLE 4.1 Soil Permeability Descriptions and Ratings Used in Soil Surveys

Descriptor	*Water Movement (cm/hour)*
Very slow	<0.15
Slow	0.15 to 0.51
Moderately Slow	0.51 to 1.5
Moderate	1.5 to 5.1
Moderately Rapid	5.1 to 15.2
Rapid	15.2 to 50.8
Very Rapid	>50.8

SOURCE: Whitney (1992).

in many forest management activities, repeated traffic with lighter-weight vehicles, and even repeated foot traffic on hiking trails, in campgrounds, or in picnic areas.

The breakdown of soil structure and the associated increase in bulk density alters the pore size distribution in the soil, changing the relative proportions of micropores and macropores. The rate at which water enters the soil is called the *infiltration rate.* The infiltration rate varies under different conditions, and the maximum possible infiltration rate is called the *infiltration capacity.* The attractive forces of water for solid surfaces are less in large pores than in small ones, so a shift from macropores to micropores may cause a change in the infiltration capacity of the soil. The infiltration rate is a soil property that is measured in the field and is usually reported in centimeters per hour (cm/hr) or inches per hour (in./hr). In undisturbed forests, the infiltration capacity is seldom less than the rainfall intensity, so most water infiltrates into the soil. When disturbed by cultivation, road building, or trail use, the soil in those areas may become less permeable and, as a result, the infiltration rate may decrease, increasing the tendency of water to move over the surface of the soil as *overland flow,* rather than infiltrating into the soil. This may cause erosion problems and may also decrease the amount of water stored in the soil.

The loss of soil structure can also decrease the water-holding capacity of sandy soils because of the loss of micropores; however, compaction is of less concern on sandy soils where the large particles have good bearing strength even when wet. When fine-textured soils lose structure or are compacted, they lose macropore space, which simultaneously decreases aeration and drainage while increasing water-holding capacity. Fine-textured soils have the lowest bearing strength, and because macropores are already limited, further reduction in macropore space may result in the soil's being imperfectly drained. Fine-textured soils that are wet may suffer the most compaction damage, because the water softens the soil structures and acts as a lubricant, allowing soil particles to slide more freely in the soil when weight is applied. High soil moisture can also make soils slippery when they are wet. This reduces vehicle traction, resulting in disturbance of surface structure and mixing of the surface soil horizons.

SOIL CHEMICAL PROPERTIES

Soil chemical properties are important to consider in assessing and describing forest soils because of their relationship to plant productivity. There are several key properties related to land use; the first is soil *pH,* which is defined as the negative logarithm of the hydrogen ion concentration in the soil:

$$pH = -\log [H^+] \qquad [4.5]$$

where $[H^+]$ is usually measured as the ionic activity of H^+ and represents the H^+ concentration, measured in centimoles of H^+ per kilogram of soil (cmol/kg), in the soil solution. Soil pH is a key measure of soil chemistry reflecting the *acidity* (high hydrogen ion concentration) or *alkalinity* (low hydrogen ion concentration or a soil dominated by hydroxyl $[OH^-]$ ions) of the soil. Since the negative logarithm of the hydrogen ion concentration is used, a lower numerical value indicates greater concentration or activity of H^+ ions relative to other ions and, thus, higher acidity. Moreover, on a logarithmic scale, the hydrogen ion concentration is ten times greater in a soil with a pH of 5.5 versus a soil with a pH of 6.5, and 100 times greater in a soil with a pH of 4.5 versus a soil with a pH of 6.5. The pH of most forest soils in eastern North America are within the range of 4.5 to 6.5. Values range from about 3 for extremely acidic peat soils to 11 for very alkaline mineral soils. A value of 7 is neutral. Soil pH is important largely because it is closely related to other soil chemical properties.

Cation exchange capacity (CEC) is a measure of the total amount of cations that may be adsorbed on the surface of soil particles; it is usually expressed in centimoles per kilogram of soil (cmol/kg). As such, it is a surrogate measure of a soil's ability to hold nutrients or positively charged organic compounds, such as some pesticides. Because of mineralogical differences and the relationship between soil particle size and surface area, cation exchange capacity is relatively low for sands (0 to 15 cmol/kg) and high for clays or organic matter (20 to 200 cmol/kg). Cation exchange capacity usually increases with increasing pH.

Organic matter has a high cation exchange capacity, so the amount of organic matter present in the soil is associated with the nutrient holding capacity of the soil. Management activities that affect the level of soil organic matter, therefore, affect the soil's capacity to hold nutrients. Similarly, acidic inputs to soils can lower pH and, concurrently, lead to a reduction in cation exchange capacity. Soil pH also affects the availability of nutrients to plants by determining the chemical form of nutrients in the soil. Most macronutrients (N, P, K, Ca, Mg) and micronutrients are in forms that plants can most readily absorb when the soil pH is between 6 and 7; they become less available when soil pH is outside this range.

Soil chemical properties are usually of interest in land management when fertilizers or certain pesticides are used or when such uses as waste disposal are contemplated. Under these circumstances, soil sampling and chemical analysis are usually performed to accurately assess site conditions for a specific treatment. For less intense management, most of the information required on soil chemical properties is of a descriptive nature and may be found in soil descriptions published by the Natural Resource Conservation Service in soil survey reports.

SOIL SURVEYS

Soil profiles are described in the field by soil scientists who are trained in soil description and classification. Soils are classified according to a hierarchical morphogenetic system, which is described in its entirety in Soil Taxonomy (USDA Soil Conservation Service, 1975). Morphogenetic means that the soil is described according to both its physical attributes and the processes by which it developed those attributes. Soil Taxonomy includes a key to identifying the taxonomic class of a soil based on a variety of soil properties and the presence or absence of diagnostic horizons. Diagnostic horizons that form at the surface of the soil profile are called *epipedons*. Examples of surface diagnostic horizons include horizons with organic matter mixed with mineral soil, forming a relatively thick A horizon, or *molic* epipedon. Traits such as thick organic deposits (O_i, O_e, O_a) make up the *histic* epipedons common to wetland soils. Other diagnostic surface horizons common to forested regions of eastern North America include *ochric* horizons, which are light-colored A horizons with low organic matter content, and *albic* horizons, which are light-colored E horizons resulting from the eluviation of aluminum and iron oxides and clay. There are also subsurface diagnostic horizons, such as B_s and B_{hs} horizons, which are enriched with iron and aluminum oxides and organic matter and referred to as spodic horizons; B_t horizons with illuviated clay form *argillic* horizons, which are common to many upland forest soils.

The broadest class used in describing soil type is soil order, followed by suborders, great groups, subgroups, families, and series. The smallest unit of classification, which is the unit of most concern in doing resource assessments, is the soil series. The criteria for differentiating series are the same as those used to differentiate other levels in the classification system, but the range of soil properties is restricted, so that the soils described at the series level not only look similar, but also respond similarly to management inputs. For this reason, information at the series level is often discussed in resource assessments as the basis for developing management alternatives.

In the vast majority of cases where resource assessments are conducted, the soils have already been identified and mapped in the field according to

TABLE 4.2 Tables and Interpretations Commonly Included in USDA Natural Resources Conservation Service Soil Survey Reports

Table Descriptions	*Information Included*
Climatological Data	Minimum and maximum temperatures, growing degree days, rainfall averages and extremes, snowfall, length of frost-free season, and growing season length
Soil Acreage	Total acreage and proportion of the survey area covered by various soil series
Prime Farmland	Soil series and slope classes that form the best agricultural soils of the county
Crop Yields	The productivity of various soil series for commonly grown agricultural crops of the county
Woodland Management and Productivity	Potential productivity for various tree species grown in the area, based on volume production and site index; management concerns by mapping unit, including erosion hazard, equipment limitations, seedling mortality, and windthrow hazard
Windbreaks and Environmental Plantings	Tree and shrub species suggested for planting on various mapping units
Recreational Development	Describes various soil properties, by map unit, that may limit the use of the land for campgrounds, picnic areas, playgrounds, and paths and trails
Wildlife Habitat	Evaluates mapping units for potential to provide habitat elements, such as vegetation type; evaluations for open land, woodland, and wetland wildlife habitat
Building Site Development	Identifies soil-related limitations for construction
Sanitary Facilities	Identifies soil-related limitations for septic systems, sewage lagoons, and landfills
Construction Materials	Identifies soil mapping unit potential for use as roadfill or source of sand, gravel, and topsoil
Water Management	Limitations for pond, dike or levee construction; information on drainage and irrigation potential
Engineering Index Properties	Soil texture, sand size, coarse fragments
Physical and Chemical Properties	Horizon thickness, bulk density, permeability, organic matter content, available water, pH, and erodibility

TABLE 4.2 *(Continued)*

Table Descriptions	*Information Included*
Soil and Water Features	Information on flooding, high water table, frost action, and estimates of runoff from precipitation, based on hydrologic soil groups

procedures and standards developed by the NRCS. Soil series are mapped across the landscape in what are known as *soil mapping units.* Mapping units can be made up of a single series, a series of a certain slope class, or a combination of two or more series that either have similar properties and uses or which occur in such a mixed pattern on the landscape that they cannot be easily separated at the scale to which the map is made.

The results of these classification and identification efforts are published in *soil survey* reports. For most counties in the United States, these reports are available from local USDA Natural Resource Conservation Service offices or Government Documents sections in many libraries. The soil survey reports include descriptions for all mapping units and soil series occurring in the county, physical and chemical properties of the soils, and maps describing the spatial distribution of different soil types within a landscape. In addition, much of the information is available electronically, including both maps and soil series descriptions with associated physical and chemical properties.

INTERPRETATION OF SOIL SURVEY INFORMATION

County soil survey reports provide a great deal of interpretive information for each mapping unit, including tables of properties, limitations, capabilities, and potential uses for the different soil mapping units in a given area (Table 4.2). They also group into categories those soils that are expected to behave similarly for a stated use. Each soil survey report contains a detailed discussion of the data and assumptions that were used in evaluating the soil and forming the groups. At the broadest scale is the Land Capability Classification, shown in Table 4.3. These classes, signified by Roman numerals I to VIII, indicate progressively greater limitations, ranging from soils for which there is essentially no limitation in use (Class I) to soils with limitations that preclude their commercial use (Class VIII; see Calus, 1996). Land Capability Classes I through IV are suitable for commonly cultivated crops, whereas Classes VI and VII may be used with special treatments. Class V soils are too wet for cultivation, but Classes II through VII may all have wet soil as a potential management concern. The system generalizes a soil map by grouping soils according to soil properties and landscape features. Although the classes remain the same from one survey to the next, the criteria for classifying a particular soil depend on local conditions. These classification criteria are

TABLE 4.3 Land Capability Classification Used for Soil Surveys in the United States

Land Capability Class	*Soil Limitation*
I	Soils have few limitations restricting use
II	Soils have moderate limitations that reduce the choice of plants or require moderate conservation practices
III	Soils have severe limitations that reduce the choice of plants or require special conservation practices, or both
IV	Soils have very severe limitations that reduce the choice of plants or that require special management or both
V	Soils are not likely to erode, but have other limitations that make them unsuitable for cultivation
VI	Soils have very severe limitations that make them generally unsuitable for cultivation
VII	Soils have very severe limitations that make them unsuitable for cultivation
VIII	Soils and miscellaneous areas have limitations that nearly preclude their use for commercial crop production

usually designed to indicate benefits in relation to management inputs for crop production. The classification parameters vary geographically but are usually based on such properties as soil depth, stoniness, wetness, salt content, slope, and available water-holding capacity. The limiting factors for a particular soil can usually be determined by examining the other tables in the county soil survey report.

There are also more specific soil classification systems that are used to describe soil suitability for particular uses, such as woodland management and productivity. For these, each soil series and mapping unit that is capable of supporting forest communities is rated according to its potential productivity and associated management concerns. Productivity is usually indicated for individual tree species based on the expected volume growth at biological rotation age (the age with maximum average annual production) or site index (average height of dominant and codominant trees at some designated index age). Management concerns are rated on a three-level scale: (1) slight, where soil properties are generally favorable for a given use with limitations that are minor and easily overcome; (2) moderate, where limitations may be overcome by planning, design, or special maintenance; and (3) severe, where soil properties are unfavorable for a given use and limitations can be mitigated only by costly reclamation, design, or intensive maintenance. Management concerns usually include (1) erosion hazard, based on soil erodibilty and slope; (2) equipment limitation, based on slope, wetness, surface soil texture and structure; (3) expected seedling mortality for natural and planted seedlings; and (4) windthrow hazard, based on factors affecting rooting depth such as depth to the water table or other root restricting layer.

Management often affects the soil erosion rate because of the resulting changes in soil cover, structure, and porosity. *Erosion* is the detachment and movement of soil material by water or wind. This transport results in soil loss from the source area and an accumulation of sediment in the area where the material is deposited. The *universal soil loss equation* (Wischmeier and Smith, 1978) recognizes the key factors influencing erosion. The expected soil loss, in tons per acre per year, is estimated by:

$$A = R \cdot K \cdot L \cdot S \cdot C \cdot P \qquad [4.6]$$

where A is the expected soil loss in tons per acre per year, R is a rainfall and runoff factor given in erosion index units per year, K is a factor indicating erodibility, L is a slope length factor, S is a slope steepness factor, C indicates the cover and crop management being used, and P is a factor to account for conservation practices being applied. The universal soil loss equation is rarely applied in its entirety to forest soils, although the concepts incorporated into the equation are often used in resource assessments in forested landscapes. The susceptibility of the soil to erosion (K) is estimated by the *erosion factor,* which is sometimes referred to as the *K-factor* because of the notation used in the universal soil loss equation (Wischmeier and Smith, 1978). The value of K is experimentally determined for a particular soil as the rate of soil loss as measured under standardized conditions (Dissmeyer and Foster, 1984); it indicates the erodibility of a soil under harsh conditions. The susceptibility of a soil to erosion increases with increasing values of the K-factor, which ranges from a low value of 0.02 to a high value of 0.69; more easily eroded soils have K-factors of 0.30 and higher. These values are regularly published for individual soil types in county soil survey reports. The soil properties with the greatest influence on erodibility are infiltration capacity and structural stability, both of which affect the movement of water in soil. Thus, the K-factor is largely dependent on the percentage of sand, silt, and organic matter content in the soil.

The universal soil loss equation is used to estimate soil losses from croplands based on soil properties and crop management. Dissmeyer and Foster (1984) modified the crop factor (C) in the universal soil loss equation to improve its relevance to forested soils if erosion estimates have to be quantified during a resource assessment (Dissmeyer, 1994). Because forest soils are rarely cropped intensively, many soil survey reports do not use the universal soil loss equation approach to rate erosion hazards for forestry applications. However, because of the equipment used in management and the concerns about slope, soil-mapping units are rated using the K-factor combined with the slope to estimate erosion hazard on a scale of slight, moderate, and severe. Soils on steeper slopes with high erodibility factors represent a much greater concern than soils that occur on flat areas and have low erodibility. If forest cover is removed and the protective forest floor is disturbed, erosion may be severe until the site is revegetated.

Soils are also placed in wind erodibility groups based on properties that affect their resistance to wind erosion after cultivation. Sandy-textured soils that lack structure are the most easily eroded by wind, unless these soils have a high proportion of coarse fragments on the surface or a high moisture content. Loams and clays pose less of a problem if management practices to control minor erosion are employed. Windblown erosion is of concern in forested landscapes when dry soils are devegetated and then exposed to windy conditions. Dunes along lake margins constitute one example of such sensitive features.

With such a large amount of information, the question often arises: How much soil information is necessary to include in a resource assessment? Each interpretive table in a soil survey report provides important information. Some of the interpretive tables include detailed technical information, some apply to the development of management alternatives, and others serve to identify areas that may have potential for economic development. The soil properties play a large role in determining both the types of activities that may be implemented within the landscape and the most acceptable methods for implementing particular activities. The choice of information to include in an assessment process depends on the management goals and objectives and the management alternatives being considered. The soil survey interpretive tables that address the major concerns in forest land management have been used as examples here; particular interpretive information is discussed in later chapters for specific landscape components.

COARSE WOODY DEBRIS

Coarse woody debris is the term used to describe larger, dead woody material lying on or near the forest floor. This includes branches, logs, and wood chunks that have fallen, as well as snags that are still standing in the forest. Anything larger than 0.64 cm (0.25 inch) in diameter is usually considered to be coarse woody debris; smaller material is considered part of the forest floor. Coarse woody debris accumulates in forests either naturally, through tree and branch mortality and subsequent blowdown, or as logging slash or material that is felled, but not removed during a harvesting operation. Large coarse woody debris that is highly decayed and incorporated into the soil is sometimes termed *soil wood* (Figure 4.3).

Coarse woody debris is important in forests for a number of reasons. It represents the fuel load on the ground that may be easily combustible in a wildfire or a prescribed fire, which is a particular concern in some fire-susceptible forest types. But coarse woody debris has also become recognized for its ecological role in providing habitat for microorganisms and insects, a substrate for lichens and mosses, a preferred seedbed and rooting medium for some

FIGURE 4.3 Soil wood is often apparent only after disturbance, such as clear cutting followed by soil scarification.

tree species (Figure 4.4), and habitat for reptiles, small mammals, and some cavity-nesting birds. Thus, the amount of coarse woody debris in an ecosystem may be of great concern in managing forested landscapes, especially when there are management goals to increase wildlife habitat, restore an ecosystem to an old-growth condition, or manage the fuel load through prescribed burning or harvest.

Because of the increased recognition of the importance of coarse woody material in forested ecosystems, management activities designed to regulate the amount of coarse woody debris may necessitate measurement of the pre- and posttreatment levels of coarse woody debris in the ecosystem. The *line intercept sampling* method as developed by Warren and Olsen (1964) and extended by Van Wagner (1968) was designed to estimate the volume or biomass of logging slash and the fuel loads in forest ecosystems. In line intercept sampling, one or more transects are established by laying a tape along the forest floor. Whenever the tape crosses woody debris greater than

(a)

FIGURE 4.4 Coarse woody debris provides an important microsite for the regeneration of some tree species: (a) eastern hemlock trees growing from coarse woody debris in the Grand Island Research Natural Area, Alger County, Michigan, and (b) a 17 cm diameter yellow birch that regenerated in an old white pine stump (the stump has since rotted away).

a predetermined minimum diameter, the diameter of the woody debris is measured. Van Wagner (1968) developed the following equation to estimate the volume of coarse woody debris in m³/hectare:

$$V = \frac{(\pi^2) \cdot \left(\sum_{i=1}^{l} d_i^2\right)}{8 \cdot L} = (1.234) \cdot \frac{\sum_{i=1}^{l} d_i^2}{L} \qquad [4.7]$$

where V is the estimated volume of coarse woody debris (m³/hectare), d_i is the diameter (cm) of the ith piece of coarse woody debris intersected by the transect, and L is the length of the transect (m). Note that this method does not require any length measurements of the pieces of coarse woody debris that are intersected by the transect. Moreover, if a piece of debris is crooked and intersects the transect more than once, it is counted each time it intersects the transect and the diameter is measured and recorded for every intersection.

Figure 4.5 illustrates an example transect that intersects coarse woody debris four times over a 15 m distance. The estimated volume of coarse woody debris per hectare is given by:

FIGURE 4 *(Continued)*

$$V = \frac{(\pi^2) \cdot (7.6^2 + 4.4^2 + 2.5^2 + 1.0^2)}{(8) \cdot (15)}$$
$$= (1.234) \cdot \frac{84.37}{15} \qquad [4.8]$$
$$= 6.9\ \text{m}^3/\text{hectare}$$

The estimated dry weight of coarse woody debris per hectare may be estimated if an average specific gravity can be assumed. The estimated dry weight in tons/hectare is given by:

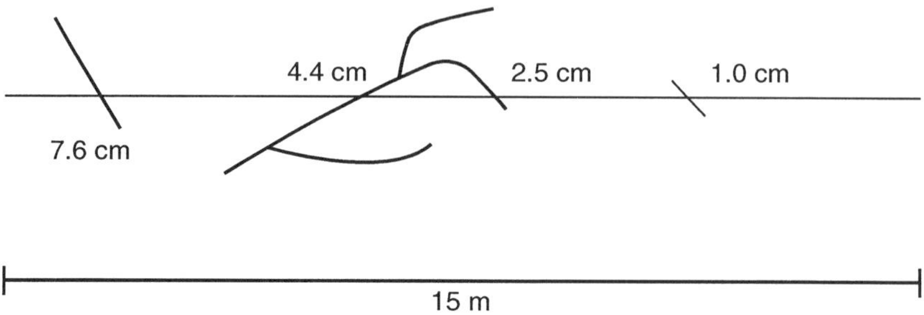

FIGURE 4.5 An application of line transect sampling and a sample transect in a forested ecosystem.

$$W = V \cdot S \qquad [4.9]$$

where W is the estimated dry weight in metric tons per hectare, V is the volume of coarse woody debris estimated using Equation [4.7], and S is the average dry weight (g/cm^3) of the coarse woody debris on the site. For the example transect in Figure 4.5, if the average dry weight of the coarse woody material on the site is 0.4 g/cm^3, then the estimated dry weight of coarse woody debris in the ecosystem is given by:

$$\begin{aligned} W &= (6.9) \cdot (0.4) \\ &= 2.8 \text{ tons/hectare} \end{aligned} \qquad [4.10]$$

There were several assumptions made in the derivation of Equation [4.7]: (1) all pieces of coarse woody debris are cylindrical; (2) all pieces of coarse woody debris are horizontal; and (3) the pieces of debris are oriented in random directions. All of these assumptions are often violated in practice. Pickford and Hazard (1978) examined a more realistic case in which the pieces of debris were assumed to be tapered rather than cylindrical. The effects of nonhorizontal debris were examined by Brown and Roussopoulos (1974) and the consequences of nonrandom orientation were examined by Bell and associates (1996). The latter investigators concluded that Equation [4.7] led to unbiased estimates of the volume of coarse woody debris per hectare if, instead of establishing transects of single straight lines, sample units consisting of three straight line segments arranged in an equilateral triangular pattern are used.

Brown (1974) noted that coarse woody debris could be segregated by diameter class, which simplifies tally methods during field sampling. He included a correction factor for nonhorizontal debris in the following equation for estimating the dry weight in English tons per acre of coarse woody debris; this equation provides the standard method for estimating downed woody material in the western United States:

$$W_i = \frac{(11.64) \cdot n \cdot d_i^2 \cdot a \cdot c \cdot s}{N \cdot L} \qquad [4.11]$$

where W_i is the estimated dry weight of coarse woody debris in English tons per acre for all material in the ith size class, the constant (11.64) is a combination of several unit conversion factors, n is the number of pieces of material in the ith size class intersecting the transect, d_i is the diameter midpoint of the ith size class, a is a correction for nonhorizontal angle, which may vary by size class of material, c is a slope correction if the transect is not on level ground, S is the specific gravity of material in the ith size class, and N is the number of transects of length L established in the ecosystem of interest. The correction factor for the non-horizontal angle of debris is simply the secant of the orientation angle, and a slope correction factor is used to adjust ground distance to horizontal distance:

$$c = \sqrt{1 + \left(\frac{\text{percent slope}}{100}\right)^2} \qquad [4.12]$$

Nested transects may be used to estimate the volume or dry weight of material of different sizes. If, for example, a total transect length of 15 m is used, 0 to 1.0 cm diameter coarse woody debris may be counted only in the first meter, 1.0 to 3.0 cm diameter coarse woody debris may be counted in the first 2 m, 3.0 to 7.0 cm diameter coarse woody debris may be counted in the first 4 m, and coarse woody debris greater than 7.0 cm in diameter may be measured over the entire 15 m transect. Note that the diameters of material in the largest size class must be all be measured and, in Equation [4.11], the quantity ($n \cdot d_i^2$) must be replaced by the sum of the squared individual diameters in order to estimate the dry weight of material in the largest size class.

Eventually, wood deposited on the forest floor is incorporated into the soil. This soil wood is still identifiable as wood for many years before it is extensively decomposed. Soil wood provides additional water-holding capacity to the soil and is a host to many soil organisms; tree roots are often concentrated in decomposing soil wood. Both the volume and the weight of soil wood may be estimated using line transect sampling with some minor modification. Owing to the extent of decomposition, for example, soil wood can no longer be assumed to have a circular cross section. To account for this, the width and depth of the material must be measured at each intersection with the transect. These measurements are substituted for the diameter measurements in Equation [4.7]:

$$V = \frac{(\pi^2) \cdot \left[\sum_{i=1}^{l} \left(\frac{\text{width}_i^2 + \text{depth}_i^2}{2}\right)\right]}{8 \cdot \text{L}} = (1.234) \frac{\sum_{i=1}^{l} \left(\frac{\text{width}_i^2 + \text{depth}_i^2}{2}\right)}{L} \qquad [4.13]$$

where $width_i$ and $depth_i$ are the width (cm) and depth (cm), respectively, of the ith piece of soil wood encountered along the transect.

In conducting an inventory of soil wood, it is often best to segregate the material into white and brown rot systems. Brown rot fungi decompose only cellulose, leaving residue that has a high lignin content. White rot fungi, on the other hand, decompose both cellulose and lignin, leaving a white residue. Brown rot systems are usually associated with conifers and oaks, whereas white rots decay other hardwoods. Brown rotted material stays in the soil for very long periods, as evidenced by many of the old hemlock and white pine stumps in the northern temperate forest of the Lake States, New England, and southern Canada that date to harvesting activity in the 1800s.

To estimate the weight of soil wood in an ecosystem, Equation [4.11] may again be used, but with modification because rotten wood cannot be assumed to have a circular cross section:

$$W = \frac{\sum_{i=1}^{I} \left[(11.64) \cdot \left(\frac{\text{width}_i^2 + \text{depth}_i^2}{2} \right) \cdot a \cdot c \cdot s_i \right]}{N \cdot L} \qquad [4.14]$$

Note that the specific gravity of soil wood may vary more than the specific gravity of sound wood on the soil surface. The specific gravity of soil wood may vary according to the extent of decay, which is largely a function of the length of time since the material has fallen, and the type of decay. An average specific gravity may be determined for different classes of soil wood if necessary to simplify the procedures.

SUMMARY

Soils are complex, but predictable, natural bodies scattered across the landscape. To a large extent, their physical and chemical properties control plant growth, or primary productivity, of the ecosystem. This, in turn, affects many of the potential uses of the landscape. Soil physical properties also control the movement of water within an ecosystem and between ecosystems across the landscape. For these reasons, the maintenance or improvement of soil productivity is a necessary consideration in developing or evaluating forest management alternatives. Soil survey reports contain descriptive as well as interpretive information that is useful in planning and assessing potential management activities. Finally, when management goals include increasing wildlife habitat, restoration of an ecosystem to an old-growth condition, or management of the fuel load through prescribed burning or harvesting, it may be necessary to assess the coarse woody material on the forest floor or the amount of woody material incorporated into the mineral soil.

APPLICATIONS

4.1 How are soil structure and drainage class related? Can changes in soil structure caused by management activity affect drainage class? What is the relationship between soil structure and soil chemical properties, such as cation exchange capacity?

4.2 Obtain a USDA Natural Resources Conservation Service soil survey report from a county near you. For four different soil mapping units, identify the slope class and the soil series located within the units. What percentage of each unit consists of each soil series? What are common uses of the sites within the different mapping units?

4.3 For a soil series within one of the mapping units identified in item 4.2, describe the typical soil profile. What horizons are typically present? What is the textural class of each horizon? What is the typical soil structure of the surface horizons? What is the typical drainage class of the series?

4.4 For a soil mapping unit identified in item 4.2 above as being a woodland soil, identify the forest tree species that may be expected to have the greatest productivity on soils within the mapping unit. What is the erodibility of the soils within the mapping unit? Are there any seasonal restrictions on equipment use within the mapping unit?

4.5 Identify a square-mile survey section (640 acres). Produce a map of the soil mapping units within the area. What are some possible management concerns associated with each mapping unit? Are certain mapping units better suited to agricultural crop production than others? Are some typical woodland soils?

4.6 Use Brown's (1974) equation (Equation [4.11]) to estimate the tons/acre of coarse woody material in a forest near you. Use line intercept sampling and establish three 50-foot-long transects in the forest to measure coarse woody debris. Assume an average specific gravity of 0.4.

Chapter

5

Climate

Climate is the long-term pattern of temperature, precipitation, wind, solar radiation, and other ambient conditions. The short-term fluctuations in these conditions are collectively referred to as *weather*. Understanding the climatic conditions in a landscape is critical to understanding many of the major factors affecting wildlife habitat and the growing conditions of plants. Many models of plant succession, for example, implicitly or explicitly assume that climatic factors determine the geographic range limits of many species (i.e., Botkin, Janak, and Wallace, 1972; Shugart, 1984); Root (1988) demonstrated that climatic conditions control the distribution of many avian species. Elevational gradients in plant community composition and structure in mountainous regions are largely dependent on elevational gradients in climatic conditions; in the Great Lakes region, the presence of large bodies of water moderates temperature and humidity extremes and greatly affects snowfall, which in turn affects both plant and animal communities. Climate, to a large extent, determines which species can survive and reproduce in a given area over an extended period of time. Year-to-year fluctuations in growth and reproductive success are largely due to shorter-term differences in weather. It is important to understand both the range and the normal variability in weather conditions in order to evaluate the risks associated with establishing a certain agricultural crop or favoring a certain tree species, protecting a floodplain to ensure water quality and reduce erosion (see Chapter 6), designing a plan to protect the habitat of an endangered species, or projecting future forest growth and productivity.

The science of climatology investigates the various processes associated with climate, atmospheric dynamics, and feedback mechanisms between the atmosphere, oceans, and land. The details of these processes are beyond the scope of this text, but they certainly have an impact on the results of landscape management activities. A general understanding of these factors can result in better informed land management decisions. Detailed descriptions of the principles of climatology and meteorology, or the study of weather, can be found in Critchfield (1974) and McIlveen (1992), among others. In most resource assessment exercises, the effect of ambient conditions on flora and fauna are of interest. Ecophysiology is the study of the ecological consequences of differing ambient conditions as mediated by the physiological responses of plants and animals. Texts such as Landsberg (1986) and Kozlowski, Kramer, and Pallardy (1991) discuss the principles involved in understanding and predicting these responses.

In developing and interpreting resource assessment reports, it is important to understand the numerous variables used to describe climate and weather, how to use and interpret this information, and where to find descriptive information for a given area. The two most important climatic factors are temperature and moisture, though a number of different variables have been used to describe climatic conditions. In many situations, the management decision-making process will not require a detailed description of local climate and there will only be a minimal discussion of climate and weather in the assessment document. For some management goals and objectives, climate and weather conditions will have to be examined more extensively during resource assessment in order to support the management planning effort.

TEMPERATURE

Temperature, moisture, and light are among the most important ambient conditions affecting the physiological responses of plants and animals. The National Weather Service, within the National Oceanic and Atmospheric Administration in the Department of Commerce, maintains a national network of weather monitoring stations throughout the United States and Environment Canada maintains a similar network in Canada. The primary data from these networks are the basis for the description of climate at different locations and, coupled with remote sensing information, form the basis for weather forecasting in the United States. Temperature and precipitation are monitored at every station, and some stations also monitor additional variables. The primary data are maintained by the National Climate Data Records Center in Boulder, Colorado; printed copies of data from individual stations are usually available within a few months of observation at libraries that are repositories of United States government documents. Electronic copies of all historic daily records from every station are available from commercial sources; these data are updated yearly. Summaries of the primary climatic data are available in map form for the United States (Baldwin, 1983) and in greater detail for individual states (e.g., Eichenlaub et al., 1990) or regions.

Climatic atlases are also becoming available in electronic form (e.g., Ollinger, 1995), which facilitates entry into geographic information systems (see Chapter 13) and integration with other sources of information.

Temperature is a measure of the average kinetic energy of a body or system. This energy is important because it is literally the power that drives many ecosystem processes, ranging from plant growth to the hydrologic cycle. The most common variables used to describe the temperature regime of a given area are the *minimum, average,* and *maximum daily temperatures.* These are often presented as average daily values for a month, but shorter summary periods are sometimes used. Longer time intervals are sometimes used as well; the averages for May through September may be combined, for instance, to estimate average conditions during the growing season, or the information may be summarized to coincide with phenotypic events such as growth initiation or leaf abscission. Since the primary daily data are easily obtained, it is relatively easy to develop customized reports for a given landscape if there is a need for doing so.

Figure 5.1 illustrates the daily minimum, average, and maximum temperatures by month for the Houghton, Michigan, station observed from 1951 to 1980. Thirty-year mean values such as these are commonly used to describe

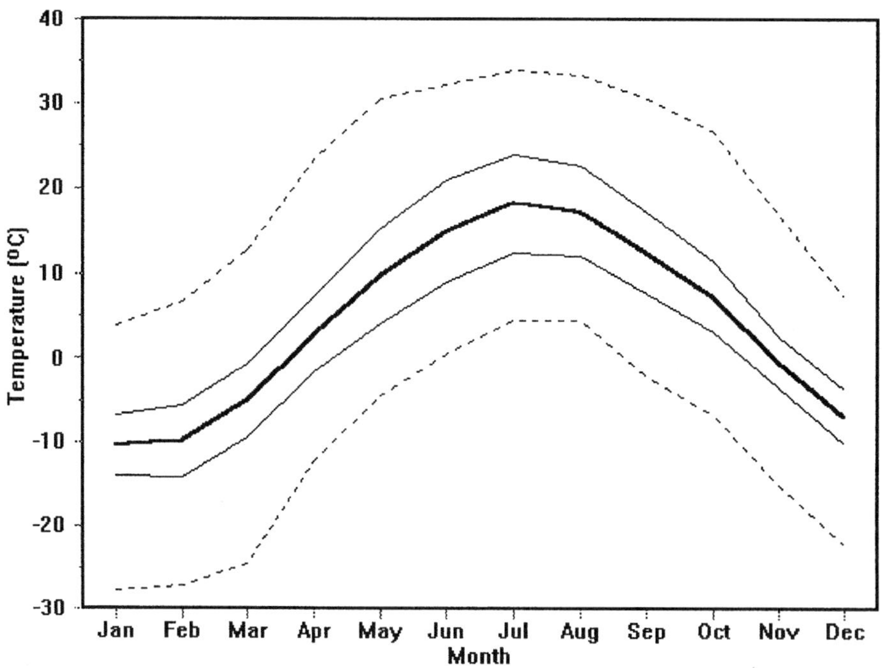

FIGURE 5.1 Thirty-year (1951–1980) average monthly temperatures for Houghton, Michigan (*heavy line*), average monthly minimum and maximum temperatures (*solid lines*), and monthly temperature extremes exceeded two years in ten (*dotted lines*).

climate. In addition to the National Weather Service reports, these data area often also included in other sources, such as the Natural Resource Conservation Service county soil survey reports (see Chapter 4).

The long-term minimum, average, and maximum temperatures provide a good indication of expected growing conditions and are often used as a basis for management decisions. The USDA Plant Hardiness Zones, for example, which are commonly consulted in the selection of plant cultivars for use in a particular location, are based on minimum January temperature isotherms (Cathey, 1990). The average daily values, however, do not indicate the degree of year-to-year variability in weather patterns that may occur. Since plants and animals are limited by extreme conditions, rather than mean conditions, it is often important to understand the variability in the average daily minimum and maximum temperatures. One method of indicating this variability is to present temperature extremes that are expected to be exceeded only every few years. In Figure 5.1, for example, the minimum and maximum daily temperature extremes that are expected to be exceeded in only two out of every ten years are included. These values help evaluate the risks of extreme conditions that may affect plant and animal life.

A third method used to present and evaluate temperature extremes is based on the number of days on which certain thresholds are exceeded. At the Houghton, Michigan, station, again, Schwenner (1991) noted that the daily minimum temperature is greater than 0°C on at least 102 days in nine out of ten years. There are 132 days when this threshold is exceeded in five out of ten years, and, in one out of ten years, there are 155 days when the minimum daily temperature is greater than 0°C. These values can be combined (Figure 5.2) to estimate the likelihood of the daily minimum temperature exceeding 0°C for differing numbers of days for a given year. Sometimes these numbers of days are referred to as the length of a *frost-free period* or the expected *growing season length*. These types of climatic summaries are helpful in describing and evaluating opportunities for recreational development or planning the operation of seasonal facilities. Another method to present similar information uses the *expected freeze dates*. Schwenner (1991) also notes, again for the station at Houghton, Michigan, that May 21 is the average date of the last day in spring when the minimum temperature is below 0°C in five out of ten years; the average date in fall when the daily minimum temperature drops below 0°C is October 1 in five out of ten years. This type of information may be used in evaluating the risks of frost damage to young plants established on different planting dates or in determining likely windows of opportunity for planting or herbicide application. Clearly, the climatic information to consider in a resource assessment depends on the particular issues and management goals in a given landscape.

A final method to describe temperature uses a measure of cumulative temperature called a *degree day*. Degree days are calculated by summing the numbers of degrees per day by which a given threshold temperature is exceeded over a time interval, such as a week, a month, or a year. An example

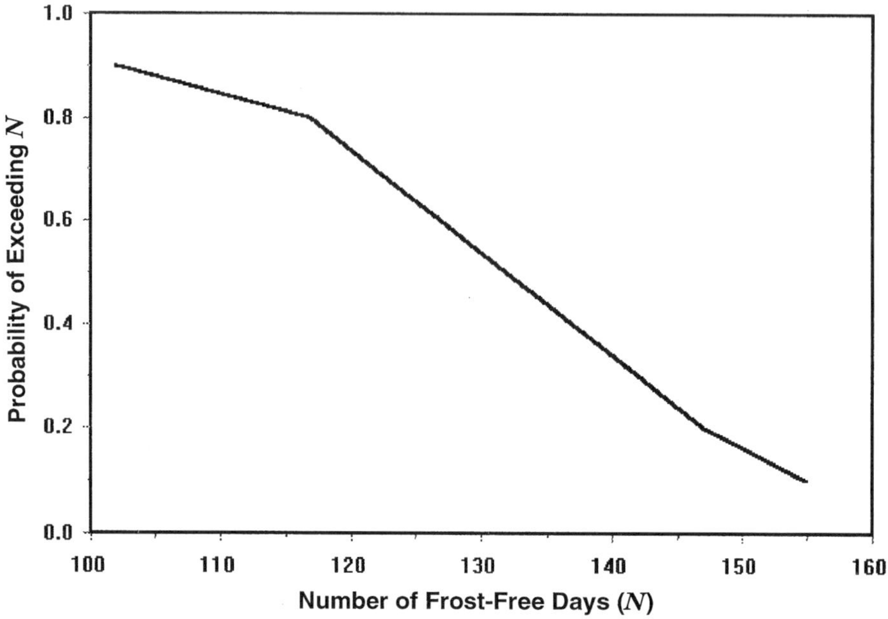

FIGURE 5.2 Probabilities of the frost-free period exceeding a given number of days; based on 30-year (1951–1980) daily minimum temperature records for Houghton, Michigan.

of this concept is *growing degree days.* The calculation of growing degree days for a time interval is based on the length of the interval and the difference between the average temperature and a threshold temperature over the time interval:

$$\begin{aligned} GDD_K &= (T_A - K) \cdot n \qquad && \text{if } T_A > K \\ &= 0 && \text{if } T_A \leq K \end{aligned} \quad [5.1]$$

where GDD_K is the growing degree days calculated for a threshold temperature K, T_A is the average daily temperature over the time interval, and n is the number of days in the interval. The threshold temperature, K, is selected to represent the minimum temperature at which plant physiological or growth processes can occur. The choice of K depends on the characteristics of the major species or processes of concern; common values of K are 4.4°C (40°F), 7.2°C (45°F), and 10°C (50°F). For example, if the month of July, which has 31 days, has an average daily temperature of 20°C, the growing degree days calculated using a base of 4.4°C are:

$$\begin{aligned} GDD_{4.4} &= (20 - 4.4) \cdot 31 \\ &= 483.6 \approx 484 \text{ growing degree days (4.4°C basis)} \end{aligned} \quad [5.2]$$

The same month would have had 310 growing degree days calculated using a 10°C basis. Growing degree days have been related to seasonal growth pattern (e.g., Perala, 1985; Jones, et al. 1991) and the occurrence of phenological events such as the timing of budbreak (e.g., Scalabrelli and Couvillon, 1986), the initiation of flowering, or the onset of senescence.

Heat sums are also used in the estimation of heating or cooling requirements in the planning of construction projects, such as the development of a greenhouse facility. *Cooling degree days* are similar to growing degree days, except that they are usually calculated on an 18.3°C (65°F) basis in the United States. Cooling degree days may be used to evaluate air conditioning requirements. A related measure is *heating degree days*. Heating degree days are like growing degree days or cooling degree days, except that they are based on temperatures that fall below a threshold, rather than temperatures that rise above a threshold. Heating degree days are given by:

$$\begin{aligned} HDD &= (18.3 - T_A) \cdot n && \text{if } T_A < 18.3°\text{C} \\ &= 0 && \text{if } T_A \geq 18.3°\text{C} \end{aligned} \qquad [5.3]$$

where HDD is the heating degree days accumulated over a time interval, T_A is the average temperature over the time interval, and n is the length of the interval in days. As noted previously, 18.3°C (65°F) is generally used as the basis of both heating and cooling degree days in the United States; if a different basis is used, the value of 18.3 in Equation [5.3] is simply replaced with the value of the basis that is used.

Even though temperature is a relatively simple quantity to measure, there are many different ways to summarize temperature that may be relevant to resource assessment and land management decision making. The appropriate measures to include depend on the management goals and the issues and concerns being addressed during the assessment process. Simple average daily temperatures or average daily extremes are probably the easiest to interpret, but growing season lengths or average frost dates are probably more important in selecting plant species or planning planting efforts. Heat sums may be used to evaluate seasonal growth or to select species to plant, but they may also be used to examine energy requirements for residential development and some types of recreational facilities.

MOISTURE

The second major factor affecting flora and fauna is moisture. As in the case of temperature, there are numerous measures of moisture and moisture availability that may be of use in evaluating management alternatives. The basic climatic variable describing the atmospheric moisture input to an area is *precipitation*. In the United States, precipitation is commonly measured using a standard 8-inch rain gauge (Hewlett, 1982). Precipitation is expressed in mm or inches; snowfall or snow depth is often reported on a water-equivalent basis, which is the depth of water that will remain when the snow melts. This is used because the water content of snow varies greatly with temperature; snow falling at or near freezing temperature has a much higher water content in comparison with snow falling at a lower temperature.

As with temperature, extreme values of monthly precipitation may be used to define the range of precipitation amounts that an area is likely to receive at different times of the year. These extremes are usually expressed as lower and upper amounts, which are likely to be surpassed with some frequency, such as two years out of ten. These values provide an objective definition of what actually constitutes an unusually dry or an unusually wet period. The risk of dry periods that could impact plant or forest regeneration efforts can be quantitatively expressed and considered in the planning process in the same way that probabilities of different growing season lengths may be expressed and considered (Figure 5.2). Similarly, the likelihood of extremely wet periods that may preclude equipment operations within an area may be considered and, if it is judged to be necessary, road construction or stream crossings may be altered to accommodate wetter-than-normal conditions that may be likely to occur over the expected life of the road. Schwenner (1991) gave the average monthly precipitation values for Houghton, Michigan, and the extreme monthly precipitation amounts that have been surpassed in two of ten years (Figure 5.3). As with temperature, these values are often estimated from the 30-year National Weather Service summaries.

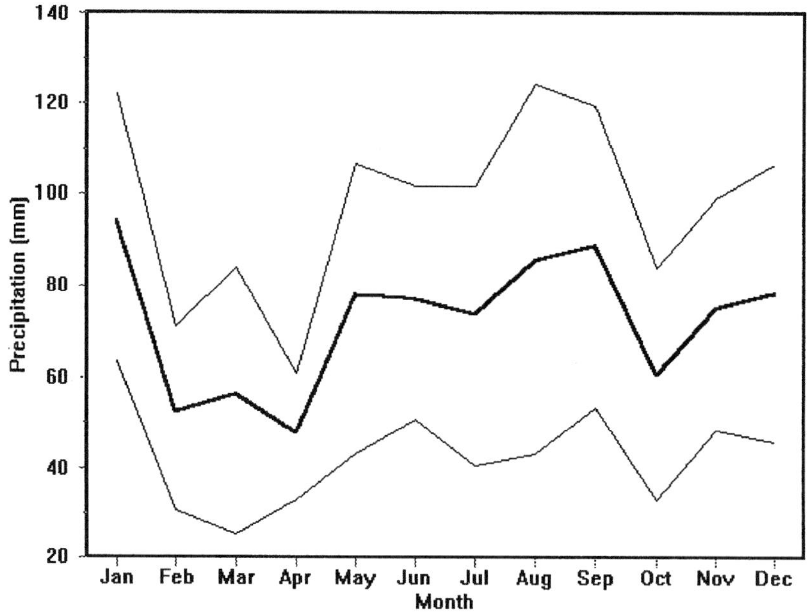

FIGURE 5.3 Thirty-year (1951–1980) average monthly precipitation values (*heavy line*) and extreme values (*solid lines*) exceeded two years in ten for Houghton, Michigan.

The precipitation during some portion of the year is often an important consideration in management planning. Growing season precipitation, rather than total annual precipitation, may be more important in controlling plant productivity. The definition of the time period constituting the growing season varies from place to place; in the Lake States summaries of climatic variables for the growing season are usually based on the months of May through September. The amount of snowfall (Figure 5.4) or the average depth of the snowpack may be important variables to consider in designing buildings to withstand snowload, or in the development of ski or snowmobile trails, in different regions. Snow depth is also important because it may limit wildlife movement and food availability for some animals, including important game species such as white-tailed deer. The depth of the snowpack in February or March defines the large flux of water that will pass through a landscape when the snow melts. Since melting occurs in a much shorter time period than accumulation, the snowmelt period is often the time that maximum water flux occurs, as do associated high water levels in streams and lakes. In other words, several months of precipitation may pass through a landscape in a two- to three-week period during the spring. Obviously, a flux of water this large can impact equipment operations, stream crossings, planting, and many other land management activities.

The amount of precipitation falling in a rainfall event ranges from a barely measurable trace to several cm. Incoming precipitation is intercepted by the

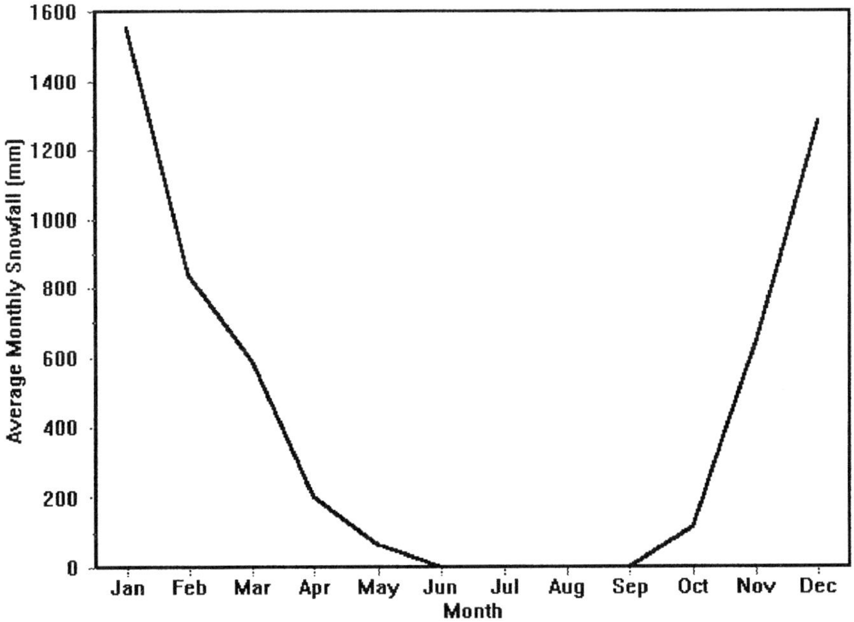

FIGURE 5.4 Thirty-year (1951–1980) average monthly snowfall values for Houghton, Michigan.

plant cover and, until the interception capacity of the canopy is exceeded, very little precipitation reaches the ground. Liechty and associates (1990), for example, collected throughfall, which is precipitation that passes through the canopy and reaches the forest floor, in a sugar maple-dominated temperate deciduous forest. They found that essentially no water reached the forest floor until the amount of precipitation exceeded about 15 mm.

The *intensity* of a precipitation event is the rate at which precipitation is delivered to the ground, which is usually expressed in millimeters per hour (mm/hour). Precipitation intensity is relatively difficult to measure and extremely variable. Often, the number of days on which precipitation exceeds some threshold amount is used as an indicator of the average intensity of rainfall events. Commonly reported values in the United States are the number of days on which precipitation exceeds 2.5 mm (0.1 inch) or 6.25 mm (0.25 inch). This concept can be extended in order to identify the maximum precipitation amounts that can be expected during rainfall events within a given area. The intensity of precipitation events is of concern in defining maximum levels of streamflow in hydrological evaluations and watershed studies (Chapter 6). The likelihood of extreme precipitation events is usually expressed through the concept of a return period, which is the inverse of the probability of the occurrence of an event:

$$T_r = \frac{1}{P_o} \qquad [5.4]$$

where T_r is the expected length of time between occurrences of an extreme event, usually expressed in years, and P_o is the probability of an event of specified size occurring during a given year. For a precipitation event that is so large that there is only a 1% chance that it will occur in a given year, the probability of occurrence is 0.01 and the return period of the event is given by:

$$\begin{aligned} T_r &= \frac{1}{0.01} \\ &= 100 \text{ years} \end{aligned} \qquad [5.5]$$

A 100-year storm, therefore, is a rainfall event with such a large amount of precipitation that it is expected to be exceeded only once a century. Building codes may require construction along lakes and rivers to be above the 100-year floodplain, which is a related concept concerning the high water mark of a body of water that is expected to be exceeded only once every 100 years.

For most of North America, the meteorological records do not extend for a sufficient time period to obtain a true estimate of the likelihood of really unusual events. Hewlett and Nutter (1969) give the following equation for use in estimating T_r using existing precipitation records:

$$T_r = \frac{n + 1}{m} \qquad [5.6]$$

where n is the number of years in the record and m is the rank of the event being considered. If there is a 20-year precipitation record for a given location and the third-largest precipitation event in the record deposited 50 mm of water in 24 hours, the estimated return period for a storm of this size is:

$$\begin{aligned} T_r &= \frac{20 + 1}{3} \\ &= 7 \text{ years} \end{aligned} \qquad [5.7]$$

Conversely, the probability of an event of this size occurring in any given year may be estimated using Equation [5.4], which implies that the likelihood of the event's occurring in any given year is the inverse of the return period. For the example, the probability of experiencing a storm event depositing greater than 50 mm of precipitation within a 24-hour period is 0.14 for any given year. The probability of experiencing an event of a given size over an extended period of k years may be calculated from the estimated value of P_o:

$$P_k = 1 - (1 - P_0)^k \qquad [5.8]$$

where P_k is the probability of an event of the given size occurring during a k-year time period and P_0 is the probability of an event of the given size occurring in a one-year time period. For the example, the probability of experiencing an event depositing 50 mm of precipitation within a 24-hour period during a single year is 0.14, which implies that the probability of experiencing an event depositing 50 mm of precipitation in a 24-hour period sometime within a ten-year interval is:

$$\begin{aligned} P_{10} &= 1 - (1 - 0.14)^{10} \\ &= 1 - (0.86)^{10} \\ &= 0.78 \text{ or } 78\% \end{aligned} \qquad [5.9]$$

Greater amounts of moisture are required by plants and animals as temperature increases. There are several measures of moisture that combine the effects of temperature with moisture. At a subset of the weather stations in the national network, the National Weather Service records pan evaporation, which is the amount of water that evaporates from a standardized four-foot diameter pan that is one-foot deep. Pan evaporation estimates evaporation from a free water surface, which is the maximum amount of evaporation that occurs. Pan evaporation largely depends on temperature, but wind and relative humidity also play a role in the amount of evaporation that occurs, because these factors govern the moisture in the layer of air closest to the evaporating surface. Evaporation increases at higher wind speeds and lower relative humidity values.

Pan evaporation is a standardized observation, but it does not consider the moisture demands of plants. Thornthwaite and Mather (1955) presented a procedure for estimating the *potential evapotranspiration,* which is the amount of water returned to the atmosphere in vapor form as a result of direct evaporation and transpiration by plants. Often referred to as *Thornthwaite's procedure,* the calculation of potential evapotranspiration produces a derived variable that is based on an assumption of fully vegetated conditions. The calculations, which are discussed in detail in Chapter 6, depend on mean temperature and some derived constants. Potential evapotranspiration is often used to estimate plant water consumption and is also used in the estimation of water balance within a watershed (see Chapter 6).

If pan evaporation or potential evapotranspiration is greater than precipitation, which is almost always true over the length of the growing season in temperate zones, the difference is often referred to as a *moisture deficit.* The moisture deficit is also expressed in mm; it indicates the combination of the amount of additional precipitation that could have evaporated into the atmosphere and the amount of additional water that plants could have utilized. As such, the difference between precipitation and potential evapotranspiration is often used as an indication of the degree to which plant growth and production may be limited by the amount of available water. The moisture deficit may be expressed on a daily, weekly, or, most commonly, monthly basis by subtracting pan evaporation or potential evapotranspiration from precipitation over the desired time intervals; cumulative values are sometimes derived for the entire growing season.

Palmer (1965) developed another index of moisture deficit, referred to as the *Palmer drought severity index.* This index has been adopted by the National Weather Service, which has produced an atlas showing monthly values of the index for the continental United States (Karl and Knight, 1985). The Palmer drought severity index is based on complex calculations involving potential evapotranspiration, precipitation, and minimum and maximum air temperatures. The index considers past conditions as well as current conditions, so it is cumulative to some degree, and compares the values from one season with long-term average conditions. The index has the value of zero when conditions are equivalent to the long-term mean conditions; it has values greater than zero when conditions are wetter than normal and values less than zero when conditions are drier than normal. As emphasized by Zahner and Myers (1986) and Zahner and Grier (1990), the Palmer drought severity index is an indicator of average regional conditions. The index does not consider site-specific soil water-holding capacity or soil texture; the drought conditions at a specific site may be greater or less than the regional conditions indicated by the Palmer drought severity index. Sandy-textured soils, for example, with low water-holding capacity, may exhibit drought conditions to a greater extent than silt-loam soils supporting similar vegetation. The index is indicative of the relative drought severity of different years. Thus, in a resource assessment, if occasional drought severity is important to consider, historical values of the index (Karl and Knight, 1985) may be coupled with past growth records or dendrochronological data to describe the range of drought conditions and growth responses to be expected within a given landscape.

LIGHT

Energy from the sun is the driving force for all climate and weather patterns on earth. The simplest measure of sunlight is merely the average number of sunny days per month or, conversely, the average number of days with cloud cover. These variables are reported in climatic atlases and are often used in marketing recreational facilities, but the information is usually not of much value in resource assessment exercises unless the potential for resort development is being considered.

The earth receives solar energy in the form of ultraviolet, visible, and infrared light with wavelengths ranging from about 30 to 3,000 nm. Incoming solar radiation, or solar *insolation,* is captured by plants during photosynthesis, but plants do not utilize the entire energy spectrum produced by the sun. Plants utilize visible light with wavelengths from 380 nm to 710 nm; this is referred to as *photosynthetically active radiation,* or PAR. Incoming solar radiation varies seasonally and daily with changes in cloud cover (Figure 5.5).

Plant canopies intercept incoming solar radiation, so there is less light below the canopy than above. Herbaceous plants living on the forest floor

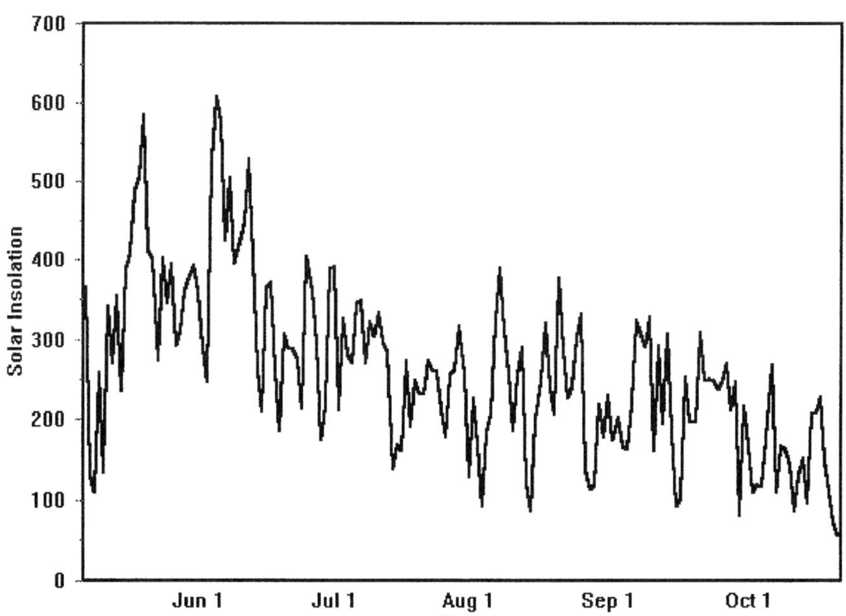

FIGURE 5.5 Daily solar insolation received during an example growing season at Crystal Falls, Michigan.

must survive in an environment with much lower light levels than plants that do not live beneath a canopy. The available light decreases as the leaf area above a point increases (see Chapter 7). This results in decreasing amounts of available light at lesser heights beneath the canopy of a forest. The vertical light profile is referred to as a *light extinction curve* (Figure 5.6).

The light extinction curve can be mathematically described by the Beer-Lambert law (Landsberg, 1986), which relates the amount of photosynthetically active radiation received at a point to the amount of incoming radiation at the top of the canopy and the projected leaf area above the point:

$$Q_i = Q_0 \, e^{-k \, LAI_i} \qquad [5.10]$$

where Q_i is the photosynthetically active radiation at height i above the ground (but below the canopy), Q_0 is the total incoming photosynthetically active radiation, k is the light extinction coefficient, and LAI_i is the projected leaf area (m^2/m^2) above height i. The value of the light extinction coefficient, k, varies by vegetation type. Pierce and Running (1988) give a value of 0.52 for coniferous forests, whereas Jarvis and Leverenz (1983) discuss the applicability of the Beer-Lambert law and give ranges of estimated values of k for a variety of vegetation types. Burton et al. (1991a) found that the value of k varied with solar angle in temperate deciduous forests:

$$k = \frac{0.437}{\sin(\theta)} \qquad [5.11]$$

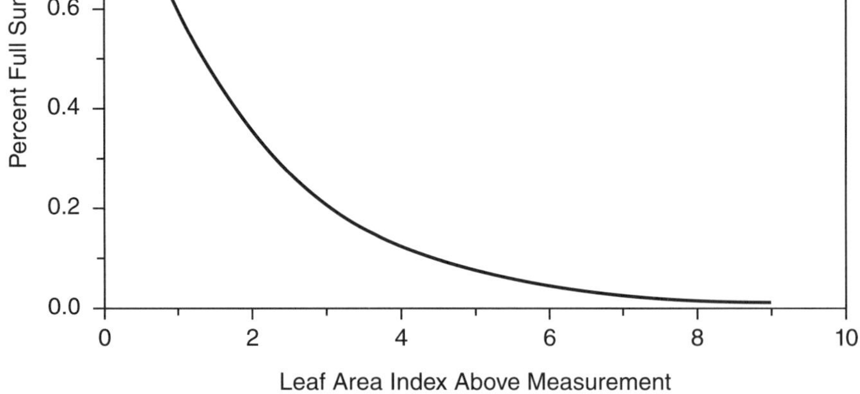

FIGURE 5.6 Typical light extinction curve for a temperate deciduous forest (adapted from Burton et al., 1991a).

where θ is the solar elevation angle. The light extinction curve can be used to assess changes in the intercepted photosynthetically active radiation within an ecosystem in response to changes in leaf area resulting from harvesting, insect defoliation, or other activity. Figure 5.6 illustrates the light extinction curve for sugar maple-dominated forests as reported by Burton et al. (1991a). A full canopy may have a leaf area index exceeding 8 m^2/m^2, which allows only 1.6 % of incoming photosynthetically active radiation to reach the forest floor. An insect defoliation that consumes 20 % of the foliar biomass will reduce the leaf area index to 6.4 m^2/m^2, but this results in only a small increase in the amount of photosynthetically active radiation reaching the forest floor (3.5%). In fact, leaf area would have to decrease to about 4.4 m^2/m^2, a reduction of 45%, before as much as 10% of the incoming photosynthetically active radiation would be expected to reach the forest floor. Similarly, removal of 45% of the canopy through harvesting will result in 10% of the incoming photosynthetically active radiation reaching the forest floor. During the assessment process, the light extinction curve may be used to examine some of the changes in ground-level microclimate, which may cause changes in plant succession or reproductive success of some species, as a result of management activity.

The light extinction curve can also be used to estimate the projected leaf area in a plant canopy. An instrument called a ceptometer is used to measure the incoming solar radiation (Q_0) and the radiation at ground level beneath the canopy (Q_i). These two values are then used to estimate the projected leaf area above the point where the measurement is made:

$$LAI = \frac{-\ ln \left[\frac{Q_i}{Q_0}\right]}{k} \qquad [5.12]$$

If the photosynthetically active radiation at the forest floor is 8% of the incoming PAR, and if the appropriate value of k is 0.50, then the estimated leaf area index in the plant canopy may be calculated using Equation [5.12]:

$$LAI = \frac{-\ ln\ [0.08]}{0.50}$$

$$= 5.1\ m^2/m^2 \qquad [5.13]$$

WIND

Wind and fire were the two major disturbance agents in eastern North American forests prior to the widespread land-use change associated with European settlement (Lorimer, 1977; Frelich and Lorimer, 1991). Notes made during the U.S. Public Land Survey (see Chapter 7) are the primary source of information on pre-European-settlement forests. In their daily work, the surveyors noted areas that had clearly been burned or that had suffered a severe blowdown. Catastrophic fire and wind disturbances occurred in all forest types, although fire was the more common disturbance factor on some site types and wind was the more common factor on others. Human suppression activity has largely removed fire from the eastern North American landscape, leaving wind as the most common natural catastrophic disturbance factor operating in eastern forested landscapes today.

Land management activity cannot suppress wind in the same sense that fire has been suppressed as a disturbance agent, but it is possible to evaluate the risk of wind damage and to undertake management activities that reduce such risks. Generally, the likelihood of wind damage increases after any partial cutting, because the gaps created by tree removal allow greater windspeeds within the canopy. As noted by Kozlowski, Kramer, and Pallardy (1991), larger trees tend to be more susceptible to wind damage following thinning or the creation of openings than do smaller trees (King, 1986). Cremer et al. (1982) found that stem form was a good indicator of the likelihood of wind damage in Australian radiata pine; trees that had height-to-diameter ratios greater than 100 were much more likely to suffer wind damage than trees with smaller height-to-diameter ratios.

The amount of damage caused by winds is the result of the speed, the duration, and the direction from which the winds originate. National Weather Service records that describe the historic patterns of windspeed and direction are available for permanent weather stations. A *wind rose* is a diagram used to illustrate the directional frequency of wind for a given location (Figure 5.7). Although winds strong enough to cause substantial damage can originate from any direction, openings in the vegetation can be oriented to minimize the likelihood that strong winds from the prevailing wind direction will cause

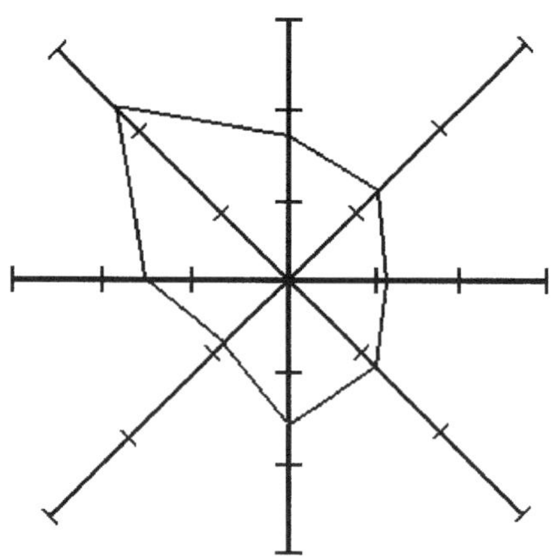

FIGURE 5.7 A wind rose displays the average numbers of days per year that the wind is from different compass directions.

damage to the remaining vegetation. As noted previously, the problem is that openings in the vegetation allow increased windspeed in the canopy; trees still standing on the downwind side of the opening, for example, may be subject to the full force of wind following creation of the opening. Moreover, opening dense forests through thinning creates opportunities for wind damage to occur if the thinning is too severe and results in an increase of windspeed within the stand. Remaining trees would never have been subject to winds of this type because they would have been sheltered by the surrounding vegetation before thinning or the creation of the opening; the management activity reducing the surrounding tree density will increase the windspeed experienced by the remaining trees. This may result in windthrow or breakage if the remaining trees are unable to sustain the new levels of stress they experience. Trees growing on shallow soils with shallow rooting zones, owing to the presence of bedrock, a fragipan, or some other physical factor, are particularly susceptible to windthrow following management activity (Figure 5.8a); trees that have developed in very dense stands will be most susceptible to blowdown or breakage caused by wind damage (Figure 5.8b). In many forests, the presence of pit and mound topography (Figure 5.9) is an excellent indicator of the susceptibility to windthrow.

Management activities can be designed to minimize wind damage on susceptible soils or on exposed physiographic features that tend to receive higher winds. Avoiding the creation of distinct boundaries between open and forested areas, especially perpendicular to the prevailing wind direction, can reduce the risk of wind damage to the remaining vegetation. The creation of small openings, less than one tree height in width, can also help to moderate wind speeds in the canopy.

The susceptibility of areas within a landscape to wind damage may be assessed in several ways. The first is to examine interpretive information associated with available soil survey descriptions to identify soil types that are susceptible to wind damage because of shallow rooting depth (see Chapter 4). Another indication of susceptibility to wind damage may be found in the General Land Office survey notes; if areas are known to have suffered catastrophic damage in the past, then there is a risk that such events will be repeated. Finally, physical evidence within the landscape, such as broken trees or tip-up mounds created when trees are blown over and their roots "tip up" out of the soil (Figure 5.10), indicates that the creation of openings or a reduction in forest density may result in increased rates of wind damage to the surrounding stand. The presence of pit and mound topography is a strong indication that there have been blowdowns in the past (Figure 5.9).

All of the planning to reduce the risk of wind damage will not protect against truly severe windstorms such as gale-force winds, tornadoes, or hurricanes. Climatic atlases often indicate severe storm frequencies so that the risk of such events can be evaluated. Little can be done to prevent, or even reduce, damage if such catastrophic storms do occur.

(*a*)

(*b*)

FIGURE 5.8 Examples of wind damage to forest stands: (*a*) a catastrophic blowdown in a forested wetland and (*b*) an isolated windthrow within a forest stand.

FIGURE 5.9 Pit and mound topography is indicative of past wind damage on shallow soils.

FIGURE 5.10 A tip-up mound in the Dukes Experimental Forest, Marquette County, Michigan.

CLIMATE CLASSIFICATION

As indicated in the preceding sections of this chapter, it is clear that there are an immense number of ways to summarize the four basic environmental factors of temperature, moisture, light, and wind. There are also complex interrelationships among the variables; in cool temperate regions, for instance, precipitation is often associated with lower temperatures, lower light levels, and higher winds. The ambient factors themselves are complex and difficult to summarize, and their interactions with flora and fauna are often even more difficult to evaluate. Even in the face of these complexities, species and vegetative associations occur repeatedly throughout a region. Obviously, wherever a species is found, the ambient environmental conditions must allow the species to survive and obtain resources. Across a regional landscape, the repeated occurrence of vegetative associations indicate that there are many areas where the requirements of a group of species are simultaneously satisfied. There have been many attempts to develop climate classification systems that identify geographic areas where, in aggregate, the combinations of climatic conditions result in a similar bioclimate with a similar ability to support a particular group of species.

A simple example of a climatic classification system is the United States Department of Agriculture frost hardiness zones (Cathey, 1990). These zones, which are presented in map form, are based on isotherms of minimum winter temperature. This relatively simple summarization reflects differences in growing-season length, summer temperatures, and other associated climatic conditions. The different frost hardiness zones are useful in identifying specific cultivars that are suitable for a given area, as well as in identifying approximate planting times. Similarly, maps based on the expected last frost date are also used in cultivar selection and in the planning of planting efforts. The most common examples of this type of simple climatic classification system are the maps of approximate planting dates found on the backs of vegetable seed packages. This type of information may be of value in comparing the suitability of potential sources of planting stock to the conditions of a particular site; conditions at the geographic source of the planting stock may be compared with those of the site to evaluate the likelihood of plant survival.

There are many more complex climate classification systems. These differ in their intended purpose and in their suitability for different uses. Gates (1972) provides a summary of many climate classification schemes and their relevance for different purposes. One of the most globally applicable schemes was developed by Holdridge (1947). He used average total annual precipitation and the ratio of annual potential evapotranspiration to total annual precipitation to define the *Holdridge life zones,* which describe overall bioclimatic conditions. There are, for instance, several types of rain forests (tropical, subtropical, and so on) included in Holdridge's classification, but all receive at least 1,000 mm of total annual precipitation and have an annual potential evapotranspiration : total precipitation ratio of less than 0.25.

Holdridge's classification is based on annual climate variables and is designed to be a global bioclimatic classification system. Many other authors have attempted to define climate classification systems that are more detailed and applicable to specific areas (e.g., Briggs and Lemin, 1992). An advantage of this approach is that the many nuances of local conditions can be explicitly considered; Rauscher (1984), for instance, developed a classification scheme for the Great Lakes region that better reflects the climate-modifying effects of the Great Lakes. Other authors have attempted to link climate with physiography to develop classification systems that better represent regional ecological conditions. Albert, Denton, and Barnes (1986), for example, developed a regional ecosystem classification system for Michigan that is based on climate, geology, and physiography. At a larger scale, Bailey's (1994) map of ecoregions of the United States (see Chapter 7) uses climate, geology, and physiography to define large regions that are, in a general sense, similar in bioclimatic conditions. McNab and Avers (1994) describe efforts to subdivide these large regions into a hierarchy of smaller-scale classifications.

In many resource assessment efforts, it may be of value to summarize climatic conditions by placing the landscape, or different portions of the landscape, in the context of one of these local classification systems. There is often a great deal of interpretive information associated with the different classifications. This information may include dominant vegetation types, possible successional paths, growing season lengths, and so on; the details vary for different classification systems. Management activity cannot change regional climate, but knowledge of regional climatic conditions and their biological relevance may be of value in developing and evaluating management alternatives.

CLIMATE CHANGE

Increasing levels of atmospheric greenhouse gases have raised concerns about consequential changes in global climate. Projected consequences of atmospheric greenhouse gas accumulation include increasing growing season temperature and decreasing moisture availability over much of North America (Manabe and Wetherald, 1986; Schneider, 1989; Hengeveld, 1990). Many studies have predicted that such changes, if they do indeed occur, will result in changes in the boundaries of the geographic ranges of many individual species (e.g., Gates, 1990; Overpeck, Bartlein, and Webb, 1991) and an associated disruption of ecological community structure (e.g., Jones, Reed, and Desanker, 1994). A good summary of the possible ecological impacts of global climate change is provided by Gates (1993). It is possible to obtain a larger-scale perspective of ecological responses to climate in North America by examining the evidence for vegetative and faunal community development following the end of the last glacial period; Davis (1989) provides an overview of this subject, and Pielou (1991) provides a very good nontechnical description of the ecological changes since the end of the last glacial period in North America.

Estimates of climatic response to increasing greenhouse gases are obtained using global circulation models (e.g., Hansen et al., 1988; Schlesinger and Zhao, 1988); these models to not produce estimates of climatic response at small geographic scales. Smaller-scale interpolations of the projections of the major global circulation models were developed for a 0.5° latitudinal and longitudinal grid by Leemans and Cramer (1990). Even at this scale, the projections are not precise for small tracts or individual watersheds. Global circulation models continue to improve and the precision of projections for smaller areas continues to increase, but the direct consequences of such changes at the scale typically considered during resource assessment projects can only, at best, be approximated at this time and in the foreseeable future.

The most that should be done during a resource assessment exercise is to complete analyses such as those undertaken by Jones, Reed, and Desanker (1994) or Reed, Jones, and Desanker (1994). Both of these studies examined the distribution of species or regional ecosystems within the landscape under present climatic conditions. They then examined future conditions for given locations as indicated by projections from global circulation models (Leemans and Cramer, 1990), found where similar ambient conditions currently occur, and compared vegetation from those locations with current vegetation at the sites of interest. Even these types of analyses warrant consideration only when management concerns extend over very long time intervals or when management goals specifically require consideration of possible future climate changes.

SUMMARY

Climate acts in concert with geology to control the distribution of vegetation on the earth's surface. The major climatic factors controlling the functioning of biological and ecological processes are temperature, moisture, light, and wind. Primary climatic data are available from regional monitoring networks; these data are summarized in a variety of different ways, depending on the information needed to support decision making. The great variety of information can be overwhelming, and a number of different systems have been proposed for delineating areas where the different combinations of climatic conditions result in roughly similar bioclimatic effects. These classification systems may be of great utility in resource assessment exercises and in providing background information needed for the development and evaluation of management alternatives. Climate is not static, and there is a great deal of concern regarding the effects of increasing atmospheric concentrations of greenhouse gases on future climate, but it is very difficult to obtain localized predictions and this information is rarely included in resource assessment exercises. The exception is when such information is specifically required in order to support the achievement of a particular management goal or when the planning period is extremely long.

APPLICATIONS

5.1 Obtain the National Weather Service daily or monthly weather records for a permanent monitoring station. Prepare figures similar to Figures 5.1, 5.3, and 5.4 describing long-term average monthly minimum, average, and maximum temperature, extreme temperatures, average monthly and extreme precipitation, and monthly snowfall for the station.

5.2 Obtain a national, regional, or local (state) climatic atlas. Make a list of the climatic variables described in the atlas. How might each of these be of use in the resource assessment process? Give an example of a landscape management question that might utilize information concerning the different climatic variables.

5.3 Obtain several consecutive years of average monthly temperatures for a National Weather Service permanent monitoring station. Use these to calculate the growing degree days on both a 4.4°C (40°F) and a 10°C (50°F) basis. How much do these values differ from year to year?

5.4 Develop the light extinction curve for a temperate deciduous forest with a light extinction coefficient of 0.50. Develop a graph with the horizontal axis representing leaf area index above the measurement point and the vertical axis representing the proportion of incoming photosynthetically active radiation reaching the measurement point. For a forest with a leaf area index of 4.5 m^2/m^2, what proportion of incoming photosynthetically active radiation reaches the forest floor? How does this change if there is an insect defoliation that consumes 25% of the foliage?

5.5 Obtain several years of precipitation data from a National Weather Service permanent monitoring station. Use these records to estimate the return period for the most extreme precipitation event in the records. What is the probability that an event of this magnitude will be experienced in a given year? What is the probability that an event of this magnitude will be experienced in a 20-year time period?

5.6 Obtain a description of a climate classification system applicable to your region. What variables does it utilize to describe different climatic zones? Does it utilize geological or physiographic features to further define bioclimatic regions? What sort of interpretive information is available from the classification system?

Chapter

6

Hydrology

Water falling to the earth's surface as precipitation is utilized by plants and animals and moves through terrestrial and aquatic ecosystems prior to returning to the atmosphere as vapor to begin the cycle anew. *Hydrology* is the study of water on the surface of the earth, including its interactions with the environment. Although water is abundant and covers two-thirds of the earth's surface, most of the earth's water is salty and a substantial fraction of the fresh water is polluted (U.S. Environmental Protection Agency, 1994). About 3% of the total water is fresh water, and about 78% of that is in the form of polar ice and snow. This leaves only a small fraction of the earth's total water available for human use; water stored in aquifers, soil, freshwater lakes, and streams makes up an estimated 0.62% of the total. All water for drinking, cooking, washing, and other human needs must come from this small portion of the total. Yet one recent estimate is that humans use about 18,000 cubic kilometers of fresh water per year, or 26% of total terrestrial evapotranspiration, leaving 51,400 km^3/year to meet the water needs of all other terrestrial species and natural communities (Postel, Daily, and Ehrlich, 1996).

Water is redistributed over the landscape by transport through the atmosphere. This is a solar-powered system mediated by the processes of evaporation and precipitation. Water shapes the landscape and influences the development and pattern of soil and plant communities. Water distributed across forested landscapes is significant to human survival, as it represents a major portion of the global resource that is available for human use. But even more significant, this water is also subject to the influences of management activities that can alter the vegetation, affecting transpiration and interception. *Transpiration* is a loss of water to the atmosphere through plant stomata, and *interception* is a loss of water to the atmosphere through the evaporation of water adsorbed on plant surfaces such as leaves, twigs, and branches. Management activity can alter physical soil properties affecting *infiltration,* the movement of water into the surface of the soil, and *storage,* the retention of water in the soil pores. Because the soil filters water and reduces pollution, land management activities influence both soil water quality and quantity, eventually influencing the amount and quality of groundwater and surface water in lakes and streams.

Although not intended to present a comprehensive review of hydrology or watershed management, this chapter explores some of the key processes and measures of water runoff and landscape structure that are important to resource managers in the assessment process. Those unfamiliar with forest or wildland hydrology, or seeking more detailed information, may refer to texts such as Hewlett (1982), Brooks et al. (1991), and Black (1991).

WATERSHED CHARACTERISTICS

A *watershed* is a sloped area of land that collects precipitation; drainage from the area passes a single point, usually referred to as the *mouth.* The term *watershed* is synonymous with *drainage basin* and *catchment,* although the last term is usually used in reference to smaller watersheds. Watersheds may vary in size from a few hectares, for small streams, to many thousands of hectares for large systems such as the Mississippi River basin. These areas are fairly well defined hydrologic units in that precipitation inputs and water outflows can be estimated with some accuracy, and they are often considered to be good natural units for management planning. No matter what its size, it is also important to remember that each watershed is part of an even larger system.

Drainage basins are separated from each other by *watershed divides,* which are physical barriers to water movement. Literally, water that falls on one side of a divide flows into a different watershed than that which falls on the other side. Surface divides between watersheds are located by finding the high elevation levels between watersheds. These are most easily determined from topographic maps in steep, mountainous terrain (Figure 6.1); it is more difficult to delineate watershed boundaries in areas with less relief. The determination of watershed divides may be facilitated by the use of a digital elevation model in a geographic information system (Chapter 13). Subsurface divides are underground barriers to water movement that may be due to variation in deposited material in glaciated areas, tectonic uplift, degradation, or faulting of underlying geologic strata. The presence of a subsurface divide may be indicated by the inability to quantitatively link watershed inputs and outputs. Subsurface divides are not always easily determined and may be

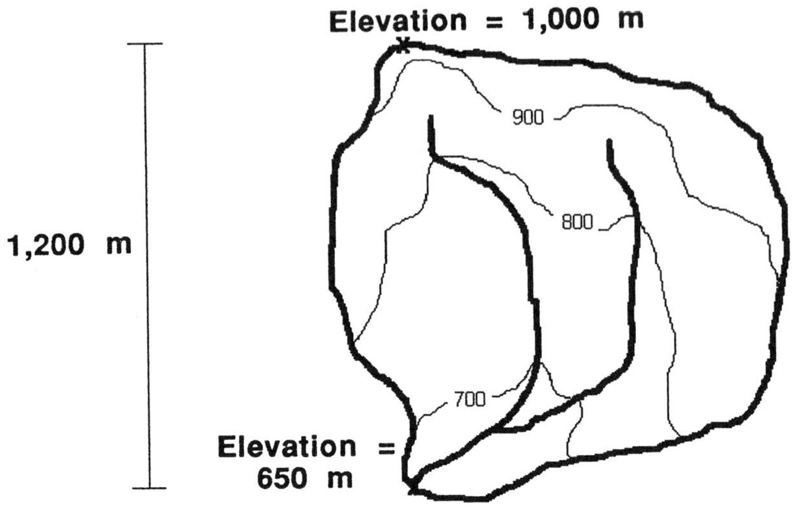

FIGURE 6.1 Topographic map of an example watershed illustrating watershed relief.

identified only with the help of intense geologic exploration. This expense is usually not warranted in most resource assessments unless the presence of a subsurface divide is indicated and the hydrologic relationships are critical to fulfilling the management goals.

Watersheds can be described in a variety of ways. The *shape* can be described as *ovoid, circular,* or *rectangular* (Figure 6.2). The shape is largely an expression of the underlying geologic structure, but the geologic structure is intertwined with shape; the shape affects the hydrology, but the hydrology is also partly responsible for the shape. Over time, there is a tendency for watersheds to become elongated; the combined effects of streams that downcut channels and erode banks result in headward erosion, which gradually advances upstream. These are the same processes by which streams go from

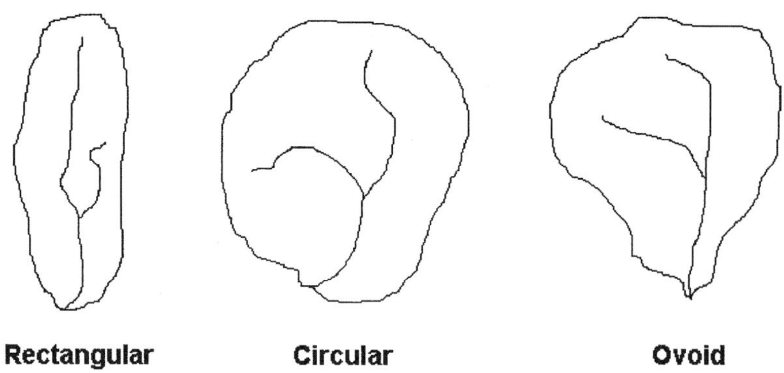

FIGURE 6.2 Rectangular-, circular-, and ovoid-shaped watersheds.

youth, characterized by steep, erratic channels, to old age, exhibiting gentle gradients and sinuous paths.

Watershed shape is important because it controls the time it takes for water to enter a stream after it falls as precipitation. On more compact, circular watersheds, water has a relatively short distance to travel to reach a stream and to leave the watershed at its mouth. On a watershed of the same size, but with a rectangular shape, water will have a longer distance to travel, on average, to reach the mouth. The dimensions of watersheds may be measured once the watershed divides have been located on topographic maps. The perimeter is the linear distance along the watershed divides delineating the watershed; perimeter is measured in linear units of meters, kilometers, feet, or miles. Once the perimeter is measured, the area of the watershed may be determined using a variety of methods, ranging from manual dot grids to digital calculation within a geographic information system (Chapter 13). Area is measured in square units corresponding to the perimeter measurement (square meters, hectares, square kilometers, square feet, acres, or square miles). Watershed shape was quantitatively described by Gravelius, as cited by Black (1991), using a *compactness coefficient:*

$$K_c = \frac{0.28\ P}{\sqrt{A}} \qquad [6.1]$$

where K_c is the compactness coefficient, P is the length of the perimeter of the watershed, and A is the area of the watershed. When watershed area is expressed in squared units of perimeter (e.g., square kilometers if perimeter is measured in kilometers), K_c will equal 1 for a circular watershed. Increasing rectangularity results in increasing values of K_c.

Several other measures are useful in describing the slope of a watershed. *Basin relief* refers to the difference in elevation between the mouth of the watershed and points on the perimeter. *Maximum basin relief* is the difference in elevation between the mouth and the highest point on the perimeter:

$$\text{Maximum basin relief} = \text{Elevation}_{\text{maximum}} - \text{Elevation}_{\text{mouth}} \qquad [6.2]$$

In Figure 6.1, the maximum basin relief is the difference between the elevation at the upper perimeter of the watershed (1,000 m) and the elevation at the mouth (650 m), which is 350 m. Maximum basin relief divided by the horizontal distance from the mouth to the highest point on the perimeter gives the *relief ratio* or *drainage basin slope:*

$$\text{Relief ratio} = \frac{\text{Maximum basin relief}}{\text{Distance from mouth to highest point}} \qquad [6.3]$$

In Figure 6.1, the maximum basin relief is 350 m and the distance from the mouth to the highest point on the perimeter is about 1,200 m, giving a relief ratio of:

$$\text{Relief ratio} = \frac{350}{1200} = 0.29 \quad [6.4]$$

These slope measures provide a manager with an idea of how water movement might be affected by various treatments. Steep slopes, as compared with flat areas, usually result in more overland flow simply because water has less opportunity to infiltrate. Slope also influences the trafficability of equipment and erosion hazard (see Chapter 4).

The *water balance* of a watershed is often used to describe the relationships between water inputs (precipitation) and outputs (yield) in drainage basins. The simplest balance for a watershed can be stated as:

$$P_g = E_t + Q + \Delta S \quad [6.5]$$

where P_g is the gross precipitation on the area, E_t is evapotranspiration, Q is streamflow or yield, and ΔS is change in storage. The gross precipitation in a watershed is the total volume of all precipitation falling anywhere within the watershed; this is usually estimated by multiplying the area of the watershed by the average amount of precipitation over the entire watershed. Gross precipitation is measured in cubic meters, area is usually expressed in hectares (10,000 m^2), and precipitation is expressed in millimeters. This leads to the following expression for estimating gross precipitation:

$$P_g = (\text{area} \cdot 10{,}000) \cdot \left(\frac{\text{precipitation}}{1{,}000}\right) \quad [6.6]$$

where *area* is watershed area in hectares and *precipitation* is average precipitation within the watershed in millimeters. The constants in the equation convert the measures to square meters or meters, respectively; if area or precipitation is expressed in another unit, the constants in the equation must be adjusted accordingly. If a 27-hectare watershed receives an average annual precipitation of 850 millimeters, the gross precipitation in the watershed is:

$$P_g = (27 \cdot 10{,}000) \cdot \left(\frac{850}{1{,}000}\right) = 229{,}500 \text{ m}^3 \quad [6.7]$$

The average annual precipitation in a watershed may be estimated by sampling using rain gages at fixed points within the watershed, annual climate information from one or more local weather stations, or regional average annual climatic information. In English units, volume of water is often expressed in acre-feet, which is equivalent to the volume of water covering one acre to a depth of one foot. This is equivalent to 43,560 cubic feet of water.

Since precipitation is usually expressed in inches, the gross precipitation in a watershed may be obtained as follows:

$$P_g = \text{area} \cdot \left(\frac{\text{precipitation}}{12}\right) \qquad [6.8]$$

where gross precipitation is in acre-feet, *area* is the watershed area in acres, and *precipitation* is the average precipitation within the watershed in inches. If an average annual total of 40 inches of precipitation falls within a 90-acre watershed, the gross precipitation is:

$$\begin{aligned} P_g &= 90 \cdot \left(\frac{40}{12}\right) \\ &= 300 \text{ acre-feet} \end{aligned} \qquad [6.9]$$

Evapotranspiration is a more difficult quantity to estimate than gross precipitation. In fact, evapotranspiration cannot be directly measured on a large scale, but must be estimated. One way to estimate potential evapotranspiration is to use the Thornthwaite Equation (Thornthwaite and Mather, 1955; Chow, 1964; Mitsch and Gosselink, 1993), which assumes a fully vegetated watershed:

$$ET_i = 16\left(\frac{10\ T_i}{I}\right)^a \qquad [6.10]$$

where ET_i is the potential evapotranspiration for month *i*, expressed in mm/month, T_i is the mean monthly temperature in °C, *I* is the local heat index given by:

$$I = \sum_{i=1}^{12}\left(\frac{T_i}{5}\right)^{1.514} \qquad [6.11]$$

and *a* is an exponent given by:

$$a = (0.0675 \cdot I^3 - 77.1 \cdot I^2 + 17{,}920 \cdot I + 492{,}390) \cdot 10^{-6} \qquad [6.12]$$

Table 6.1 contains average monthly temperatures for an example watershed. The value of *I* for use in the Thornthwaite Equation is found using Equation [6.11]:

$$\begin{aligned} I &= 0 + 0 + 0 + \left(\frac{3}{5}\right)^{1.514} + \left(\frac{9}{5}\right)^{1.514} + \ldots + \left(\frac{5}{5}\right)^{1.514} + 0 \\ &= 31.1 \end{aligned} \qquad [6.13]$$

Note that months when the average temperature is less than 0 °C are entered as zeros into the calculation of *I*. The value for the exponent in the example data is found using Equation [6.12]:

TABLE 6.1 Potential Evapotranspiration Calculated Using Thornthwaite's Equation (Thornthwaite and Mather, 1955) for an Example Watershed

Month	*Average Temperature (°C)*	*Potential Evapotranspiration (mm)**
January	−12	0
February	−10	0
March	−5	0
April	3	15
May	9	45
June	14	70
July	17	84
August	18	89
September	16	79
October	11	55
November	5	25
December	−2	0
Total		462

* $ET_i = 16\left(\frac{10\,T_i}{I}\right)^a$

where ET_i is the potential evapotranspiration and T_i is the average temperature in month i; I and a are given by:

$$I = \sum_{i=1}^{12}\left(\frac{T_i}{5}\right)^{1.514}$$

$$a = (0.0675 \cdot I^3 - 77.1 \cdot I^2 + 17{,}920 \cdot I + 492{,}390) \cdot 10^{-6}$$

$$\begin{aligned} a &= (0.0675 \cdot (31.1)^3 - 77.1 \cdot (31.1)^2 \\ &\quad + 17{,}920 \cdot 31.1 + 492{,}390) \cdot 10^{-6} \\ &= 0.977 \end{aligned} \quad [6.14]$$

With values of I and a, the potential evapotranspiration, in mm/month, can be calculated for each month using Equation [6.10]. The potential evapotranspiration for months when the average temperature is less than 0°C is assumed to be zero; the potential evapotranspiration for April in the example is calculated using the average temperature for April and the values of I and a from Equations [6.13] and [6.14], respectively:

$$\begin{aligned} ET_{\text{April}} &= 16 \cdot \left(\frac{10 \cdot 3}{31.1}\right)^{0.977} \\ &= 15.45 \approx 15 \text{ mm} \end{aligned} \quad [6.15]$$

In this example, summing the potential evapotranspiration for all months gives a total of 462 mm for the year. The gross potential evapotranspiration for the area is calculated in the same fashion as gross precipitation in Equation [6.6]:

$$E_t = (\text{area} \cdot 10{,}000) \cdot \left(\frac{\sum_{i=1}^{12} ET_i}{1{,}000} \right) \qquad [6.16]$$

where area is measured in hectares. For a 27-hectare watershed with 462 mm of potential evapotranspiration per year, the gross potential evapotranspiration is:

$$E_t = (27 \cdot 10{,}000) \cdot \left(\frac{462}{1{,}000} \right)$$
$$= 124{,}740 \text{ m}^3 \qquad [6.17]$$

Returning to the watershed water balance relationship (Equation [6.5]), the water balance for this example watershed can be estimated using the estimated values of gross precipitation (Equation [6.7]) and gross evapotranspiration (Equation [6.17]):

$$229{,}500 = 124{,}740 + Q + \Delta S \qquad [6.18]$$

which implies that:

$$Q + \Delta S = 104{,}760 \text{ m}^3 \qquad [6.18a]$$

If the annual change in water storage in the watershed is assumed to be zero, then the annual volume of streamflow from the watershed, assuming that the watershed is fully vegetated, is estimated to be 104,760 cubic meters of water. The Thornthwaite Equation assumes fully vegetated conditions, and there is no adjustment for different types of vegetation or different types of management treatments. The equation may be used to get a rough estimate of potential evapotranspiration before and after treatment. The expected reduction in evapotranspiration for the treated portion of the watershed must be weighted by the area of the watershed to be treated:

$$\sum_{i=1}^{12} ET_i' = \left(\sum_{i=1}^{12} ET_i \right) \left[1 - R \cdot \left(\frac{\text{area treated}}{\text{total area}} \right) \right] \qquad [6.19]$$

where ET_i' is the estimated potential evapotranspiration for the watershed following treatment and R is the expected reduction in evapotranspiration for the treated portion of the watershed.

If a management treatment that is expected to result in a 30% decrease in evapotranspiration is applied over the entire example watershed, then the potential annual evapotranspiration following harvest would simply be 30% less, or 323 mm for the year:

$$\sum_{i=1}^{12} ET_i' = 462 \cdot [1 - (0.30 \cdot 1)]$$
$$= 323 \text{ mm} \qquad [6.20]$$

Equation [6.17] can be used to estimate the gross evapotranspiration following treatment, which is expected to be 87,210 m^3 of water. Estimated streamflow from the watershed following treatment, again assuming no change in net water storage, is:

$$Q = 229{,}500 - 87{,}210$$
$$= 144{,}290 \text{ m}^3 \qquad [6.21]$$

If the same treatment is applied to 8 of the 27 hectares in the watershed, the average potential evapotranspiration for the watershed following treatment is:

$$\sum_{i=1}^{12} ET_i' = 462 \cdot \left[1 - \left(0.30 \cdot \left(\frac{8}{27}\right)\right)\right]$$
$$= 421 \text{ mm} \qquad [6.22]$$

Again using Equation [6.17], this implies 113,670 m^3 of gross evapotranspiration in the example watershed; the estimated annual streamflow following this treatment is 115,830 m^3. This compares to estimated streamflow values of 104,760 m^3 if the treatment is not implemented and 144,290 m^3 if the entire watershed is treated. These should be considered only rough estimates; there are many assumptions made in deriving these values. As noted earlier, Thornthwaite's Equation assumes fully vegetated conditions and that all vegetation types will have equal evapotranspiration given equal temperature. This is generally not true, but the equation has been proven to give reasonable approximations of potential evapotranspiration under a wide range of physical and climatic conditions.

Another consideration is that these estimates also assume that there is no change in soil storage following treatment. Road building and the clearing of vegetation may lead to increased runoff, so there is less time for precipitation to penetrate into the soil. Many recommended management practices to maintain water quality are designed to minimize changes in water storage caused by increased runoff or soil compaction. Vegetation removal results in less evapotranspiration, but whether the additional water enters into storage or becomes streamflow depends, in large part, on the soil characteristics and infiltration capacity, the proportion of the watershed treated, and the effect that the treatment has on soil surface properties.

Direct estimation of streamflow requires installation of equipment to estimate *stage height,* which is the depth of the streamwater in a calibrated streambed. The streambed profile (Figure 6.3a) is determined at the point of measurement of stage height; knowledge of stage height is then used to determine the cross-sectional area of the water in the stream at any point in time (Figure 6.3b). A stage height recorder may be permanently installed in the streambed to provide continuous observation of stage height. This alone

$$A = (V_1 + V_2 + V_3 + V_4 + V_5) \cdot L$$

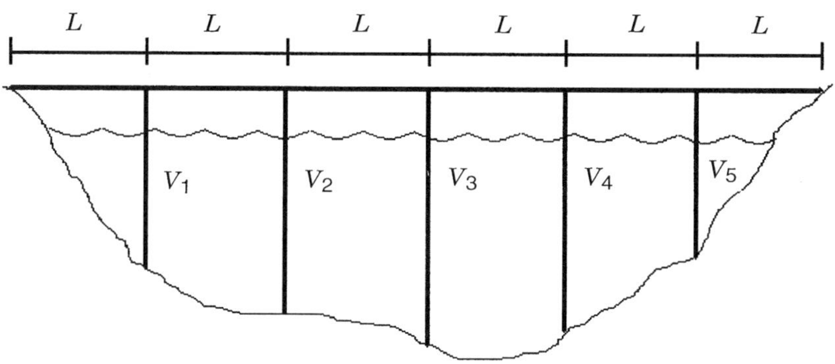

(a)

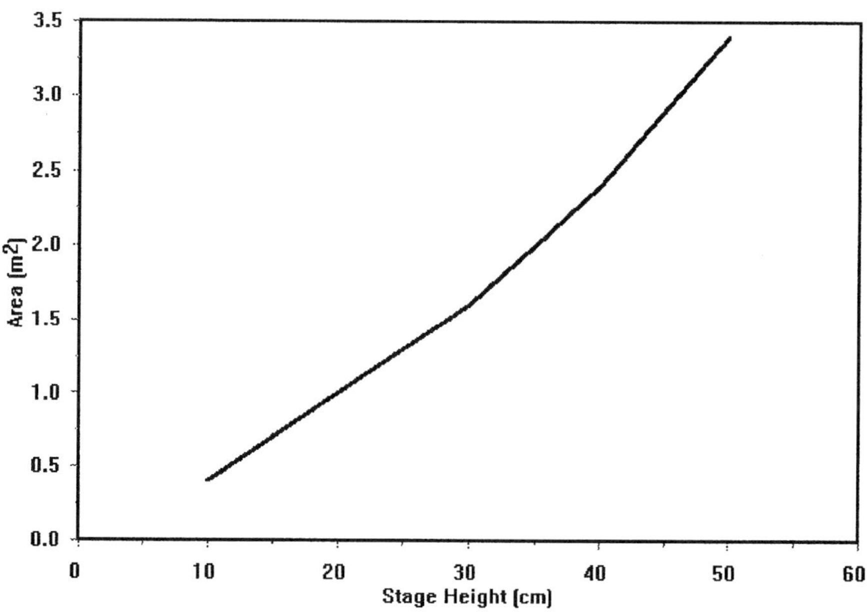

(b)

FIGURE 6.3 Measurement of streambed characteristics: (*a*) streambed profile and (*b*) relationship between cross-sectional area and stage height.

is not sufficient to determine the volume of water in the stream; it is also necessary to know the rate of flow, which is affected by the depth of the water. A current meter is used to determine the average velocity of water in the stream. The volume of water passing the measurement point during a specified time interval can be calculated by multiplying the average stage height times the average rate of flow during the time period:

$$Q_i = A_i \cdot V_i \tag{6.23}$$

where Q_i is the streamflow in cubic meters per second (m^3/s) during the *i*th time period, A_i is the average cross-sectional area (m^2) of water in the stream during the time period, and V_i is the average velocity (m/s) of water during the time period. The total volume of streamflow during a time period may be calculated by multiplying the estimated average streamflow from Equation [6.23] by the length of the time period in seconds.

For a streambed such as that illustrated in Figure 6.3a, measurement of the stage height is used to estimate the cross-sectional area of the water, as in Figure 6.3b. For the streambed illustrated in Figure 6.3a, the depth of the water is 0.4 m, which corresponds to a stream cross-sectional area of 2.4 m^2 (Figure 6.3b). If the average stage height over a one-hour time period is 0.4 m and the average water velocity during the hour is 0.2 m/s, the streamflow during the one-hour period may be calculated using Equation [6.23]:

$$\begin{aligned} Q_i &= 2.4 \cdot 0.2 \\ &= 0.48 \text{ m}^3/\text{s} \end{aligned} \tag{6.24}$$

The total streamflow volume over a one-hour time period may be estimated by multiplying the average velocity of 0.48 m^3/s by 3,600, which is the number of seconds in an hour, to obtain an hourly volume of 1,728 m^3 of water.

By measuring flow rates at different stage heights, a stage height-discharge relationship can be developed for the stream (Figure 6.4). The water velocity in a streambed is generally highly correlated with stage height. The exact relationship varies with the stream profile, but following a calibration period, a stage-discharge relationship such as that illustrated in Figure 6.4 can be developed for a stream. After this relationship is developed, it is no longer necessary to measure velocity in order to estimate discharge rate.

Once the stream profile is measured and the instrumentation is installed, the watershed is referred to as a *gaged* watershed. Once the relationships in Figures 6.3 and 6.4 are known, only knowledge of average stage height during a specified time period is required to estimate streamflow over that time period.

It requires much time and effort to establish and maintain a gaged watershed. This may be justified in some landscapes, particularly those where water yield is an important management consideration. In most instances, however, the time and expense required to maintain a gaged watershed are not justified. In these cases, it is still possible to estimate streamflow during a resource

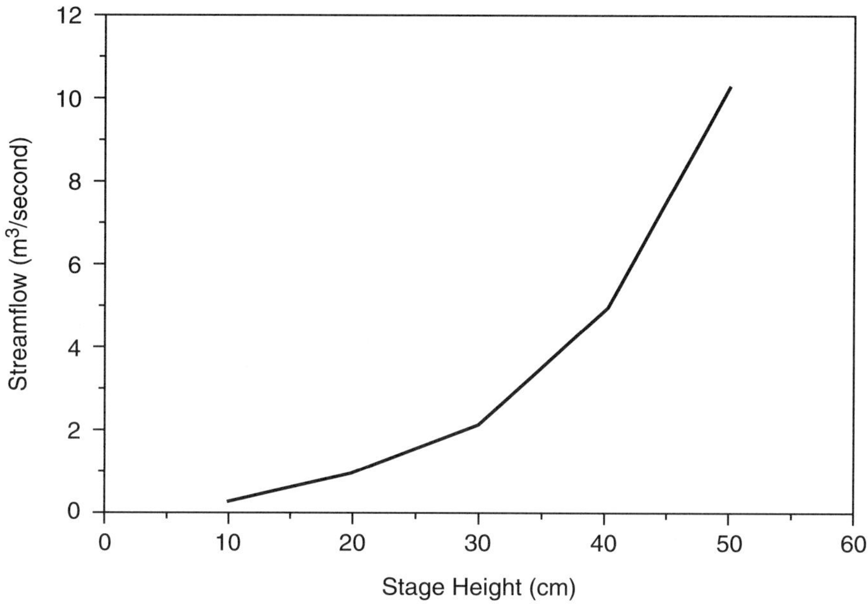

FIGURE 6.4 Stage-discharge relationship for the example stream profile in Figure 6.3.

assessment exercise. This method utilizes the *Manning Equation,* which may be used if there is some estimate of the stream slope and roughness of the streambed (Mitsch and Gosselink, 1993):

$$Q_i = \frac{(A_i) \cdot (R_i^{2/3}) \cdot (\sqrt{S})}{n} \qquad [6.25]$$

where Q_i is streamflow in m^3/s, A_i is the cross-sectional area of the stream in m^2, R_i is the hydraulic radius, which is the stream cross-sectional area divided by the wetted perimeter of the stream, S is the energy gradient, approximated by the channel slope, and n is a roughness coefficient, termed the Manning coefficient, which varies by the type of stream (Table 6.2). The wetted perime-

TABLE 6.2 Roughness Coefficients for Use in the Manning Equation to Determine Stream Flow in Ungaged Streams (after Mitsch and Gosselink (1993), Chow (1964), and Lee (1980))

Stream Condition	*Manning Coefficient,* n
Straightened earth canals	0.020
Winding natural streams with some plant growth	0.035
Mountain streams with rocky streambeds	0.040–0.050
Winding natural streams with high plant growth	0.042–0.052
Sluggish streams with high plant growth	0.065
Very sluggish streams with high plant growth	0.112

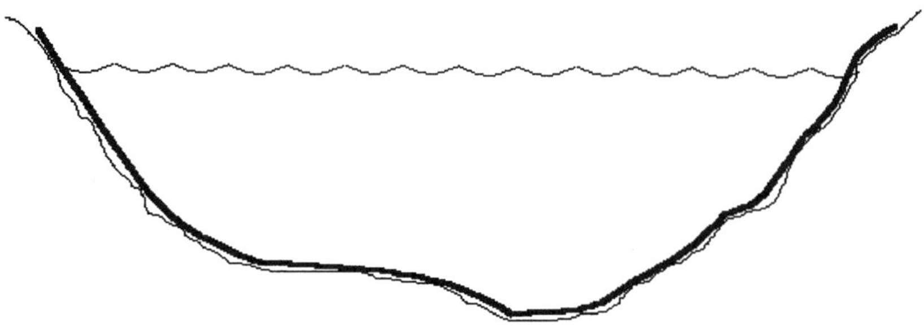

FIGURE 6.5 Measurement of the wetted perimeter of a stream.

ter of the stream is measured by laying a tape across the bottom of the streambed from the waterline on each bank (Figure 6.5). The hydraulic radius at time i is calculated by dividing the cross-sectional area of the water in the streambed (Figure 6.3b) by the wetted perimeter (Figure 6.5):

$$R_i = \frac{A_i}{P_i} \quad [6.26]$$

The hydraulic radius is expressed in meters or feet. Figure 6.5 illustrates the measurement of the wetted perimeter for the stream illustrated in Figure 6.3. The cross-sectional area of the stream (2.4 m^2 in Figure 6.3b) is divided by the wetted perimeter (1.8 m for the stream in Figure 6.5), to estimate the hydraulic radius for this stream:

$$\begin{aligned} R_i &= \frac{2.4}{1.8} \\ &= 1.5 \text{ m} \end{aligned} \quad [6.27]$$

The energy gradient, S, is approximated by the channel slope. As shown in Figure 6.6, the channel slope for some length of stream is estimated by dividing the change in elevation of the stream by the length of the stream section:

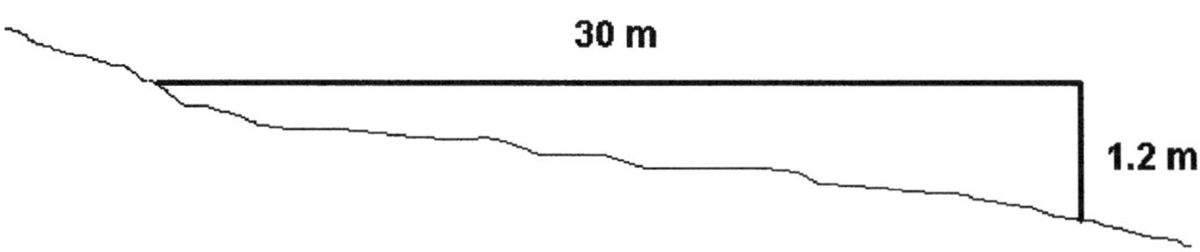

FIGURE 6.6 Measurement of the channel slope for an example length of stream.

$$S = \frac{\Delta E_L}{L} \qquad [6.28]$$

where S is the channel slope, which is expressed in m/m, ΔE_L is the change in elevation (m) over the section of stream, and L is the length of the section (m). Note that the change in elevation is the relative difference in elevation over the length of the stream, which may be easily measured; absolute elevation measurements are not required. The channel slope may be measured directly, using a level, which gives a percent slope reading, but if the slope is measured in percent it must be divided by 100 to obtain the energy gradient value required by the Manning Equation. The channel slope may be assumed to be constant over reasonable time periods; it does not change unless there is a change in the relative elevations of the two ends of the stream section. For the length of stream illustrated in Figure 6.6, the channel slope is estimated by dividing the relative change in elevation (1.2 m) by the length of the stream section (30 m):

$$\begin{aligned} S &= \frac{1.2}{30} \\ &= 0.04 \text{ m/m} \qquad [6.29] \end{aligned}$$

The cross-sectional area, hydraulic ratio, and channel slope can be combined in the Manning Equation (Equation [6.25]) to estimate streamflow volumes for different types of streams. Using the Manning coefficient from Table 6.2 with the preceding example, and assuming that it is a natural winding stream with some plant growth, the estimated streamflow volume is given by:

$$\begin{aligned} Q_i &= \frac{(2.4)\ (1.5^{2/3})\ (\sqrt{0.04})}{0.035} \\ &= 18.0 \text{ m}^3\text{/s} \qquad [6.30] \end{aligned}$$

Because the Manning Equation uses only physical measurements of the stream channel, it can be used to estimate the volumes of high and low flows from measures of the wetted perimeter at the high water mark and at any level below it. Usually, measures obtained during spring snowmelt and runoff in the northern United States and Canada reflect the period of high flow because of rain-on-snow events. Most harvest or thinning treatments temporarily lead to increased flows for several years after treatment. Higher flows are a result of decreased interception loss because of decreased canopy surface area, and decreased evapotranspiration loss because of less vegetation on the site. Although not always necessary in describing a stream in a resource assessment, this approach may be useful in weighing the effects of alternative treatments if the effects of those treatments may be adequately predicted from existing knowledge of the treated watershed and the effect of the treatment on soil surface properties.

STREAM DESCRIPTORS

Stream patterns can be described as dendritic, trellis, rectangular, or radial (Figure 6.7). The most common pattern in the Lake States region is dendritic, which, as the name suggests, has tributaries that form a pattern reminiscent of tree branches. *Dendritic* streams develop on soil and geologic material of relatively equal resistance to erosion. Other patterns occur in areas where there is unequal resistance to erosion. *Radial* patterns typically occur on volcanoes, *rectangular* patterns have tributaries that more or less follow right angles and are common along geologic fault lines, and *trellis* patterns have short, closely spaced tributaries and occur in more mountainous areas with uplifted rock strata.

Duration of flow is another stream descriptor. *Perennial streams* flow throughout the year, with flows ceasing only during drought years. *Intermittent streams* flow only during a rainy season or during a period of snowmelt in a well-defined channel. *Ephemeral streams* flow after a storm or snowmelt,

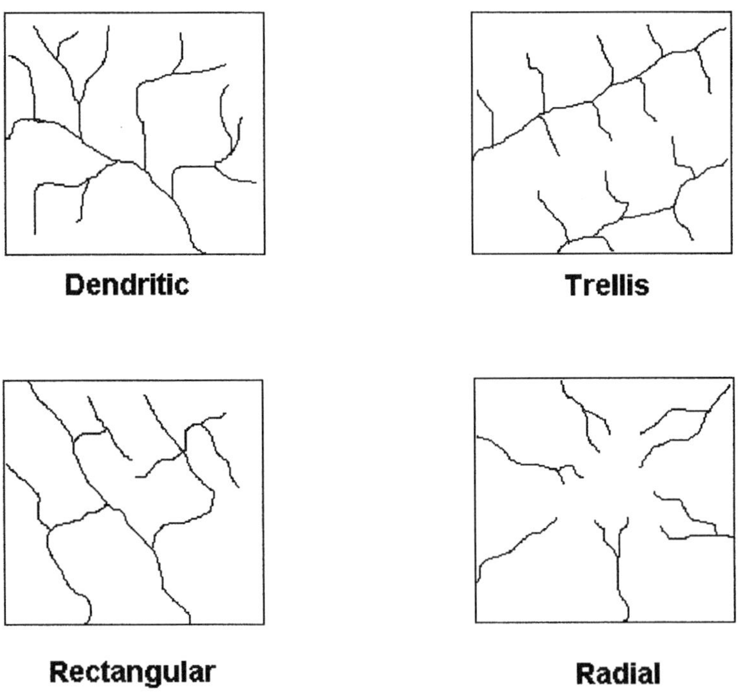

FIGURE 6.7 Dendritic, trellis, rectangular, and radial stream patterns.

but lack a defined channel. Ephemeral, intermittent, and perennial streams are equally important in the resource assessment process, because all need special consideration in designing treatments to maintain water quality and yield. In eastern North America, ephemeral and intermittent streams are often at the headwaters of a drainage basin, so inappropriate disturbance may influence water quality and habitat both in the immediate area of disturbance and at an extensive distance downstream.

Stream order measures the amount of branching of a stream; it also provides insight into watershed size and complexity. A *first-order stream* is an unbranched stream and is the smallest unit. The confluence of two first-order streams results in the downstream section or *reach* becoming a second-order stream. If another first-order stream joins the second-order stream, the resultant downstream reach is still a second-order stream, but if two second-order streams join together, a third-order stream is formed, and so on (Figure 6.8). First-order streams are usually small, but are important because of their location at the headwaters of the drainage system. Because of their small size, it is also more likely that forest management treatments will cover more, or perhaps all, of the corresponding watershed area, which may then result in significant changes in water behavior.

Stormflow is water that rapidly enters a stream following a precipitation event. Stormflow is the combination of precipitation falling directly into the stream, subsurface flow through the soil, and surface flow that does not infiltrate into the soil but, instead, rapidly moves across the surface to the stream channel. *Hydrologic response* is the proportion of total precipitation or streamflow that enters the stream with sufficient speed or volume following a particular storm event that it can be identified as stormflow. Hydrologic

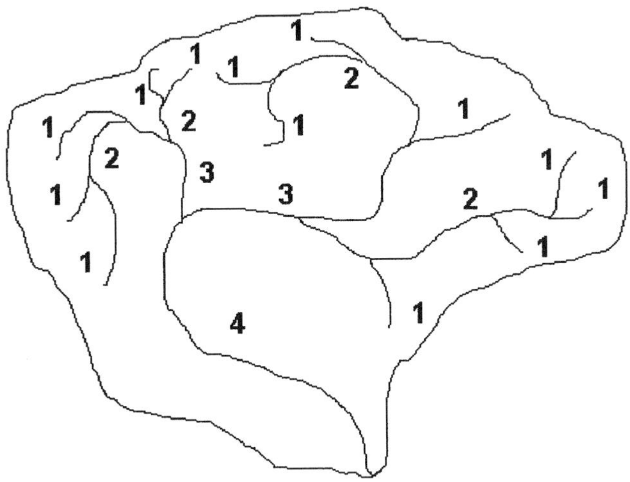

FIGURE 6.8 The relationships among first-, second-, third-, and fourth-order streams.

response on gaged watersheds, where there are one or more stage height recorders to measure streamflow, can be estimated using annual data:

$$R_{P_g} = \frac{Q_s}{P_g} \qquad [6.31]$$

where R_{P_g} is the hydrologic response based on gross precipitation, Q_s is streamflow attributable to storms (stormflow), and P_g is gross precipitation. If a watershed with a gaged stream has a gross precipitation of 100,000 m^3 per year, and if an estimated 15,000 m^3 of streamflow can be identified as stormflow, then the hydrologic response based on gross precipitation is:

$$\begin{aligned} R_{P_g} &= \frac{15{,}000}{100{,}000} \\ &= 0.15 \end{aligned} \qquad [6.32]$$

This implies that an average of 15% of the gross precipitation received in the watershed during a rainfall event rapidly enters the stream as stormflow. Less than 15% of the water falling in a small event will enter the stream as stormflow, whereas more than 15% of the water falling during a major precipitation event will become stormflow. The estimated hydrologic response from Equation [6.31] is the average proportion over all of the precipitation events occurring within a year.

Hydrologic response may also be expressed as a percentage of the total annual streamflow:

$$R_y = \frac{Q_s}{Q} \qquad [6.33]$$

where Q_s is streamflow identifiable as stormflow and Q is total annual streamflow, or yield. If a watershed has 50,000 m^3 of annual streamflow, and if 15,000 m^3 of that can be identified as stormflow following precipitation events, the hydrologic response based on yield is:

$$\begin{aligned} R_y &= \frac{15{,}000}{50{,}000} \\ &= 0.30 \end{aligned} \qquad [6.34]$$

The calculated hydrologic response based on yield (Equation [6.33]) will always be greater than that based on gross precipitation (Equation [6.31]), because streamflow (Q) is less than gross precipitation (P_g). Either calculated hydrologic response variable is indicative of the amount of time between the beginning of a precipitation event (or the onset of snowmelt) and its detection as an increased volume of streamflow by a stage height recorder. A high hydrologic response indicates a large proportion of stormflow, which is sometimes referred to as a *flashy* response. A flash flood occurs if a large amount of water enters a stream following a storm event and causes a very rapid increase in streamwater volume owing to the associated stormflow. A low value of hydrologic response indicates that water falling on the watershed takes a long time to reach the stream and that little of it appears as stormflow.

A measure that combines stream and watershed descriptors is *drainage density,* which is defined as the total length of stream within a watershed divided by the total area of the basin. The resulting quantity is expressed as kilometers of stream per square kilometer of watershed (km/km^2); English units are miles of stream per square mile of watershed (mi/mi^2). Drainage density is often used as a surrogate measure of hydrologic response in watersheds that are not gaged. Generally, a lower drainage density implies that there are fewer streams to carry water, which then results in a lower hydrologic response because more water tends to infiltrate the soil and groundwater. Another watershed characteristic associated with a low hydrologic response rate is the presence of permeable soils, which permits the infiltration of precipitation. In permeable soils, precipitation moves to the stream by either interflow through the soil or percolation to the groundwater and then into the stream as baseflow. In any event, these pathways of water movement take far longer to reach a stream than overland surface flow and are difficult to distinguish on a stage height recorder as stormflow. Areas with high hydrologic response generally have low infiltration rates. This may be because of steep slopes, impermeable soils, sparse vegetation, or easily erodible soil. Thus, the drainage density may be used to infer other ecosystem characteristics and, when combined with vegetation and soil information, also serves as a good indicator of watershed characteristics in a resource assessment. Soil physical properties associated with infiltration and percolation may be altered during land management activities. These effects are usually only temporary, but it is often important to evaluate whether proposed management alternatives will significantly increase or decrease water yield.

WETLAND DESCRIPTORS

The *water table* is the top of the saturated zone in the soil; depth to the water table is a key factor affecting both the hydrologic properties of a watershed and the vegetation of an area. Land management is most impacted when the water table is within a few feet of the soil surface. This can happen if the site under consideration is in a low topographic position, as is the case with the *riparian* areas bordering lakes and streams (Figure 6.9). Riparian zones constitute the interface between a lake or stream and surrounding drier areas. They are important for buffering and filtration of overland and subsurface flow as water moves to stream channels or lakes. They are also important habitat areas for many species and often support different vegetative communities than those found in drier areas that are just a few meters distant. The width of a riparian area depends on the slope, the soil physical properties,

FIGURE 6.9 Riparian zones border lakes and streams and are the interface between terrestrial and aquatic ecosystems.

and the characteristics of the surrounding areas. In the assessment process, areas that function as riparian zones can usually be identified by geomorphology, soil properties, and the composition of the vegetative community. The soil in riparian zones may have finer texture or higher organic matter content than soil in surrounding areas, for example, and the presence of a floodplain often helps delineate the extent of a riparian area. The vegetative community in riparian areas usually contains species that are tolerant of high water levels, such as alder, willow, or spruce. Species composition of the understory may be particularly affected by the conditions in riparian zones.

The water table may also be near the surface because of the presence of a soil layer such as a fragipan, which impedes drainage, or because of the presence of springs. Springs are particularly common in moist areas in the lower portions of a slope. These types of conditions should be noted in an assessment, because disturbance of such areas may result in alteration of the hydrologic characteristics of the site. These areas, similar to riparian zones, are often rich in species composition and form important wildlife habitat. Alteration of the hydrologic properties of an area with a shallow water table may affect the infiltration of water into the soil and, hence, affect water storage throughout an entire watershed. Changes in water storage result in corresponding changes in streamflow and water yield from the watershed (see Equation [6.5]).

Under some conditions, sensitive areas may be considered *jurisdictional wetlands,* which means that their management and use falls under Section 404 of the Clean Water Act of 1970 (40 CFR Part 230 and 40 CFR Part 232; U.S. Environmental Protection Agency, 1980, 1988) and a variety of state laws that were passed subsequent to the federal legislation. The law defines the waters of the United States to include essentially all continental and coastal waters and wetlands. These laws cover dredge and fill activities in wetlands and describe management activities that require a permit, as well as some special activities that are exempted from permitting requirements (Table 6.3).

The depth to the water table normally fluctuates through the year; it is usually closest to the soil surface in the spring before plants begin to transpire and generally falls through the summer. If water is at or near the surface for extended periods during the year, the site may be considered a jurisdictional wetland. The U.S. Fish and Wildlife Service (Cowardin et al., 1979) defines wetlands as

> lands transitional between terrestrial and aquatic ecosystems where the water table is usually at or near the surface or the land is covered by shallow water. For purposes of this classification, wetlands must have one or more of the following attributes: (1) at least periodically, the land supports predominantly hydrophytes, (2) the substrate is predominantly undrained hydric soil, and (3) the substrate is nonsoil and is saturated with water or covered by shallow water at some time during the growing season of each year.

A second regulatory definition used by the U.S. Environmental Protection Agency and U.S. Army Corps of Engineers, which originates from the Section 404 permit program, identifies wetlands as

> those areas that are inundated or saturated by surface or groundwater at a frequency and duration sufficient to support, and that under normal circumstances do support, a prevalence of vegetation typically adapted for life in saturated soil conditions. Wetlands generally include swamps, marshes, bogs and similar areas.

TABLE 6.3 Summary of Permitting Requirements for Management Activities in Forested Wetlands (Refer to 40 CFR Part 230 as reprinted in the *Federal Register,* Vol. 45, No. 249, December 24, 1980, for details)

Permit Requirement	*Activity*
Prohibited	Any dredge or fill activity, provided there is a practicable alternative that does not have other adverse environmental impacts
	Any dredge or fill activity that causes or contributes to significant degradation of the waters of the United States
Permit Required	Discharge of dredged or fill material that contains a toxic pollutant
	Activities whose purpose is to convert an area of waters of the United States to a use to which it was not previously subject
	Activities in which the flow or circulation of the waters of the United States may be impaired or the reach of such waters is reduced
	Activities in which there will be discernible alterations of flow or circulation
Exempt (Generally, unless activities noted earlier are involved)	Normal farming, silviculture, and ranching activities of an established operation
	Maintenance and emergency reconstruction of serviceable structures such as dams, dikes, etc.
	Construction or maintenance of farm or stock ponds or irrigation ditches, or the maintenance (but not the construction of) drainage ditches
	Construction of temporary sedimentation basins that does not include placement of fill material into the waters of the United States
	Any activity for which a state has an approved program
	Construction or maintenance of farm roads, forest roads, or temporary roads, where such roads are constructed in accordance with best management practices to ensure that flow and circulation patterns and chemical and biological characteristics of waters of the United States are not impaired, that the reach of the waters of the United States is not reduced, and that any adverse effect of the aquatic environment will be otherwise minimized

The criteria for determining and delineating a wetland are (1) the presence of hydric soils, (2) the presence of redoximorphic soil features indicating waterlogged conditions or a fluctuating water table, and (3) 50% of the dominant plants on the site are commonly found in wetlands (Lyon 1993; Federal Interagency Committee for Wetland Delineation 1989). As a general rule, if soils are flooded for more than one or two weeks a year, the area requires careful scrutiny using methods defined by the previously cited references, and many activities are permissible only after a permit is obtained from one or more state or federal regulatory agencies.

An approach to wetland identification and delineation was presented by Tiner (1993). Referred to as the Primary Indicators Method, the procedure uses the examination and classification of the overstory and understory species present within an area, including the determination of whether or not obligate or facultative wetland species are present. Soil pits are used to detect indicators such as high organic matter content or the presence of redoximorphic features. If the plant communities and soils contain these wetland indicators, the area is judged to meet the wetland identification criteria. The border of the wetland is delineated by physical inspection to determine where the soil and plant wetland indicators are present and where they are lacking.

If a published soil survey is available for an area, it will list the soils meeting hydric criteria (Chapter 4), and wetlands are often delineated on soil maps. In many resource assessment situations, this information is sufficient for planning purposes because direct manipulation of wetland areas will be avoided; after field verification, the map information allows the planner to design necessary buffer strips to protect and maintain wetland process and function. In cases in which direct management is considered, particularly if potentially permittable activities are involved, the extent of the wetland should be delineated and verified through field observation. In addition, a hydrogeomorphic classification of wetlands can be used to develop methodology to estimate the impacts of management alternatives during the assessment process (Brinson, 1993). The system provides a functional classification system for wetlands using geomorphic setting (topographic location), water source and transport (precipitation, surface flow, groundwater, or these factors in combination), and hydrodynamics (the direction and strength of flow). Brinson (1993) determined functional and ecological relationships for various depressional wetlands, extensive peatlands, riverine wetlands, and fringe wetlands along lakes and coastal areas. Any proposed management activity may be evaluated in terms of its effect in maintaining the original function of the basic wetland processes; that is, if the water relationships change, the wetland is assumed to change. As an assessment tool, the system requires the establishment of reference wetlands for comparative purposes. Reference wetlands may or may not be available in a given area, but can usually be established. However, as a planning tool, the system is useful because it guides the user in considering basic process and functions in the evaluation of management activities in and around wetlands.

Some states have developed classification systems that can provide a basis for planning. One such system has been developed for the riparian and wetland sites in the State of Montana (Hansen et al., 1995). This system uses a habitat type approach (see Chapter 7) for ecological classification and provides information on location and landform, vegetation, soils, and adjacent communities, successional information, management interpretive information, and a cross-index to other classification schemes. Obviously, such sources are very helpful in the process of conducting a resource assessment.

WATER QUALITY MEASURES

Water quality is a relative term; implicit in any discussion is the intended use of the water. For example, water may be suitable for irrigation but not for "higher" uses such as bathing or drinking. The U.S. Environmental Protection Agency sets criteria and standards for water quality which, if not exceeded, are expected to support aquatic systems suitable for the higher uses of water. Polluted waters are defined as those degraded below their natural level by human activities. Although they do not provide an exhaustive list of possible pollutants, the following sections briefly describe the classification and measurement of the most common water pollution concerns encountered in planning management activities to be implemented within forested landscapes.

Dissolved oxygen is gaseous oxygen that is essentially trapped in the lattice of loosely fitting water molecules. It is essential to support fish and insect populations and is critical in organic matter decomposition. Dissolved oxygen levels are strongly controlled by temperature; generally, the colder the water, the higher the dissolved oxygen levels. Forest management practices have the potential to affect water temperature in a number of ways (Brown, 1985). For example, a harvest that results in greater direct sunlight reaching the surface of a stream will cause temperature to rise, generally decreasing the dissolved oxygen levels. Similarly, removal of vegetation may result in higher surface soil temperature. As moisture moves through the soil and to the stream, it is warmed above pretreatment levels and will contain less dissolved oxygen. The magnitude of this warming is difficult to predict, as it is dependent on the climate, soil properties, and residence time of the water in the soil. Usually, riparian buffer strips or streamside management zones are maintained in undisturbed states along water courses in order to shade streams and minimize the impact of temperature changes following treatment.

Solids are material that is suspended or dissolved in water. The amount of solids can be determined by measuring material left in a vessel after evaporation in a drying oven. *Total suspended solids* is the portion of the solids that cannot pass through a filter; it is retained on the filter surface. This material is commonly referred to as *sediment* and consists of particulate matter, including colloids, that may be either mineral or organic material. These suspended solids may affect the aesthetic quality of water by making it appear cloudy or murky. It may also decrease the light available to aquatic plants, which reduces primary productivity. They may cover the stream bottom, which destroys the benthic community, or they may act as an abrasive by mechanically removing plants and insects from the benthic community. Suspended solid concentration is commonly estimated using a variety of commercially available turbidimeters or monophotometers, which measure light transmission through a sample; the less light transmitted, the higher the nonfilterable residue content.

Total dissolved solids is the portion of total solids that passes through a filter. These are not colloidal; they are organic and mineral materials present in dissolved forms and include dissolved organic carbon, ions, carbonates, sulfates, and chlorides. These are commonly measured in total by evaporating a filtered sample or by measuring the electrical conductance of the sample. In addition, the concentration of individual ions and compounds may be measured using gas chromatographs, carbon-nitrogen analyzers, atomic absorption spectrophotometers, or inductively coupled plasma spectrophotometers. The effect of elevated dissolved solids on aquatic ecosystems and water quality largely depends on the specific compound. As a group, these materials are responsible for the taste of water. High levels can also affect use; for example, high levels of arsenic, copper, or lead can, to varying degrees, have an impact on the health of a variety of organisms. Levels of NO_3 in excess of 10 ppm may cause a condition called methmoglobenemia, which can affect oxygen uptake by mammals. Concentrations of some compounds can be used as indicators of disturbance following management activity; dissolved organic carbon, for instance, can be used to monitor the degree of wetland disturbance. More information specifically related to forest management activity is given by Brown (1985).

Residue becomes a problem in water as a result of *erosion,* which is the dissolution, detachment, and movement of mineral and organic material by water, wind, and gravity (Chapter 4). *Sedimentation* is the deposition of eroded material in aquatic ecosystems. In most undisturbed forested ecosystems, these processes are negligible over time spans of days or years. They usually become a concern when ecosystems are disturbed, whether the disturbance is due to natural phenomena or to management activity. In most management activities, *buffer* or *filter strips,* which are undisturbed corridors, are used along streams and lakeshore areas to minimize the deposition of eroded materials into the aquatic ecosystem. Activities such as road building require a great deal of care in planning and execution because roads can be corridors of water movement; there can also be direct physical disturbance at stream crossings and increased erosion resulting from streambank disturbance. As noted in the following section, relatively minor design modifications and proper construction techniques can often minimize the negative impacts of road construction and maintenance on water quality.

TABLE 6.4 State of Michigan Recommended Best Management Practices for Road Location, Construction, and Maintenance, along with Timber Harvesting Guidelines, for the Protection of Water Quality During Forest Management Activities

Activity Guidelines
Road Location and Design
• Roads and trails should be located where soil disturbance and water crossings will be minimized.
• Where feasible, roads should be constructed with a slight grade of 1 to 2%; slopes greater than 10% should be avoided, and slopes greater than 5% should be avoided on erodible soils (Chapter 4).
• Adequate cross drains should be provided when wetland crossings cannot be avoided, to ensure that both surface and subsurface water is equalized on both sides of the road.
• Roads should be located away from lakes, ponds, creeks, and low areas to maximize filter areas and minimize siltation potential or alteration of soil drainage.
• Water crossings on streams should be in a straight stretch of the channel and as close to a 90° angle to the streambed as possible.
• Water crossing locations should be chosen where the streambank is most cohesive and where the approaching road will have a low slope in order to minimize disturbance of the streambank.
• If the road has a long slope to the crossing, broad-based dips (Figure 6.10a) should be used to keep overland flow from depositing sediment directly into the stream.
Road Construction
• Surfacing the road in the vicinity of a stream should be considered if it is anticipated that sediment cannot be controlled by broad-based dips or open-topped culverts.
• Where fish migration is critical, bridges should be used instead of culverts. Portable bridges are desirable in many situations, because they can be put in with a minimum of disturbance and removed easily, so that costly maintenance is avoided.
• Crossings should be designed to accommodate 25- to 50-year flood frequency to minimize the possibility of washouts.
• Construction should be done in periods of low flow.
• Treated wood used at crossings should not contain excessive surface chemicals or oils when installed.
Road Maintenance
• Roads that are not closed to public vehicular traffice should be maintained.
• Clean debris and windfalls from roadside drainage ditches prior to periods of high flow.
• Minimize traffic roads during the spring runoff period and other wet periods to avoid excessive road damage.

TABLE 6.4 *(Continued)*

Activity Guidelines
• Roads that are open for use should be inspected regularly to make sure that erosion control features are not compromised.
• Repairs should be made as soon as feasible. These may include grading, to fill ruts that can channel water, and minimizing berms along the edges of roads that can trap water on a road and impede normal drainage.
• Low-volume roads should have waterbars installed on slopes and be closed with a berm, where feasible, to minimize damage from recreational vehicles and maintenance costs (Figure 6.10b).
Timber Harvesting
• Skid trails should always be designated to be on the highest and driest ground, avoiding wet areas.
• Use as few skid trails as possible to minimize widespread disturbance and keep skid trail slopes less than 10%.
• Skid trails should cross streams only at designated areas, usually where a crossing with a bridge or culvert has been installed.
• Avoid skidding in riparian areas; if trees are cut in riparian zones, logs should be winched out to minimize disturbance to soil.

SOURCE: Michigan Department of Natural Resources (1994b).

BEST MANAGEMENT PRACTICES

Best management practices, or BMPs, for maintaining water quality and ecosystem integrity have been developed by natural resource management agencies in most states. BMPs are also specified in Section 404 of the Clean Water Act of 1970 under the silvicultural exemption. These guidelines are used in the design and analysis of management alternatives in order to minimize adverse water quality impacts of management activity. Most BMPs deal with road location and construction because improperly constructed roads may have large impacts on water quality and aquatic or wetland habitat.

Most states have common elements in their BMPs regarding road location and construction. These are mostly commonsense guidelines, designed to minimize soil disturbance, overland water movement, and subsequent erosion. In some states these guidelines are mandatory and require the approval of plans, issuance of permits, and inspection before, during, and after construction. In other states, these guidelines are voluntary, though there may be some mandatory permitting requirements for some situations (Chapter 10). Recommended or required BMPs can be quite explicit in some instances, but there are some differences among the requirements for different states. Land managers must be familiar with both the regulatory features and the details of guidelines applicable in their state. Table 6.4 lists some guidelines for road location, construction, and maintenance, and Figure 6.10 illustrates some water control structures that are taken from the State of Michigan's recommended best management practices (Michigan Department of Natural Resources, 1994b); many features of the Michigan guidelines are similar to those of other states.

As noted earlier, BMP guidelines may be either voluntary or mandatory, depending on the issuing state. In some states, they are enforced with financial or criminal penalties. In many states, there are permitting procedures that must be followed when stream crossings are installed or when construction crosses wetland areas. Most states publish their water quality BMP guidelines (e.g., Michigan Department of Natural Resources, 1994b), and many hold either formal or informal training sessions to assist landowners, land managers, and contractors with compliance.

Since many forested areas are already roaded, much of the emphasis in planning and assessment is necessarily placed on the examination of existing road systems to ensure that they meet, or can be upgraded to meet, state best management practice standards. Most states currently have audit procedures

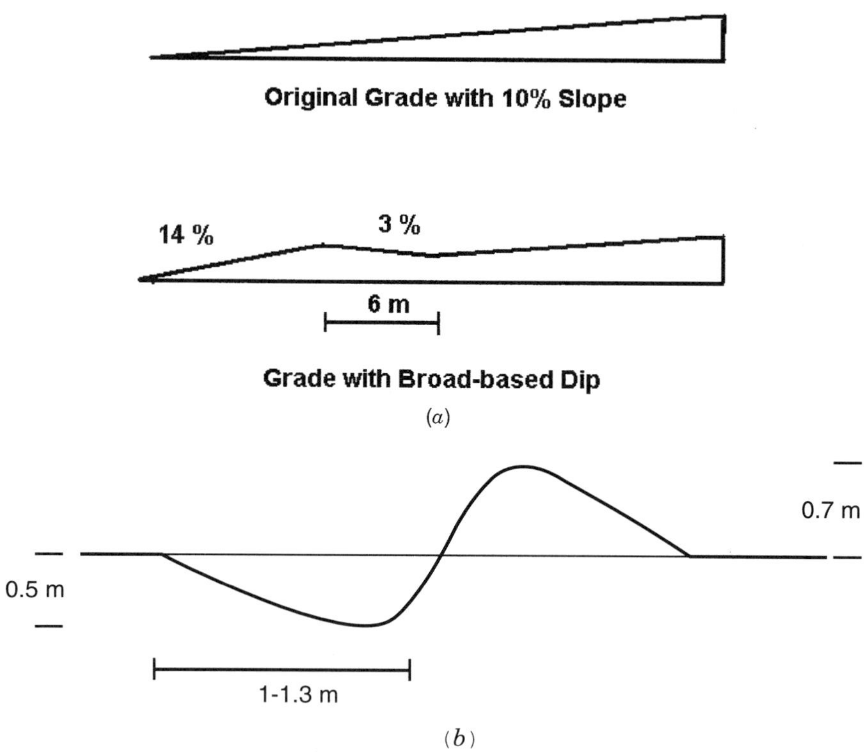

FIGURE 6.10 Water control structures recommended by Michigan's voluntary best management practices to maintain water quality along forest roads: (*a*) cross section of a broad-based dip and (*b*) cross section of a waterbar (adapted from Michigan Department of Natural Resources, 1994b).

for the evaluation of BMP use and effectiveness in recently treated areas. These audit procedures may also be used in an assessment to evaluate the need for treatment of problem areas and may serve as a checklist for regulatory compliance.

In most instances, the audit procedures have been developed from guidelines developed by Dissmeyer (1994). When audits are conducted, they usually require the talents of an interdisciplinary team that rates the application of best management practices in the field. On an individual site, an audit team first determines whether a particular best management practice (Table 6.4) was applicable to the site. Second, the audit team determines whether the practice was applied correctly, usually using a rating scale that ranges from *gross neglect* to *meeting* and *exceeding the best management practice standards.* Finally, the team rates the degree to which the applied practice was effective in meeting the goal of maintaining water quality, using a scale ranging from *direct and prolonged* (longer than one year) *impact* to *adequate* or *improved protection of soil and water resources.* Utilization of the same procedures prior to treatment in assessment and management planning will help to ensure regulatory compliance and the protection of soil and water resources.

SUMMARY

The availability of high-quality water in sufficient quantity is a necessity for all living things. Consideration of possible impacts on water yield and quality is important in the evaluation of management alternatives within forested landscapes. Changes in water yield following management activity may be particularly important, especially in watersheds with large hydrologic responses. Areas bordering lakes and streams are buffers that prevent sediment and material moving in surface and groundwater from entering surface water. They are important in providing habitat areas for many species, as are wetlands. Management activities should be designed to protect the integrity and function of riparian and wetland areas. States have developed either mandatory or voluntary best management practices that allow management activity to proceed while protecting water quality; it is important for managers to familiarize themselves with the legal and regulatory situation in their area to ensure compliance.

APPLICATIONS

6.1 Obtain a USGS quadrangle map; outline the major watersheds by examining the contour lines. Calculate the maximum basin relief and the relief ratio for a watershed outlined on the map. How can you tell which way is upstream in watersheds that are not entirely within the map boundaries?

6.2 Calculate the gross precipitation in the units specified:

(a) In cubic meters for a 6 km^2 watershed that receives 900 mm of annual precipitation
(b) In acre-feet for a 12-square-mile watershed that receives 32 inches of precipitation

6.3 Use Thornthwaite's Equation to calculate the average annual potential evapotranspiration for a watershed with the following average monthly temperatures:

January	13.4°F
February	16.1°F
March	26.6°F
April	41.6°F
May	54.1°F
June	64.0°F
July	68.4°F
August	66.4°F
September	57.3°F
October	47.8°F
November	33.0°F
December	18.6°F

6.4 If there is a watershed with gross precipitation of 200 acre-feet of water per year and 125 acre-feet of water lost through evapotranspiration per year, what is the water yield of this watershed? What is the expected water yield if a management treatment reduces the vegetation by 20% on 33% of the watershed area?

6.5 What is the hydrologic response of a watershed with 200 acre-feet of gross precipitation per year and 150 acre-feet of streamflow? What is the hydrologic response following implementation of a management activity that reduces evapotranspiration by 15%? What are some of the implications of this change?

6.6 Are there water quality best management practices for your state? Are they mandatory or voluntary? Are there any permitting requirements applicable to vegetation management activity? What about the construction of stream crossings or woods roads? Is there an audit checklist that may also be used for assessment purposes?

PART

III

Biotic Resources

Chapter

7

Vegetative Community

Qualitative and quantitative descriptions of the vegetative community provide insights into the abiotic potential of a site to support different types of vegetation, including species not currently present in the community. The vegetative community may also indicate potential site productivity, information concerning disturbance history, basic information concerning ecological function, and current successional stage. Observations may be made on the entire vegetative community or on some component of the community (e.g., the understory, the overstory, or the regeneration of overstory species). The methods described here are general and are not restricted to use in analyzing vegetative communities; many of the methods may be applied to the analysis of any type of biological community.

Resource assessment efforts usually begin with a historical treatment of the vegetative components within a landscape. Additional qualitative descriptions of existing communities, including such interpretive information as likely successional paths, may provide adequate information for land managers. In some situations, however, it may be necessary to provide more explicit descriptions of community structure during a resource assessment. An example is a case in which both disturbed and undisturbed ecosystems exist within a landscape and a goal of the assessment is to describe the effects of past disturbance on the vegetative community. Another example requiring more explicit description of the vegetative community is an instance when it is necessary to locate comparable ecosystems and describe baseline information that will be used in future monitoring to evaluate the impacts of planned management activity. If a management goal concerns biodiversity, quantitative measures of community structure will be required from the resource assessment effort. Therefore, some of the methods described in this chapter are appropriate for use in specific situations, but will not be included in other resource assessment efforts. There are many texts providing detailed treatment of plant ecology and vegetative community analyses (e.g., Grieg-Smith, 1983; Kent and Coker, 1992). In addition, Pielou (1977) and Ludwig and Reynolds (1988) describe many statistical methods applicable to ecological problems of various types, beyond analyses concerning just the structure of the vegetative community.

HISTORIC INFORMATION

There is a great deal of information concerning the state of North American forests at the time of European settlement. For the original 13 colonies along the Atlantic Coast and other areas heavily settled prior to 1790, it may be difficult to find spatially explicit pre-European-settlement information for a given landscape. For much of the rest of the United States, though, there is a great deal of spatially explicit information in the U.S. Public Land Survey and in General Land Office (GLO) records. The utility of this information for ecological analyses has been explored by many; a good summary of the typical problems associated with these data and some methods to address these problems are given by Bourdo (1956).

Information contained in the GLO records include witness tree descriptions for each section corner. Witness trees are trees the surveyors used to mark the locations of corners; the trees were usually blazed and the distance and bearing to the corner was recorded in the field notes (Table 7.1). These descriptions include the size and species of four trees at each section corner, resulting in an approximate one-mile by one-mile grid of samples. In addition, the surveyors described the general cover types they passed through, listed the major species, and provided comments regarding the soil conditions and the general landscape. In fact, these original surveys may in some ways be considered early resource assessments, designed to provide information relative to potential agricultural or industrial development. Table 7.1 contains information transcribed by the authors from the original notes for the survey of one section line in Houghton County, Michigan.

Clearly, the General Land Office records contain a great deal of spatially explicit information about the original overstory vegetation. These records have been used to investigate original cover (e.g., Finley, 1976, and many others) and other topics such as natural disturbance frequency (e.g., Lorimer, 1977, 1980, 1985; Frelich and Lorimer, 1991), but they also provide information that may be used to describe the original natural forest in a resource assessment effort. This information provides a context for evaluating the current forest, the extent of disturbance in the current forest and surrounding landscape, and the potential natural vegetation at different locations within

TABLE 7.1 Information Transcribed by the Authors From the Original Notes for the 1846 Survey of the Section Line Between Sections 5 and 6, T56N, R32W, Houghton County, Michigan; the Numeric Values to the Left Are Distances in Chains and Links From the Section Corner

North True line between Sections 5 and 6

Land Descending N.E. Variation 3°00′W

15.50 Spruce 18 inches diameter

20.00 Summit of Ridge course N.E.

34.40 W. Pine 12 inches diameter

39.10 Stream 10 L. course East

40.00 Set Post quarter section corner
Y. Birch 28 [inches] S 80 W 21

Variation 4°E. Fir 6 [inches] N 60 E 29

58.75 Birch 10 inches diameter

62.50 Stream 3 L. course East

72.95 Sugar [maple] 10 inches diameter

73.50 Boulders

78.00 Ridge course East

80.01 Intersected line 18 links East of Post
set Post Corner of sections 5 and 6

Land poor 2nd Rate rolling, soil Sand and gravel,

Timber Hemlock Sugar [maple] Birch W. Pine Spruce and Cedar

a landscape. GLO records are usually available in depositories of surveying information, including county courthouse records and state archives.

There are other sources from which detailed records of forest overstory condition may be obtained. These include wills, business records, and photographic records. As a rule, such sources of information are more difficult to locate and, although they provide a great deal of information, are not systematically available in the sense that survey records are available. For a given landscape, there may be a prior landscape-level resource assessment, though this level of detail is probably available only from very recent efforts. It is more likely that there are existing ownership records, management plans, and treatment or sales records for individual parcels within a landscape. Such information may be incomplete, and the information may have been collected at different times using different standards. Any available information, however, can provide important perspectives that may be of value in assessing current landscape conditions.

QUALITATIVE DESCRIPTIONS OF VEGETATION

Qualitative measures of the vegetative community may be quite simple descriptions of the type of vegetation on a site, such as grassland, deciduous forest, fen, and so on, or they may be sophisticated classifications of vegetative communities. The mechanisms and processes involved in the formation of vegetative communities have been much debated, dating back at least to the classic presentations of Clements (1904, 1905, 1916) and Gleason (1917, 1926, 1939). Today it is generally accepted that individual species respond to the abiotic environment in their own way. Each species also modifies the environment and indirectly impacts the ability of other species to reproduce and grow. Communities consist of interacting species, and collections of species tend to occur together repeatedly in landscapes where similar environmental conditions are encountered. These assemblies are termed *vegetative associations* or, sometimes, *habitat series* or *habitat types*.

The development of systems to classify vegetative associations is not uniform across North America; in the Intermountain region of the west and in the Great Lakes region there are well-developed systems to classify vegetative associations, and these are commonly used by resource managers as a basis for making management decisions. These systems may provide extensive interpretive information concerning the expected productivity of numerous possible overstory species, likely successional paths, the likelihood of threatened or endangered species occurrence, and the suitability of an area for different land uses, such as agriculture or commercial forest production (see Figure 7.1 and Table 7.2). In some areas, these systems are tightly linked with soil surveys or with systems for classifying physiographic features of the landscape. In almost all systems, the vegetative association is assumed to be dependent on the abiotic environment and relatively independent of disturbance or site history. In other words, early and late successional stages of a particular association are assumed to be recognizable, and vegetative association is assumed to be invariant over time.

Field methods for assessing vegetative association differ according to the classification system being utilized. For some systems, a visual inspection to develop a listing of the species present in an area, and, possibly, their relative frequencies, is made by a trained observer. This list is compared with standard lists describing the ideal communities for each association; in some cases there may be a dichotomous key based on species presence or absence to assist in the field identification of vegetative association. Combined with the observer's assessment of the impacts of such factors as disturbance, the species list provides the basis for classifying the area as one of the vegetative associations present in the region. These classification methods, although not rigorously quantitative, are usually repeatable by trained observers in a given region. Success in applying these methods depends on the observer's ability to recognize different vegetative associations.

Systems based on more rigorous sampling procedures have been developed, partially to reduce the reliance on observer skill. These methods may require the establishment of quadrats, or fixed area plots, and assessment of species presence/absence, relative frequency, or relative cover among plots. Rules or keys can be used by the observer that are based on both species lists and

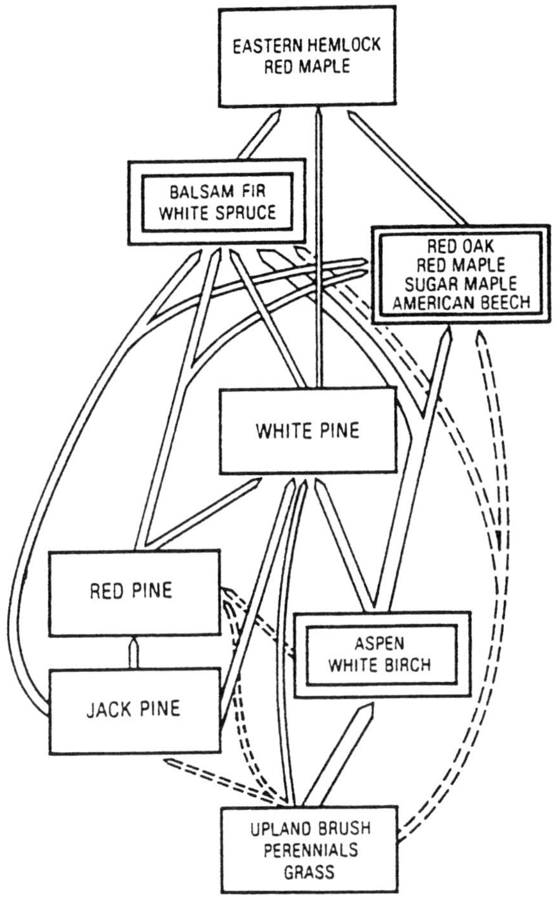

FIGURE 7.1 Probable successional stages of the *Tsuga-Maianthemum-Vaccinium* habitat type as defined by Coffman et al. (1984). The thicker the arrows, the faster the rate of succession from one stage to the next; the dashed arrows indicate possible, but unlikely, paths. Double-lined boxes represent the most common successional stages in the Upper Peninsula of Michigan; the wider the spacing between lines, the more common the successional stage.

relative density measures to identify the particular vegetative association in a given area.

It is difficult to be more explicit concerning sampling methods because of the variety of vegetative classification systems that exist. It is very important that the users of particular classification systems carefully study the field procedures used to collect the information that was utilized in the development of the system. A particular classification system implicitly assumes certain sampling methods. Failure to follow similar field procedures can lead to misclassification and misinterpretation of this type of information.

TABLE 7.2 Example of Interpretive Information for the *Tsuga-Maianthemum-Vaccinium* Habitat Type in Upper Michigan and Northeastern Wisconsin as Described by Coffman et al. (1984).[a]

Species	Site Index (m at Age 50)	Mean Annual Growth (m^3/ha)	Rotation Age (Years)	Successional Stability
Jack Pine	(22–24)[b]	(4.9–6.3)	40	Very Low
Red Pine	22–24	7.7–10.1	40	Low
White Pine	–[c]	–	–	Moderate
Aspen	(20–24)[d]	(3.8–8.3)	45–55	Very Low
White Birch	–	–	–	Very Low
White Spruce	15–19	–	–	Moderately High
Balsam Fir	–	–	–	High
Red Oak	(19–21)	–	–	Moderately High
Red Maple	(15–17)	(1.4–2.2)	–	High

[a] Used with permission.
[b] Values in parentheses are based on preliminary information.
[c] A (–) indicates that no value was given for that species by Coffman et al. (1984).
[d] Aspen is highly variable owing to genetic variability and high dependency on stand treatment at the time of regeneration.

Systems for classifying vegetation association vary in their availability and sophistication by region. Such systems usually utilize the names of common species or genera to denote the different associations. In the Intermountain region, therefore, the Douglas fir habitat series is recognizable based on the potential overstory species that could occupy the site in the absence of disturbance. Even within regions there may be various vegetation classification systems used by different management organizations.

It is certainly possible to develop classification systems designed for particular regions or for particular purposes, though this is not a trivial effort. Pregitzer and Barnes (1984), Carleton, Jones, and Pierpont, (1985), Spies and Barnes (1985), Hix (1988), and Fincher and Smith (1993) provide examples of different approaches that may be used to develop vegetative classification systems. It is beyond the scope of this text to further describe the various approaches for developing such systems; Grieg-Smith (1983) and Kent and Coker (1992) provide good summaries of the range of statistical and field methods used in such efforts.

The U.S. Forest Service is developing a comprehensive national system of ecological classification, referred to as the National Hierarchical Framework (McNab and Avers, 1994), it is designed to provide a structure for ecosystem classification that extends from the global scale to the local level within a landscape (Table 7.3; Figure 7.2). Within regions, ecological landtypes within the national framework can be linked to local systems of classifying vegetative association for which interpretive information is already available.

TABLE 7.3 Classification Units Within the National Hierarchical Framework

Analysis Scale	*Ecological Units*	*General Spatial Scale*
Ecoregion		
Global	Domain	10^4–10^9 ha
Continental	Division	
Regional	Province	
Subregion	Section Subsection	10^3–10^4 ha
Landscape	Landtype Association	10^2–10^3 ha
Land Unit	Landtype Landtype Phase	10^0–10^2 ha

Source: McNab and Avers (1994).

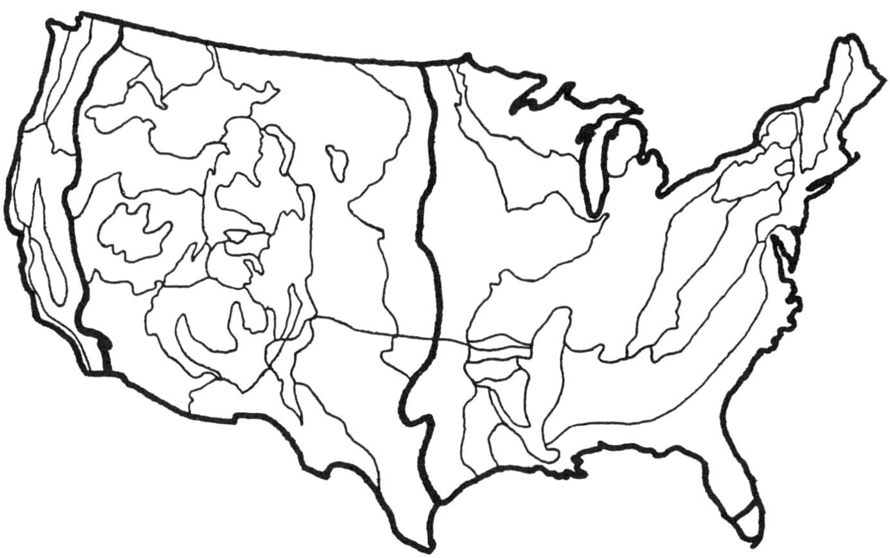

FIGURE 7.2 Ecoregions and subregions of the United States (adapted from Bailey (1994)).

Keys et al. (1995) developed a map of ecological units of the eastern United States that is also available in electronic form that is compatible with common geographic information system software (see Chapter 13). In some parts of the United States, the system is fully developed down through the descriptions of landtype phases. In northern Michigan and Wisconsin, the National Hierarchical Framework is linked with the soil survey and an existing habitat type classification system (Coffman et al., 1984). Application of this integrated system was initiated in the National Forests of the two states and is now being extended across the region (Figure 7.3).

The combined system links soils, physiography, and habitat type within the context of the national framework. The value of linking with existing local systems is that interpretive information, such as the estimated productivity of different overstory species, that has already been developed for soil mapping

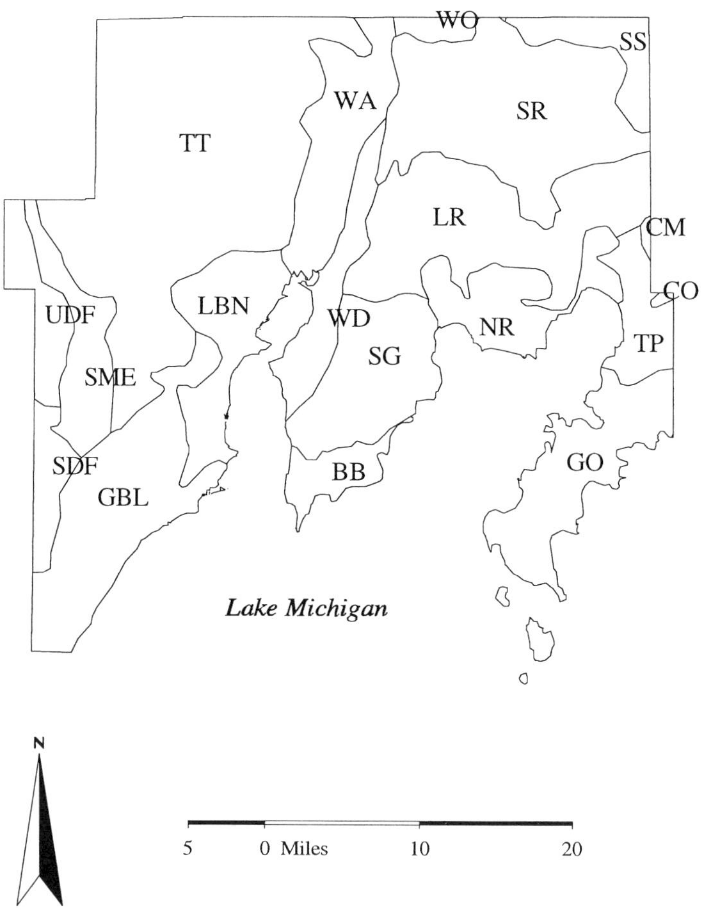

FIGURE 7.3 Ecological landtype associations delineated for Delta County, Michigan (adapted from U.S. Forest Service data); the different letters represent names associated with the different landtype associations.

units and habitat types in the region may be utilized in the combined system. It is then a matter of refining the estimates for the ranges of conditions associated with each landtype, rather than beginning anew in developing interpretive information for each landtype and landtype phase.

QUANTITATIVE MEASURES OF COMMUNITY COMPOSITION AND STRUCTURE

Quantitative measures are designed to provide repeatable descriptions of community composition and structure. Aspects of community structure that can be measured include the presence, density, and abundance of individual species, as well as overall species richness, evenness, and diversity within the community. *Species richness* refers to the number of species occupying an area; areas with greater numbers of species are said to have greater richness. *Species evenness* refers to the relative abundance of different species in a community; communities in which the species have approximately equal relative frequencies are said to have a high degree of evenness (Pielou, 1977). *Species diversity* refers to both the number of species in an area and their relative frequencies; diversity is high when the number of species is high and their relative abundance is about equal.

Management questions examine the proportion of the landscape that should be maintained in each successional stage and the best method of mixing the successional stages across the landscape (Boyce, 1995). For large contiguous blocks of common ownership, this is a difficult question. For landscapes where the pattern of land ownership is highly fractionated, methods and examples of coordinating management activity among landowners in order to maximize landscape diversity are in their infancy, though instances in which such coordination is being attempted are becoming more numerous. Campbell and Kittredge (1996), for example, outline an effort to coordinate landscape management activity among small private landowners in western Massachusetts.

At the ecosystem level, methods of quantifying community structure are relatively well defined. In landscape-level resource assessment efforts, it is usually not possible to conduct a complete inventory or census of the species present in a community; some method of sampling is required. In almost all applications, numerous fixed area plots, or *quadrats,* are established in a random or systematic fashion (see Chapters 8 and 15). Observations are made of individual species' presence or density in the quadrats and the numbers averaged to obtain an estimate for the overall population. To assess community composition, the species present at each plot are tallied and the results across all sampled plots are combined to assess richness, diversity, and evenness. Indices that are appropriate for use in sampled populations are emphasized here; more in-depth presentation and discussion of these methods may be found in Pielou (1977), Ludwig and Reynolds (1988), Kent and Coker (1992), and many other sources.

Individual Species

To assess structure, it is common to utilize a measure of the relative frequency of each species for each plot. This could take the form of an estimate of cover, such as that given by an ordinal scale (Table 7.4). For overstory species, such an estimate could be achieved by a measure of density such as the number of trees per hectare. In *regeneration surveys,* estimates of the density or frequency of seedlings of different potential overstory species are of primary interest.

The timing of management treatments may have a great impact on the success of natural regeneration (Smith, 1986). Many overstory species produce heavy seed crops on a periodic basis, usually every three to five years. Implementing management activity when there is adequate seed production or existing seedling density may be critical to the successful natural reestablishment of the overstory community following treatment. For a species to be present in the future overstory, it is often not germination that is most critical, but establishment and recruitment into larger size classes. If artificial regeneration is being considered, the amount of natural regeneration in an area is critical to the successful establishment of the desired species.

Regeneration surveys are designed to estimate the density of seedlings or saplings of potential overstory species. This may be done to determine the likely successional path of an overstory community, or it may be done to determine silvicultural options for stand treatment. In some cases, there may be legal or regulatory requirements concerning the establishment of a minimum level of regeneration following harvesting; regeneration surveys may be necessary to demonstrate compliance with these requirements. The definition

TABLE 7.4 Example of An Ordinal Scale for Recording Data on Percent Cover

Percent Cover	*Cover Code*
0–1	1
2–5	2
6–10	3
11–20	4
21–40	5
41–60	6
61–80	7
81–100	8

of *seedlings* is usually structural, such as "individuals less than 1.37 m in height," rather than physiological, whereby seedlings may be defined as individuals that still have attached cotyledons. Surveys of *advanced regeneration,* with individuals sometimes termed *saplings,* usually consider individuals greater than 1.37 m in height but less than 10 cm in diameter.

In most regeneration surveys, small circular plots (10 m^2, for example) are established throughout the area and counts are made of the seedlings and/or saplings of each species present. In cases where there are very high seedling densities, small plots (1 m^2, for example) may be used to observe seedlings, and these plots may be nested inside larger plots that are used to observe advanced regeneration. As noted by Schreuder, Gregoire, and Wood (1993), this is a relatively straightforward sampling problem.

If sample quadrats are randomly selected (see Chapter 15), the mean seedling density may be estimated by calculating the mean number of seedlings per quadrat and converting to a per unit area basis.

$$\overline{D} = \frac{\sum_{i=1}^{n} d_i}{A \cdot n} \qquad [7.1]$$

where $\overline{D}$ is the estimated number of seedlings per unit area, d_i is the number of seedlings on the ith sample plot or quadrat, A is the area of the individual sample plots or quadrats, and n is the number of sample plots or quadrats established in the population. Table 7.5 contains a set of data describing sugar maple advanced regeneration in a northern hardwood forest in northern Michigan. A total of forty 100 m^2 (0.01 ha) circular plots were established in the population; all saplings greater than 1.3 m in height, but less than 10 cm in diameter, were counted on each plot. There were a total of 283 saplings observed on the 40 plots, and the estimated number of sugar maple saplings per hectare was:

$$\begin{aligned}\overline{D} &= \frac{283}{(0.01) \cdot (40)} \\ &= 707.5 \text{ saplings per hectare}\end{aligned} \qquad [7.2]$$

The standard error of the estimated density is estimated from the standard deviation of the number of individuals per sample plot, the number of sample plots, and their size:

$$s_{\overline{D}} = \frac{s_d}{A \cdot \sqrt{n}} \qquad [7.3]$$

The standard deviation of the number of individuals per sample plot is calculated as in simple random sampling (Chapter 15):

TABLE 7.5 Observed Numbers of Sugar Maple Saplings Greater Than 1.3 m in Height, But Less Than 10 cm in Diameter, on 100 m² (0.01 Hectare) Sample Plots in a Northern Hardwood Population in Northern Michigan

Plot Number	*Sapling Count*	*I**	*Plot Number*	*Sapling Count*	*I**
1	6	1	21	8	1
2	7	1	22	9	1
3	4	0	23	7	1
4	9	1	24	6	1
5	2	0	25	4	0
6	8	1	26	0	0
7	6	1	27	16	1
8	12	1	28	5	1
9	0	0	29	7	1
10	9	1	30	10	1
11	3	0	31	6	1
12	5	1	32	8	1
13	6	1	33	5	1
14	10	1	34	9	1
15	1	0	35	11	1
16	9	1	36	14	1
17	8	1	37	8	1
18	7	1	38	5	1
19	12	1	39	6	1
20	6	1	40	9	1

* The indicator variable, I, is assigned the value 1 if the number of saplings on the 0.01 hectare plot is ≥ 5 (greater than 500 saplings/hectare), and the value 0 if the number of saplings on the 0.01 hectare plot is <5 (less than 500 saplings/hectare).

$$s_d = \sqrt{\frac{\sum_{i=1}^{n} (d_i - \bar{d})^2}{n - 1}} \qquad [7.4]$$

For the example in Table 7.5, the standard deviation of the number of saplings per quadrat is 3.4 saplings. The standard error of the estimated mean number of saplings per hectare is:

$$s_{\bar{D}} = \frac{3.4}{(0.01) \cdot (\sqrt{40})} = 54.2 \text{ saplings per hectare} \qquad [7.5]$$

Other sampling designs, particularly stratified random sampling and cluster sampling, may be applicable to regeneration surveys in some situations (see Chapter 15). Understory plants tend to have clumped spatial distributions, and special consideration may be needed in some situations. There are cases in which increasing the quadrat size may sufficiently address this problem; this is especially true when the understory environment is relatively homogeneous. In other instances, stratification by microsite may be advisable in order to obtain an adequate estimate of seedling density.

In some cases, the criteria determining the success of regeneration efforts may specify that some fraction of the area in question have a seedling density exceeding some minimum specified number per unit area. In this situation, sampling will again consist of quadrats or small plots, but the variable that is recorded is not the number of seedlings present; instead, the data recorded for each plot consist of an indicator variable that takes the value of 1 if the plot meets the density criteria and a value of 0 if it does not. The proportion of plots meeting the minimum density criteria, then, is simply the average of the indicator variables for all the sampled plots:

$$\hat{p}_D = \frac{\sum_{i=1}^{n} I_i}{n} \qquad [7.6]$$

where I_i is the indicator variable for the ith plot and n is the number of plots. For the example in Table 7.5, an indicator variable could be developed for each plot if the minimum number of saplings per plot to meet some criteria could be specified. If, for example, there was a requirement that there be a minimum of 500 saplings per hectare on at least 75% of the area before a particular overstory treatment could be implemented, the indicator variable would take the value of 1 on all of the 0.01 hectare plots that contained at least five saplings, and the value of 0 on all plots with less than five saplings. For the example, the minimum of five saplings was met on 33 of the 40 plots, and the estimated proportion of the population meeting this criteria is:

$$\begin{aligned} \hat{p}_D &= \frac{33}{40} \\ &= 0.825 \end{aligned} \qquad [7.7]$$

In this example, therefore, the proportion of the population meeting the required minimum number of saplings per hectare to implement the management treatment is 0.825, implying that the management treatment may be implemented if it is the best available procedure to meet the management goals. The standard error of the estimated proportion is estimated by substituting the indicator variables for the seedling numbers and the estimated proportion for the mean density in Equation [7.3]; a confidence interval or lower confidence limit may be estimated using the standard error and a t-value, as described in Chapter 15.

Diversity

Diversity is a difficult concept to reduce to a single quantitative measure. The term *biodiversity* is often used in the sense of presence/absence and abundance of species in a given area. It is also possible to discuss structural diversity, whereby diversity is related to the number of individuals in different height or diameter classes within an area. Buongiorno et al. (1994), for example, examined diameter class diversity in northern hardwood forests under different silvicultural regimes. Volumetric diversity refers to the distribution of biomass or, sometimes, basal area (usually of overstory species). The following discussion is in the context of species diversity. With appropriate substitution of notation and variable definitions, the same methods may be used to assess other aspects of diversity, such as structural diversity.

The observation and interpretation of these measures in single ecosystems is quite different from their measurement and interpretation at the landscape level. Whittaker (1972) described different types of diversity as a function of scale; *α-diversity* indicates the diversity within a single community or ecosystem, *β-diversity* indicates diversity along an environmental gradient, and *γ-diversity* is the larger-scale, landscape-level diversity. Species diversity is also a function of past disturbance. It is not possible to maximize diversity at the landscape level (γ-diversity) by maximizing α-diversity in every ecosystem. Different successional stages contain different assemblages of species. Maintaining a certain successional stage will maximize α-diversity for the ecosystem, but all successional stages must be present in a landscape in order to maximize landscape diversity (γ-diversity). Some communities, therefore, must be "suboptimal" (i.e., have less than their maximum possible α-diversity) in order to increase the overall γ-diversity of the landscape.

Simpson (1949) introduced the first quantitative measure of ecological diversity, now referred to as *Simpson's Index* and commonly denoted by λ. Simpson's index approximates the probability that two individuals sampled from the population will be of the same species; the value of the index ranges from 0 to 1, with less diverse communities having values closer to 1. Simpson gave the following unbiased estimator of λ for use when the data are from a sample of the population:

$$\hat{\lambda} = \sum_{i=1}^{s} \frac{n_i\,(n_i - 1)}{n\,(n - 1)} \qquad [7.8]$$

where the summation is over all species in the sample, n_i is the number of individuals of the ith species in the sample, and n is the total number of individuals in the sample. Table 7.6 contains data on overstory composition from two forested stands in northern Michigan. Based on the numbers of trees per hectare, the value of Simpson's index is 0.400 for site 1 and 0.535 for site 2. This implies that the community at site 2 is less diverse than the community at site 1.

TABLE 7.6 Data on Overstory Composition from Two Forested Sites in Northern Michigan

	Stems per ha		Biomass (kg/ha)	
Species	*Site 1*	*Site 2*	*Site 1*	*Site 2*
Bigtooth Aspen	82	44	1,131	637
Quaking Aspen	63	–	24,915	–
Sugar Maple	3	–	205	–
Red Maple	862	656	95,929	134,999
Northern Red Oak	174	209	541,895	744,029
Paper Birch	257	25	4,572	576
Red Pine	13	3	21,143	383
White Pine	3	6	77	542
Total	1,457	943	689,867	881,166

Source: Mroz et al. (1983).

The most common single measure of diversity utilized in plant community studies is the *Shannon-Weaver Index* (Shannon and Weaver, 1949). This index was not developed for application in an ecological context, but as an extension of mathematical information theory. Pielou (1977) provides an excellent discussion of the theoretical basis of the application of this index to ecological studies. The Shannon-Weaver Index is always greater than or equal to zero; it takes the value of zero when there is only one species in a community and takes its maximum value when all species are present in equal abundance. The Shannon-Weaver Index, usually denoted by H', is given by:

$$H' = -\sum_{i=1}^{s} p_i \log p_i \qquad [7.9]$$

where the summation is over all species present in a stand (s) and p_i is the proportion of the population belonging to the ith species; $p_i = N_i/N$ (where N is the total number of sampled individuals and N_i is the number of sampled individuals of the ith species) and the p_i values sum to unity. Using the number of trees per acre in Table 7.6, the value of the Shannon-Weaver diversity index is 1.236 for site 1 and 0.876 for site 2. This implies that the community at site 1 is more diverse than the community at site 2, which is the same conclusion found in using Simpson's Index. For each site, the value of p_i is calculated for each species by dividing the number of stems per acre for the species by the total number of stems per acre.

Unfortunately, the estimation of the Shannon-Weaver Index in practice is more difficult than might be expected. Utilizing $p_i = N_i/N$ for all species leads to a biased estimate of H'; the bias is a function of the unknown true total number of species in the community and the total number of individuals observed (Basharin, 1959; Pielou, 1966). Figure 7.4 illustrates the estimated values of H' for different numbers of samples in a simulated population of known structure. There is an upper asymptote that represents the approximate true value of H'. Pielou (1975, 1977) discusses a method to estimate H' and the associated sample variance that converges to the true expected value of H' as N_i increases. The method she presents is a sequential sampling procedure in which the number of sample plots is increased until the estimate of H' obtained from Equation [7.9] stabilizes and approximates an asymptote (see Figure 7.4b).

For small areas in which a complete inventory is taken, another measure of diversity that may be used is Brillouin's Index (Brillouin, 1962) which, as noted by Pielou (1977), was first applied in an ecological setting by Margalef (1958). This index, H, is given by:

$$H = \frac{1}{N} \log \frac{N!}{N_1!\,N_2!\,.\,.\,.\,N_s!} \qquad [7.10]$$

where N_i $(i = 1, 2, \ldots, s)$ is the number of individuals in the ith species and N is the total number of individuals in the population of s species. The Shannon-Weaver Index is appropriate in settings where the population is sampled; Brillouin's Index is appropriate when a complete inventory of the population is performed. Since a complete census is required to use Brillouin's Index, it is rarely applicable to surveys of large areas. It is, however, more appropriate than the Shannon-Weaver Index for describing the diversity of small, completely censused plots or quadrats.

Evenness

Many indices indicating evenness have been proposed for use in completely censused populations. Many of these are sensitive to the occurrence of rare species, which makes their utility problematic in sampled populations. One index, referred to as the *Modified Hill's Ratio,* is relatively insensitive to the occurrence of rare species (Alatalo, 1981) and may be used in sampled populations. This index is calculated using the estimated values of Simpson's Index and the Shannon-Weaver Index for the population:

$$E = \frac{\left(\frac{1}{\lambda}\right) - 1}{e^{H'} - 1} \qquad [7.11]$$

where estimates of λ and H' are obtained using Equations [7.8] and [7.9], respectively. This evenness index approaches zero if one species tends to dominate the community. For the data in Table 7.6, the value of E for site

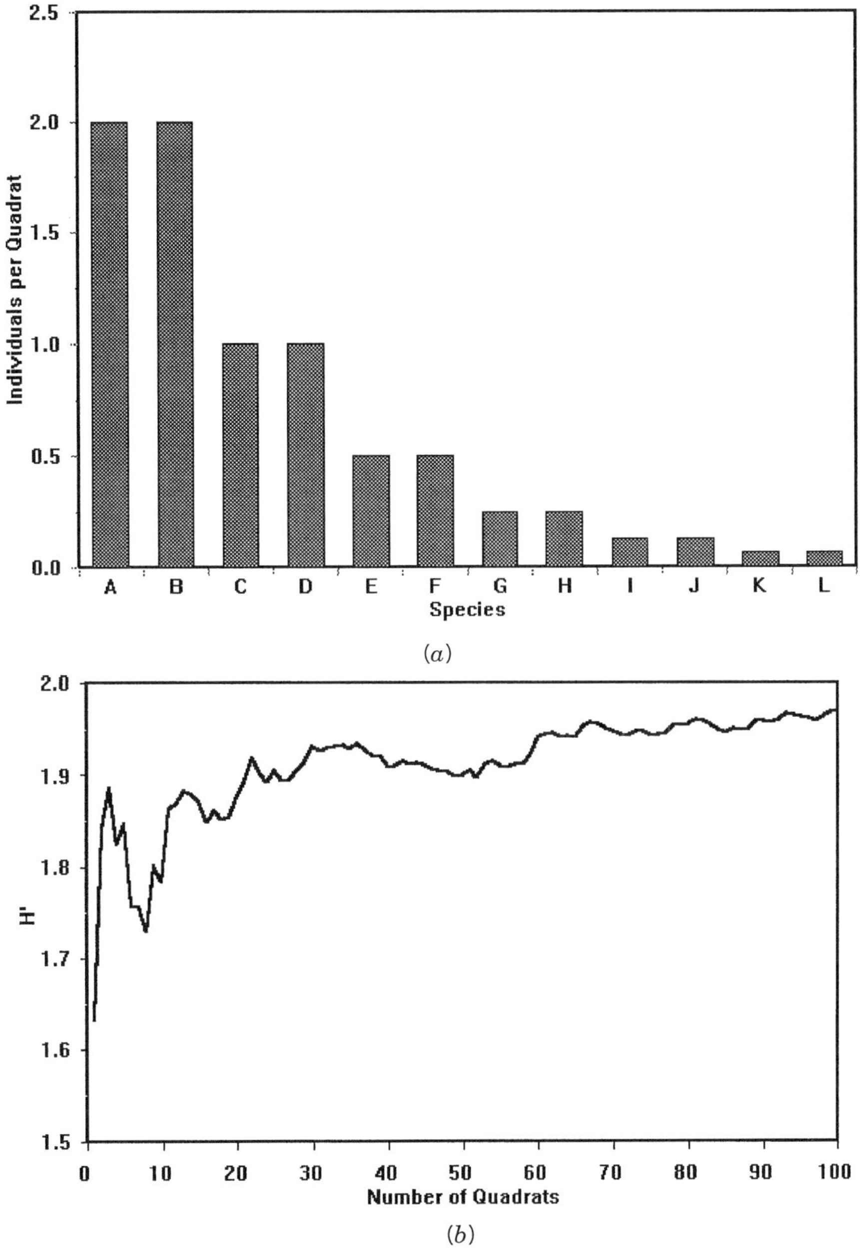

FIGURE 7.4 Calculation of community diversity for a simulated plant community: (*a*) Relative frequencies of different plant species in the simulated community; (*b*) estimates of the Shannon-Weaver diversity index for different numbers of quadrat samples simulated in the population shown in (*a*).

1 is 0.614 and the value of E for site 2 is 0.620. These values indicate that the evenness of these two communities are very similar.

Community Similarity

In some applications, it may be necessary to compare vegetative communities from two different locations and assess their similarity. One situation in which this may arise is when there are two communities on similar soil types and in similar topographic positions, but one of them has been disturbed or treated in some way in the past. An evaluation of the degree of similarity between the vegetation community on disturbed and undisturbed sites is one method of evaluating the ecological impact of the disturbance. Another situation may be one in which two treatments will be applied in close proximity and baseline information is needed in order to assess changes in the degree of similarity in vegetative communities following disturbance.

Several similarity indices have been widely applied. Mueller-Dombois and Ellenberg (1974) present a good summary of these techniques, as do Kent and Coker (1992); Ludwig and Reynolds (1988) provide a basic introductory discussion of the use and interpretation of community similarity measures. The first two discussed here utilize a list of the species present in each stand to be compared. The first index, Jaccard's Index (S_J), was developed by Jaccard (1901, 1912, 1928):

$$S_J = \frac{\text{number of common species}}{\text{total number of species}} \cdot (100) \qquad [7.12]$$

where the number of common species is the number of species appearing on both sites and the total number of species is from the merged species list. For the example data in Table 7.6, there are six species in common between the two sites and a total of eight species occurring on the two sites combined. Jaccard's Index, therefore, indicates that there is a 75% similarity in species composition between the two sites.

The second index considered here is Sorensen's Index (S_S) (Sorensen, 1948):

$$S_S = \frac{\text{number of common species}}{(S_1 + S_2)/2} \cdot 100 \qquad [7.13]$$

where the number of common species is the number of species occurring in both communities, S_1 is the number of species occurring in the first community, and S_2 is the number of species occurring in the second community. For the data in Table 7.6, there are six common species, with eight species occurring on site 1 and six species occurring on site 2. Sorensen's Index, therefore, indicates an 86% similarity in species composition between the two sites.

Neither Jaccard's nor Sorensen's index considers the evenness of community composition. A number of indices have been developed that utilize quantitative information, such as number of individuals, basal area, or biomass for each species. Gleason (1920) utilized the same formulation as Jaccard, but with quantitative data rather than simple presence/absence information to describe the similarity of community biomass or frequency data. The similarity in biomass by this method, for example, is simply the percentage of the total biomass on the two sites that is in common species. For the data in Table 7.6, Gleason's method leads to a biomass similarity index (S_G) value of 98%.

Another index utilizing quantitative information as a basis of comparison, rather than just a count of the species present in the respective populations, is Ellenberg's Index (Ellenberg, 1956):

$$S_E = \frac{C/2}{A + B + C/2} \cdot 100 \qquad [7.14]$$

where A is the sum of percent biomass unique to the first population, B is the sum of percent biomass unique to the second population, and C is the sum of percent biomass common to both populations. For the data in Table 7.6, 3.6% of the biomass at site 1 is in species occurring only on site 1, and 96.4% of the biomass at site 1 is in species common to both sites; there are no unique species at site 2. Ellenberg's Index, therefore, indicates a 96% similarity in biomass composition between the two sites:

$$\begin{aligned} S_E &= \frac{(196.4/2)}{3.6 + 0 + (196.4/2)} \cdot (100) \\ &= 96.5\% \end{aligned} \qquad [7.15]$$

Czekanowski (1913) also utilized quantitative information to express similarity between plant communities. Czekanowski's Index, which is usually referred to as the Bray-Curtis Index (Bray and Curtis, 1957), places more emphasis on the differences between the communities than does Gleason's or Ellenberg's Index. Czekanowski expressed the index as a proportion, but here it is presented as a percentage in order to facilitate comparisons with the other indices:

$$S_C = \frac{2 \sum_{i=1}^{s} \min(X_i, Y_i)}{\sum_{i=1}^{s} X_i + \sum_{i=1}^{s} Y_i} \cdot (100) \qquad [7.16]$$

where the summations are over all species occurring in either community and *min*( ) is the minimum function. For the data in Table 7.6, the value of Czekanowski's Index for biomass is given by:

$$S_C = \frac{2\,(637 + 0 + \ldots + 77)}{689{,}867 + 881{,}166} \cdot (100) = 81\% \qquad [7.17]$$

A final measure of similarity to be discussed here is not an index in the sense of a unitless measure expressed in a percentage, but is a measure of the quantitative "distance" between the two populations. The Euclidean distance between two populations may be calculated based on quantitative information. This results in a measure of difference that is expressed in the units of the information used in the calculation. For instance, if the similarity in biomass is being calculated, the Euclidean distance, the measure of distance between the populations, is expressed in tons per acre. Euclidean distance is calculated as a simple geometric distance:

$$D_{ij} = \sqrt{\sum_{i=1}^{s} (X_i - Y_i)^2} \qquad [7.18]$$

where the summation is over all species occurring in the two communities being compared. For the data in Table 7.6, the Euclidean distance in biomass is:

$$D_{ij} = \sqrt{(1{,}131 - 637)^2 + (24{,}915 - 0)^2 + \ldots + (77 - 542)^2} = 208{,}453 \text{ kg/ha} \qquad [7.19]$$

This distance is related to measures used in statistical discriminant analysis and in various clustering algorithms to define group membership. The use of Euclidean distance is not limited to the pairwise comparison of different communities; a matrix of distance values may be obtained if more than two communities are evaluated.

Ludwig and Reynolds (1988) present a method of standardizing the Euclidean distance, which they refer to as the *Relative Euclidean Distance.* Here, the equation is rescaled and expressed as a percentage so that the calculated values range from 0 to 100:

$$R_{D_{ij}} = (100)\sqrt{\frac{\sum_{i=1}^{s}\left[\frac{x_1}{\sum_{j=1}^{s} x_j} - \frac{y_i}{\sum_{j=1}^{s} y_j}\right]^2}{2}} \qquad [7.20]$$

This index will take the value of 0 if the same species are present on each site with equal relative abundances; it will take the value of 100 if there is a single species on one site and a single, different species on the second site. For the data in Table 7.6, the Relative Euclidean Distance, in terms of biomass, is:

$$R_{D_{ij}} = (100) \cdot \sqrt{\frac{\left(\frac{1,131}{689,867} - \frac{637}{881166}\right)^2 + \ . \ . \ . \ + \left(\frac{77}{689,867} - \frac{542}{881,166}\right)^2}{2}}$$
$$= 5.4 \qquad [7.21]$$

Each of these respective indices provides a different perspective on the differences in community composition between two sites. If one species is clearly dominant on both sites, then no matter what other species are present on the two sites, Gleason's, Ellenberg's, and Czekanowski's indices will indicate a great degree of similarity, whereas Jaccard's and Sorensen's indices may indicate a very low degree of similarity. Table 7.7, for example, illustrates two artificial situations in which there are considerable differences between the similarity indices discussed earlier. In the first example, two communities are dominated by a single species, but have no other species in common. The indices based on presence/absence information (Jaccard's and Sorensen's)

TABLE 7.7 Two Examples Illustrating the Behavior of the Various Indices of Similarity. Values are Biomass per Unit Area of the Different Species at Each Site

Species	*Site A*	*Site B*	*Site C*	*Site D*
A	90	90	20	10
B	1	–	10	20
C	1	–	20	10
D	1	–	10	20
E	1	–	20	10
F	1	–	10	20
G	–	1	10	–
H	–	1	–	10
I	–	1	10	–
J	–	1	–	–
K	–	1	–	–
Total	95	95	110	100

Similarity Indices (%)	Sites A–B	Sites C–D
Jaccard's (S_J)	9	67
Sorensen's (S_S)	17	80
Gleason's (S_G)	95	86
Ellenberg's (S_E)	90	75
Czekanowski's (S_C)	95	57

indicate less than 20% similarity, whereas indices based on biomass (Gleason's, Ellenberg's, and Czekanowski's) all indicate more than 90% similarity. In the second example, there are two communities with no single dominant species, but the more common species in one community are the less common in the other. In this case, the values of the similarity indices range from 57 to 86%, with the two lowest being Jaccard's and Czekanowski's. Clearly, the indications of the different indices vary with community structure. Moreover, similarity in the species present may be very different from similarity in biomass composition. The choice of the most appropriate indices for describing a particular situation depends on the nature of the problems being addressed and the underlying reason for undertaking the analysis. In most cases, calculating two or more indices is preferable to reliance on a single index in describing the relationship between two plant communities.

ECOSYSTEM SIMILARITY

Many vegetative classification systems rely on site factors in addition to descriptions of the vegetative community (see Carleton, Jones, and Pierpont (1985), Spies and Barnes (1985), Hix (1988), and Fincher and Smith (1993) for examples). There are also classification systems that do not rely on specific vegetative communities, but instead are based on climate, soils, and geology (e.g., Albert et al. (1986); see Chapter 5). The problem of assessing ecosystem similarity is even more difficult than assessing the similarity between two plant communities because there are many additional factors to consider.

One possibility for assessing the similarity between ecosystems is to utilize an approach similar to the Euclidean distance method of assessing vegetative community similarity that was discussed earlier. To use this method, a list of variables describing the most important aspects of the ecosystems to be compared must be developed. This list may contain variables describing the vegetative community (e.g., Shannon-Weaver diversity index or total basal area), soil (e.g., pH or C:N ratio), site (e.g., slope percentage or aspect), or other relevant variables as deemed important for a particular situation. The Euclidean distance between two ecosystems is calculated using Equation [7.18]. It is important to standardize the respective variables prior to calculating the Euclidean distance. This is necessary because the variables may be measured in different units or have different precision. Variables are usually standardized for each observation by subtracting their mean from each of the observed values and dividing by the standard deviation of the variable obtained over the entire data set.

$$x_{si} = \frac{x_i - \overline{x}}{s} \qquad [7.22]$$

where x_{si} is the standardized value of the ith observation of x; $\overline{x}$ is the sample mean and s is the sample standard deviation of x, respectively.

Smith and Iles (1988) presented a method to describe the similarity among ecosystems, and Burton et al. (1991) developed an integrated procedure utilizing several different methods of comparison to describe the similarity among ecosystems. Their overall goal was to identify the systems that were "most similar" among the group of systems being evaluated. Burton et al. (1991) first developed a list of variables that they perceived as describing the most important ecosystem characteristics for their purposes. These characteristics included overstory age, overstory basal area, basal area of each of eight selected overstory species, and percentage of sand in the soil B horizon. They then used cluster analysis (Anderberg, 1973), principal coordinates analysis (Digby and Kempton, 1987), and correspondence analysis (Greenacre, 1984) to identify similar sites. Their purpose was to illustrate an integrated approach in identifying similar ecosystems for utilization in an intensive study of ecological processes. Similar methods are useful in assessment and monitoring efforts when it is necessary to locate reference areas or to identify appropriate treatment and control areas so as to evaluate the longer-term consequences of management activity.

LANDSCAPE STRUCTURE

The description and assessment of landscape structure is in its infancy. Both the sophistication and the interpretation of various measures describing landscape structure are likely to increase rapidly in coming years. The various measures described in the following discussion are relatively simple, but they are very difficult to obtain without the assistance of computational routines such as those available in some geographic information systems (see Chapter 13) or in specialized software packages (e.g., McGarigal and Marks, 1995). Landscapes consist of interacting ecosystems, but the appropriate spatial extent of ecosystem delineation is dependent on the ecological processes of interest. It is implicitly assumed here that an appropriate delineation of ecological units within the landscape of interest is available.

Cullinan and Thomas (1992) compared a number of quantitative methods that examine landscape pattern and scale. They examined the use of simple patch size (Grieg-Smith, 1983), indices of nonrandomness (Grieg-Smith, 1983), spectral analysis (Hill, 1973; Ripley, 1978), fractal dimension (Milne, 1990), variance-ratio analysis (Carlile et al., 1989), and correlation analysis (Carlile et al., 1989). Riitters et al. (1995) also examined different measures of landscape pattern and structure. In general, no single method is uniformly best for all situations, and efforts to investigate landscape pattern and scale should include more than one method in the analyses.

The simplest measure of landscape structure is represented by the average size of the delineated landscape units, also referred to as *average patch size* (Grieg-Smith, 1983). In a fixed-size landscape, the average patch size is inversely related to the number of delineated landscape units, as illustrated in

Figures 7.5a and 7.5b. The same landscape may be delineated differently for different purposes, as noted in Chapter 1. The delineation of first-order watersheds may be appropriate for some purposes, whereas for others it may be more appropriate to utilize overstory community composition. It is possible to use several variables to delineate landscape units, but, in any case, the average size of the landscape units provides information concerning the typical spatial extent of the relevant processes within the landscape. In one sense, landscapes that have smaller average patch sizes may be thought of as being more complex than those with larger patches. In another sense, however, it

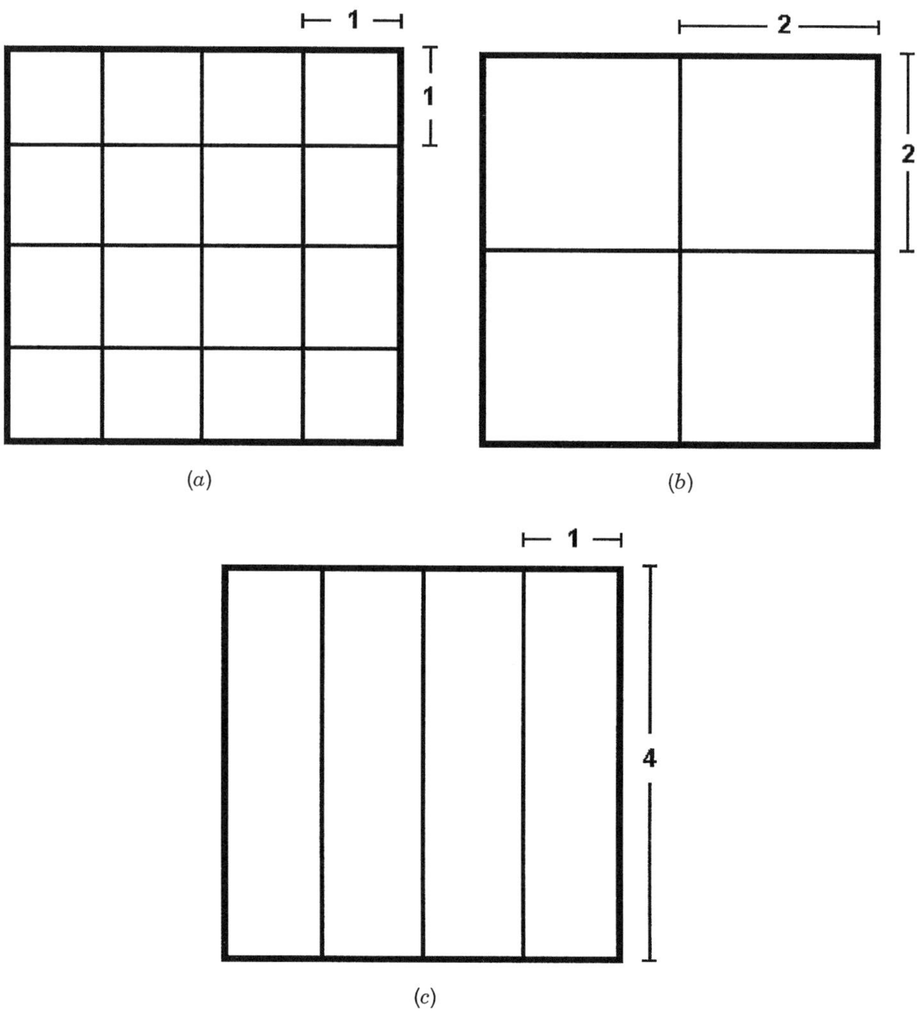

FIGURE 7.5 Artificial landscapes with different structures: (*a*) small square landscape units; (*b*) larger square landscape units; and (*c*) larger rectangular landscape units.

is probably better to think of complexity as a property of the interactions between the landscape units, rather than their number or size.

Another measure of landscape structure pertains to the length of the boundaries of landscape units, which is referred to as the *perimeter length*. This is a direct measure of "edge," which may in and of itself be of interest in assessing wildlife habitat for some species (see Chapter 9). When coupled with average area, this provides insight into the landscape unit shape. The perimeter length in a landscape increases with the number of landscape units and is related to their shape. As seen in Figures 7.5b and 7.5c, an artificial landscape made of rectangular elements has a greater perimeter length (28 units) than a landscape of square elements (24 units), even though the number of patches and the average patch size are unchanged.

A third measure of landscape structure is the average *perimeter : area ratio* of the component landscape units. The geometric shape that minimizes the perimeter : area ratio is the circle, whereas convex polygons have much greater perimeter : area ratios. Landscapes, of course, cannot be completely constructed from circular units. The perimeter : area ratio is a function of the units used to measure length and area, but is minimized when the landscape units are square. This can be seen in Figures 7.5a, 7.5b, and 7.5c. In Figure 7.5a, the perimeter : area ratio is 4 for an artificial landscape with small square elements, whereas in Figure 7.5b the perimeter : area ratio is 2 for a landscape with the same total area, but larger individual elements. In Figure 7.5c, the perimeter : area ratio is 2.5 for a landscape with rectangular elements, instead of square elements as in Figure 7.5b.

There are different implications in calculating the perimeter:area ratio for each unit and then averaging for the landscape, in contrast to calculating the average perimeter and the average area and then the ratio. These procedures are equivalent to the mean-of-ratios and ratio-of-means estimation procedures, respectively (see Chapter 15). The procedure used for sampling landscape units affects the properties of these alternative estimators of the perimeter:area ratio; a good discussion of the different implications and the advantages and disadvantages of each approach is given by Cochran (1977), with further review by Schreuder, Gregoire, and Wood (1993) and Shiver and Borders (1996). In general, the ratio-of-means estimator is the most consistent estimator of the two, which means that the precision of the resulting estimate increases as the sample size increases; this is not true in general for the mean-of-ratios estimator. The ratio-of-means estimator is found by dividing the total perimeter length in the landscape by the total landscape area.

$$\text{Perimeter : area ratio} = \frac{\sum_{i=1}^{n} P_i}{\sum_{i=1}^{n} A_i} \qquad [7.23]$$

where P_i is the perimeter and A_i is the area of the ith landscape unit; n is the number of sampled units in the landscape. Because perimeter is in linear units and area is in square units, the perimeter : area ratio is expressed in inverse linear units. The mean-of-ratios estimator is obtained by averaging the perimeter : area ratios of the individual landscape units.

$$\text{Perimeter : area ratio} = \sum_{i=1}^{n} \left[\frac{P_i}{A_i} \right] \qquad [7.24]$$

where P_i and A_i are again the perimeter and the area, respectively, of the ith landscape unit. The mean-of-ratios estimator may have a smaller variance in the special case when the probability that the landscape units are selected for sampling is proportional to their area (see Chapters 8 and 15).

Another method of describing landscape structure is to describe the *fractal dimension* of the landscape units. A good nontechnical discussion of these methods is given by Allen and Hoekstra (1992); Milne (1988, 1990) describes techniques for measuring and interpreting the fractal geometry of landscape units. Essentially, the more complex the shape of the landscape units, the greater the fractal dimension of the landscape (Figure 7.6). An agricultural matrix, with fields delineated by long, straight boundaries, has a low fractal dimension. A forested landscape, with overstory composition governed by physiographic features, drainage patterns, and soil types has a higher fractal dimension.

Fractal dimensions may vary at different scales and may provide insights into the processes operating at the landscape level. In an investigation of forested landscapes in the United States, Krummel et al. (1987) noted that at small spatial scales there was a lower fractal dimension because the structure of landscape units tended to be governed by human social and economic factors, such as property ownership patterns. At larger spatial scales, the fractal dimension increased because the shapes of the landscape units tended to be governed by natural physiographic features. Turner (1987) found a temporal feature to the fractal dimension of forested landscapes that reflected changing social and economic conditions. Investigating forested landscapes in Georgia, she found that there tended to be a low fractal dimension in the 1930s when the landscape was dominated by agricultural fields. With the

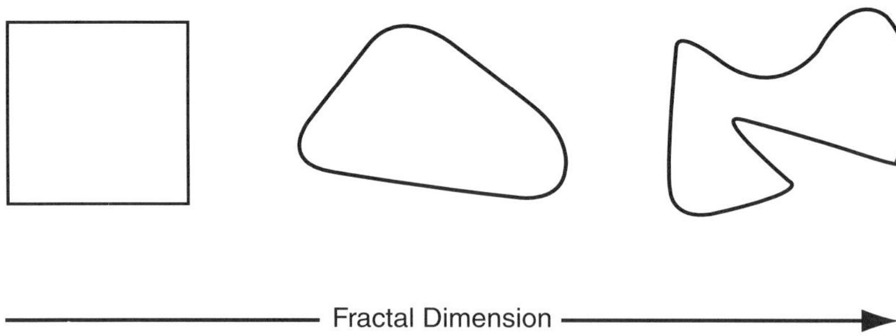

FIGURE 7.6 Increasing fractal dimension as a function of the complexity of the shape of landscape units.

abandonment of farmland following the Depression, the fractal dimension of the landscape increased over time as poorer farmland reverted back to forest.

In a resource assessment process, therefore, the fractal dimension of a landscape may provide an index of human impact on landscape structure. One goal of management in some situations may be to emulate the landscape structure found in pre-European-settlement forests. In this case, the fractal dimension of landscape pattern may provide a means of quantifying the progress made toward achieving this management goal.

SUMMARY

There are many different aspects of the vegetative community that may be described during a resource assessment. The particular information needs will change, depending on the situation. In some cases, it will be sufficient simply to describe the vegetative community in order to develop interpretive information, such as site productivity, for forest management. In other cases, it may be necessary to develop a more quantitative description of the diversity of the vegetative community or of the landscape structure. There is generally no accepted minimum set of analyses that should be conducted in all situations. Instead, the particular information needs and regulatory requirements must be formulated for each case.

APPLICATIONS

7.1 Identify a vegetative classification system applicable to your region. Describe the sampling methods necessary to classify the vegetation of a particular site. Give a list of the types of interpretive information available from the classification system.

7.2 For a section of land, obtain a copy of the original GLO survey notes and describe the pre-European-settlement forest community. Compare this with the current land use and dominant vegetative communities.

7.3 Conduct a regeneration survey in a forest stand using a systematic survey of 30 0.001 ha plots. Develop a table listing, by species, the estimated number of seedlings and saplings per hectare.

7.4 Use the biomass data in Table 7.6 to calculate Simpson's Index, the Shannon-Weaver Index, and the Modified Hill's Ratio for each stand. How do these values compare with the calculations performed in the text using the number of stems?

7.5 Using the information on the number of stems per acre in Table 7.6, calculate the values of the following similarity indices:

- **a.** Jaccard's Index
- **b.** Sorensen's Index

c. Ellenberg's Index
d. Czekanowski's Index

Compare these results with those in the text calculated from the biomass data.

7.6 Using the information in Table 7.7, calculate the values of Simpson's Index, the Shannon-Weaver Index, and the Modified Hill's Ratio for both pairs of communities. How do the differences in community structure affect each index?

7.7 What is the true value of the Shannon-Weaver Index for the community in Figure 7.4a? Note that the frequency decreases by one-half for each successive pair of less frequent species. How does this relate to the values from the simulated samples in Figure 7.4b? What does this imply about estimates of H' obtained from field samples?

7.8 Consider two landscapes of equal area, one of which consists of square vegetation units corresponding to property ownership boundaries and a second that consists of forest communities of irregular shape corresponding to the boundaries of soil taxonomic units and physiographic features. Assume that there are an equal number of units in each landscape. Which landscape has the greater:

a. Average patch size
b. Perimeter length
c. Average perimeter-area ratio
d. Fractal dimension

Chapter

8

Forest Overstory

The forest overstory consists of all vegetation exceeding the size of saplings (individuals greater than 1.37 m in height, but less than some threshold diameter, such as 10 cm). The forest overstory is the vegetation component most often manipulated during management activities. It is also the major component of the vegetation utilized for commodity production. The forest overstory, therefore, is of commercial importance; it simultaneously provides the structure and shelter used by wildlife and modifies the microclimate, affecting the regeneration success of all vegetation.

During a resource assessment, it is important to characterize the forest overstory of the different stands within a landscape. A *stand* is defined as an area of relatively homogeneous site and vegetative characteristics. For some ecological processes, stands are equivalent to ecosystems. For other processes operating at larger spatial scales, stands may be viewed as distinct vegetative communities within ecosystems. A stand is usually the unit of management manipulation. In other words, management activities are usually scheduled and planned on a stand-by-stand basis, even when landscape-level implications are considered. Stands are usually treated uniformly unless the management intent is to reduce the size or change the shape of homogeneous vegetative patches.

Characterization of the forest overstory involves the observation and measurement of living and dead stems of all individuals greater in size than saplings. Characteristics of commercial value, such as size or grade, are recorded in order to conduct a financial appraisal of the commercial value of the overstory. Even if economic return is not a primary management goal, it is usually true that most potential economic returns from management derive from the harvesting of commodities and that such returns are often utilized to fund management activities designed to promote noneconomic management goals and objectives. Almost all land managers, or individuals advising landowners of management options, must be familiar with the means of estimating commodity inventory and value so as to develop sound information to support the decision-making process. Vegetation structure and species composition must be observed in order to define the environmental conditions beneath the overstory, the state of advanced regeneration within the stand, the successional stage of the vegetation, and the habitat characteristics relevant to various wildlife species.

The measurement of forest products and individual trees is discussed in the following sections, as are the techniques for conducting inventories and developing stand-level summaries of the forest overstory structure and composition. Husch, Miller, and Beers (1983), Philip (1994), and Avery and Burkhart (1994) all provide detailed treatment of forest overstory measurement and inventory; Schreuder, Gregoire, and Wood (1993) and Shiver and Borders (1996) provide more advanced treatment of forest inventory issues than can be discussed here.

MEASURING HARVESTED PRODUCTS

The process of measuring the volume or weight of harvested forest products is referred to as *scaling*. The simplest method of scaling is *weight scaling*, whereby the raw material is weighed before it is unloaded from the truck taking it to a mill or processing yard. Raw material may be bought and sold on a weight basis, or there may be a conversion relating weight of a particular species to some other measurement unit.

A *cord*[1] is a common unit for the measure of pulpwood in the United States; it is a stack of roundwood, including solid wood, airspace, bark, and so on, that measures four feet by four feet by eight feet, or 128 cubic feet. A truck carrying eight-foot sticks of pulpwood stacked 20 feet long and eight feet high contains 10 cords:

$$\begin{aligned} \text{Cords} &= \frac{8 \text{ feet} \cdot 20 \text{ feet} \cdot 8 \text{ feet}}{128 \text{ ft}^3/\text{cord}} \\ &= \frac{1{,}280 \text{ ft}^3}{128 \text{ ft}^3/\text{cord}} \\ &= 10 \text{ cords} \end{aligned} \qquad [8.1]$$

[1] English units are used in much of this chapter in the discussion of the measurement of wood products; this is because most of the buying and selling of wood products in the United States is conducted using English units. The international units for commerce in wood products are the cubic meter for solid wood and the metric ton for chips and pulp.

As a rule of thumb, a cord of softwood weighs about 4,000 pounds (2 English tons) and a cord of hardwood weighs about 5,000 pounds (2.5 tons). The exact conversion factors depend on species, moisture content, and size of material. On average, a cord contains 70 to 75 ft^3 of solid wood.

A *board foot* is a common unit of measure of solid wood products in the United States; a board foot is a piece of solid wood measuring one inch by one foot by one foot. Board feet are used as units of measure for finished lumber and for the sale and purchase of raw sawlogs. For finished lumber, the number of board feet in any piece can be calculated by converting the dimensions to inches $\cdot$ ft^2 and calculating the cubic volume. For a nominal 2″ × 4″ piece of lumber eight feet long, the board feet are calculated as follows:

$$\begin{aligned} \text{Board feet} &= \frac{2 \text{ inches} \cdot 4 \text{ inches} \cdot 8 \text{ feet}}{12 \text{ inches/foot}} \\ &= \frac{2 \cdot 4 \cdot 8}{12} \text{ board feet} \\ &= 5.3 \text{ board feet} \end{aligned} \quad [8.2]$$

Sawlogs are usually bought and sold in units of one thousand board feet (MBF) or one million board feet (MMBF). A rough conversion factor between board feet and cords is that one cord of stacked wood equals approximately 500 board feet of solid wood. A cord contains 70 to 75 cubic feet of solid wood, but in processing logs into boards there is waste and loss owing to slabbing and saw kerf, which results in fewer board feet recovered per cubic foot of solid wood than might be expected. The exact conversion varies by size of material and species, but this is a rough guide that can be used for quick calculations in the field.

The process of scaling pulpwood or cordwood in the field involves the relatively simple application of geometric principles to estimate the volume of the stack of roundwood in question. Deductions may be made for "loose piling," excessive limbs remaining on the pieces, or for excessive amounts of dirt and debris in the pile, but otherwise the process is usually straightforward.

The scaling of sawlogs is more complicated and requires considerable expertise. The initial step is to measure the diameter and length of the log. Sawlog diameter is measured as the diameter inside the bark at the small end of the sawlog. Length is commonly measured in two-foot increments, with a usual minimum length of eight feet. An eight-foot sawlog is considered to be a log at least eight feet in length with no notches or other missing pieces on the ends. In addition to the nominal length, a *trim allowance* of at least four more inches is usually required in order to allow the log to be squared-off at the ends. An eight-foot sawlog, therefore, is a minimum of eight feet, four inches, in length.

Once the diameter and length are measured, a *log rule* is used to estimate the gross board foot content of the sawlog. Graves (1906) listed 42 log rules and Recknagel, Bentley, and Guise (1926) listed 50 different log rules then in use in different regions of the United States and Canada; in one of the most comprehensive listings, Belyea (1953) listed 115 different North American log rules. Today there are three commonly used log rules in the United States: Doyle's, the Scribner Decimal C, and the International 1/4″. These log rules differ according to the assumptions they make about log taper, the pattern of sawing boards from the log, the thicknesses of boards to be sawn from the log, and the amount of allowance for slabbing and saw kerf.

According to Avery and Burkhart (1994), the Doyle log rule was developed at about 1825 by E. Doyle. It is a formula-based log rule whereby the board foot volume of a log is given by:

$$\text{Board feet} = \frac{(D - 4)^2}{16} \cdot L \qquad [8.3]$$

where D is the diameter of the small end of the log (in inches) and L is the log length in feet. An 18-inch log, 12 feet long, has the following board foot volume, Doyle scale:

$$\begin{aligned} \text{Board feet} &= \frac{(18 - 4)^2}{16} \cdot 12 \\ &= 147 \text{ board feet} \end{aligned} \qquad [8.4]$$

The Doyle log rule has long been recognized for underestimating the volume of small-diameter logs and overestimating the volume of larger logs (Recknagel, Bentley, and Guise, 1926; see Figure 8.1). Doyle volumes of sawlogs from 8 to 40 inches in diameter and 8 to 20 feet in length are given in Table 8.1.

The second major log rule in use today was developed by J. Scribner and published prior to 1846 (Belyea, 1931). The Scribner log rule is a diagram log rule. To develop a diagram log rule, a circular cross section of a given diameter is drawn on a sheet of paper, the cross sections of boards are drawn in the circle (including allowance for slabs and saw kerf), and the board feet are calculated for each board using the drawn dimensions and the log length. The exact assumptions regarding board dimensions and sawing pattern that Scribner used have been lost. In its usual form, the Scribner log rule is rounded off to the nearest 10 board feet and the zero is dropped, ostensibly to make it easier for scalers to tally. This is usually referred to as the Scribner Decimal C log rule. Grosenbaugh (1952) developed an empirical formula for approximating the Scribner log rule:

$$\text{Board feet} = \left[0.8 \cdot (D - 1)^2 - \frac{D}{2}\right] \cdot \frac{L}{16} \qquad [8.5]$$

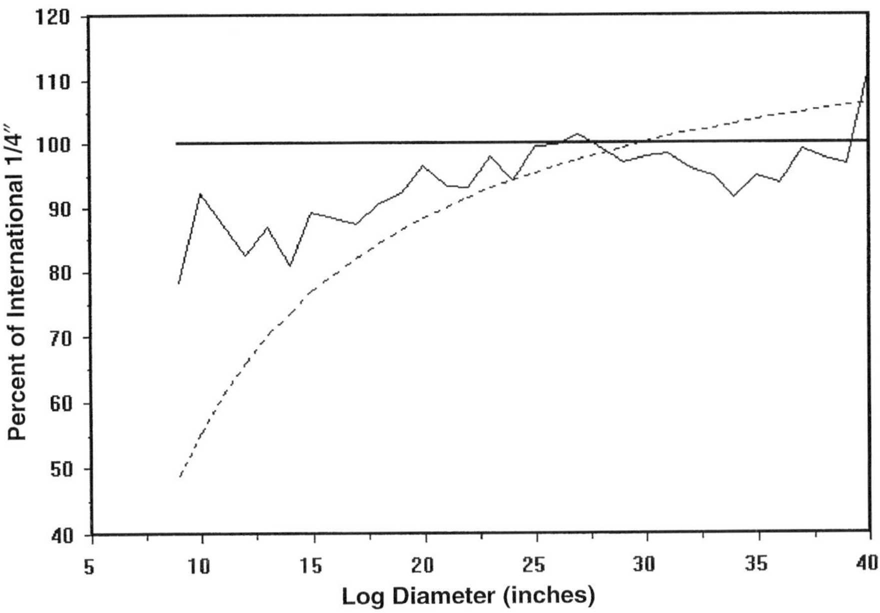

FIGURE 8.1 Comparison of the gross board foot scale from the Doyle (*dotted line*) and Scribner Decimal C (*solid line*) log rules with the International 1/4" log rule for 16-foot logs from 8 to 40 inches in diameter.

where D is the diameter of the small end of the log in inches and L is the log length in feet. Scribner Decimal C volumes calculated using Equation [8.5] are given in Table 8.2 for logs 8 to 40 inches in diameter and 8 to 20 feet in length. The original Scribner Decimal C table indicated that an 18-inch, 12-foot sawlog contains 16 Scribner Decimal C or about 160 board feet; Grosenbaugh's (1952) equation estimates 166.7 or 167 board feet in the same log. The Scribner Decimal C log rule is intermediate in estimated volume (Figure 8.1) as compared with the other two major log rules.

The International log rule is a formula-based log rule that was developed by Clark (1906). The International 1/4" log rule assumes 1/4" of saw kerf plus an additional 1/16" for shrinkage during drying. The International log rule also assumes taper in the log, a factor not considered by either the Doyle or the Scribner log rule. A fixed taper rate of 0.5 inches of diameter per 4 feet of log length is assumed. In the International log rule, volume is first computed for the 4 feet of log section at the small end, scaling diameter is then increased by 0.5 inch, volume is computed for the next 4-foot section, and this continues in the same manner until the total volume of the log is calculated. For the International 1/4" log rule, volume for each 4-foot section is calculated as:

$$\text{Board feet} = 0.905\,(0.22\,D^2 - 0.71\,D) \qquad [8.6]$$

TABLE 8.1 Gross Board Foot Volumes of Sawlogs Based on the Doyle Log Rule

Scaling Diameter (inches)	**Log Length (feet)**						
	8	*10*	*12*	*14*	*16*	*18*	*20*
				Board Feet			
8	8	10	12	14	16	18	20
9	13	16	19	22	25	28	31
10	18	23	27	32	36	41	45
11	25	31	37	43	49	55	61
12	32	40	48	56	64	72	80
13	41	51	61	71	81	91	101
14	50	63	75	88	100	113	125
15	61	76	91	106	121	136	151
16	72	90	108	126	144	162	180
17	85	106	126	148	169	190	211
18	98	123	147	172	196	221	245
19	113	141	169	197	225	253	281
20	128	160	192	224	256	288	320
21	145	181	217	253	289	325	361
22	162	203	243	284	324	365	405
23	181	226	271	316	361	406	451
24	200	250	300	350	400	450	500
25	221	276	331	386	441	496	551
26	242	303	363	424	484	545	605
27	265	331	397	463	529	595	661
28	288	360	432	504	576	648	720
29	313	391	469	547	625	703	781
30	338	423	507	592	676	761	845
31	365	456	547	638	729	820	911
32	392	490	588	686	784	882	980
33	421	526	631	736	841	946	1,051
34	450	563	675	788	900	1,013	1,125
35	481	601	721	841	961	1,081	1,201
36	512	640	768	896	1,024	1,152	1,280
37	545	681	817	953	1,089	1,225	1,361
38	578	723	867	1,012	1,156	1,301	1,445
39	613	766	919	1,072	1,225	1,378	1,531
40	648	810	972	1,134	1,296	1,458	1,620

TABLE 8.2 Gross Board Foot Volumes of Sawlogs Based on the Scribner Decimal C Log Rule as Approximated by Grosenbaugh's Formula (Equation [8.5])

Scaling Diameter (inches)	Log Length (feet)						
	8	*10*	*12*	*14*	*16*	*18*	*20*
				Board Feet			
8	2	2	3	3	4	4	4
9	2	3	4	4	5	5	6
10	3	4	4	5	6	7	7
11	4	5	6	7	7	8	9
12	5	6	7	8	9	10	11
13	5	7	8	10	11	12	14
14	6	8	10	11	13	14	16
15	7	9	11	13	15	17	19
16	9	11	13	15	17	19	22
17	10	12	15	17	20	22	25
18	11	14	17	19	22	25	28
19	12	16	19	22	25	28	31
20	14	17	21	24	28	31	35
21	15	19	23	27	31	35	39
22	17	21	26	30	34	38	43
23	19	23	28	33	38	42	47
24	21	26	31	36	41	46	51
25	22	28	34	39	45	50	56
26	24	30	37	43	49	55	61
27	26	33	40	46	53	59	66
28	28	36	43	50	57	64	71
29	31	38	46	54	61	69	77
30	33	41	49	58	66	74	82
31	35	44	53	62	70	79	88
32	38	47	56	66	75	85	94
33	40	50	60	70	80	90	100
34	43	53	64	75	85	96	107
35	45	57	68	79	91	102	113
36	48	60	72	84	96	108	120
37	51	64	76	89	102	115	127
38	54	67	81	94	108	121	135
39	57	71	85	99	114	128	142
40	60	75	90	105	120	135	150

where D is the scaling diameter in inches at the small end of the 4-foot section. Table 8.3 contains the International 1/4″ volumes for sawlogs from 8 to 40 inches in diameter and 8 to 20 feet in length. For an 18-inch, 12-foot log, volume is calculated in three 4-foot sections, increasing the diameter by 0.5 inch per section, as follows:

$$\begin{aligned}\text{Board feet} &= 0.905\ (0.22\ (18^2) - 0.71\ (18)) + \\ &\quad 0.905\ (0.22\ (18.5^2) - 0.71\ (18.5)) + \\ &\quad 0.905\ (0.22\ (19^2)) - 0.71\ (19)) \\ &= 52.9 + 56.3 + 59.7 \\ &= 168.9 \approx 169 \text{ board feet}\end{aligned} \qquad [8.7]$$

Note that the International 1/4″ volume of 169 board feet is nearly 15% larger than the Doyle volume of 147 board feet calculated for the same log (Equation [8.4]). Figure 8.1 illustrates the relationships among the board foot volumes estimated by the three log rules for 16-foot logs as a function of log diameter. The behavior of the Scribner Decimal C log rule is erratic, because it is a diagram log rule and because volumes are rounded off to the nearest 10 board feet. The Doyle log rule underestimates volumes for small logs. It estimates larger volumes than the Scribner Decimal C for logs larger than 28 inches in diameter and larger volumes than the International 1/4″ for logs greater than 30 inches in diameter.

As long as both buyer and seller understand the customary local log rule, and local prices are adjusted to account for the customary log rule, then there is a fair market and transactions are equitable for all concerned. The authors are aware of anecdotal evidence, however, of log buyers purchasing smaller sawlogs from uninformed private landowners using the Doyle scale, but with the prevailing local market prices based on the Scribner Decimal C log rule, resulting in an economic loss for the landowner.

The use of a log rule, along with the measured scaling diameter and log length, results in estimates of *gross volume* or *gross scale*. Sawlogs are bought and sold on the basis of *net volume* or *net scale*, which is gross volume minus any volume that cannot be utilized owing to defects. There are several types of *scaling defects* that result in a reduction of net volume. For hardwood sawlogs in the eastern United States, a 50% scale reduction is the maximum allowed, although up to 60% reduction is allowed by grading rule exceptions for high-quality logs of valuable species. Logs with greater scale reductions are termed *cull* logs, which means they are of such poor quality that they have no commercial value. Softwood sawlogs are allowed scaling losses of up to 2/3 of gross volume before they become cull logs.

There are three general classes of scaling defects that result in a loss of scale volume: (1) interior defects, (2) side defects, and (3) crook or sweep. *Interior defects* result from heart or stump rot, ring shake (the separation of annual rings resulting from damage during felling), heart check (substantial cracks in the center of the log), and other similar defects. *Side defects* include

TABLE 8.3 Gross Board Foot Volumes of Sawlogs Based on the International 1/4″ Log Rule

Scaling Diameter (inches)	Log Length (feet)						
	8	*10*	*12*	*14*	*16*	*18*	*20*
				Board Feet			
8	17	22	27	33	39	45	52
9	22	29	36	43	51	59	68
10	29	37	46	55	65	75	86
11	36	46	57	68	80	93	105
12	44	57	69	83	97	112	127
13	53	68	83	99	115	133	151
14	63	80	98	117	136	156	176
15	73	93	114	136	157	181	204
16	84	108	131	156	181	207	234
17	96	123	149	177	206	235	265
18	109	139	169	200	232	265	299
19	123	156	190	225	260	297	334
20	137	174	212	251	290	331	372
21	153	194	235	278	321	366	412
22	169	214	259	307	354	404	453
23	185	235	285	337	388	443	497
24	203	257	311	368	424	483	542
25	221	280	339	401	462	526	590
26	241	305	368	435	501	570	639
27	261	330	399	471	542	617	691
28	282	356	430	507	585	665	745
29	303	383	463	546	629	714	800
30	326	411	497	586	674	766	858
31	349	440	532	627	721	819	917
32	373	471	568	669	770	874	979
33	398	502	606	713	821	931	1,042
34	423	534	645	759	873	990	1,108
35	450	567	684	805	926	1,051	1,175
36	477	601	725	853	981	1,113	1,244
37	505	636	768	903	1,038	1,177	1,316
38	533	672	811	954	1,096	1,243	1,389
39	563	709	856	1,006	1,156	1,311	1,465
40	593	748	902	1,060	1,218	1,380	1,542

seams, lightning scars, catfaces (old wounds) and fire scars, sap rot, and similar defects. *Sweep* is defined as the gradual and continuous deviation from a straight centerline; *crook* is the abrupt deviation from a straight centerline in the log.

There are seven general methods for making scale deductions because of defects (Table 8.4). Each method is presented in the following paragraphs, along with an example application. It is difficult to visualize log scaling defects without actually examining real logs, and these examples do not substitute for experience. They are, however, designed to illustrate the major methods of estimating defect volumes and to provide some general concepts regarding sawlog scaling.

1. Squared-defect Method. An imaginary square or rectangle is placed around the defect, the length of the defect is estimated, and the board feet lost to the defect are calculated as follows:

$$\text{Board foot deduction} = 0.8 \cdot \frac{(W + 1) \cdot (H + 1)}{12} \cdot L_D \qquad [8.8]$$

where W is the width of the defect in inches, H is the height of the defect in inches, L_D is the length of the defect in feet, the division by 12 converts one of the inch measurements to feet (so that the units are inches · ft^2, or board feet), the 0.8 factor adjusts for saw kerf and waste, and one inch is added to

TABLE 8.4 Appropriate Scale Deduction Methods for Major Types of Scaling Defects

	Appropriate Deduction Methods	
Type of Defect	*Primary*	*Secondary*
Heart Rot	Squared Defect	Length Deduction
Stump Rot	Squared Defect	Length Deduction
Spiral Side Defect	Pie Cut	–
Straight Side Defect	Squared Defect	Pie Cut
Catface-Fire Scar	Squared Defect	Pie Cut
Sap Rot	Pie Cut	Diameter Deduction
Ring Shake	Squared Defect	Length Deduction
Pitch Seam	Squared Defect	–
Heart Check	Squared Defect	–
Frost Cracks	Squared Defect	–
Weather Checks	Diameter Deduction	–
Sweep	Sweep Deduction	–
Crook	Crook Deduction	Length Deduction
Crotch	Crotch Deduction	–

the height and width measurements of the defect to allow for sawing. A sawlog with a gross Scribner Decimal C volume of 17 (170 board feet) and a 5-inch by 5-inch hole in the butt section that is 4 feet deep, has a scaling defect of:

$$\text{Board foot deduction} = 0.8 \cdot \frac{(5 + 1) \cdot (5 + 1)}{12} \cdot 4$$
$$= 9.6 \approx 10 \text{ board feet} \qquad [8.9]$$

which is equivalent to 1 on the Scribner Decimal C scale. The net scale of this log, therefore, is the gross scale (17) minus the scaling defect (1), which equals 16 (160 board feet).

2. Pie-cut Method. Usually used with side defects, the pie-cut method encases the defect in an imaginary radial sector of the log. The proportion of the log volume lost to the defect is calculated as:

$$\text{Board foot deduction} = R \cdot \left[\frac{L_D}{L}\right] \cdot V_G \qquad [8.10]$$

where R is the proportion of the circumference of the log lost to the defect, L_D is the length of the defect, L is the length of the log, and V_G is the gross scale of the log. For a 16-foot log with a gross Scribner Decimal C scale of 17 (170 board feet), if there is a seam extending 8 feet of the length of the log and occupying 1/8 of the circumference of the log, the scaling defect is:

$$\text{Board foot deduction} = \left[\frac{1}{8}\right] \cdot \left[\frac{8}{16}\right] \cdot 17$$
$$= 1 \text{ Scribner Decimal C} \qquad [8.11]$$

so that the net scale of the log is $17 - 1 = 16$ Scribner Decimal C (160 board feet).

3. Length-deduction Method. In this method, the volume is reduced proportionally when there is a defect at one end of the log that is so severe that no sound lumber can be sawn from that section of the log. If the length of the defect is such that the sound portion of the log corresponds to a usual nominal scaling length, the scaling length of the log is reduced accordingly. If less than 8 feet of the log is sound, the entire log is usually culled, unless the sound portion of the log would qualify as some specialty product. If the length of the defect is such that the sound log length does not correspond to a usual nominal scaling length, the volume lost to the defect is calculated as:

$$\text{Board foot deduction} = \left[\frac{L_D}{L}\right] \cdot V_G \qquad [8.12]$$

where L_D is the length of the defect, L is the scaling length of the log, and V_G is the gross scale volume of the log. A 16-foot log that has a serious end defect extending 2 feet into the log, which affects the entire circumference of the log, is reduced to a 14-foot log for the purpose of defining net scale. The volume lost to the defect in a 16-foot log that has a Scribner Decimal C volume of 17 and a serious defect extending 3 feet into the log, which affects the entire circumference of the log, is calculated as:

$$\begin{aligned} \text{Board foot deduction} &= \left[\frac{3}{16}\right] \cdot 17 \\ &= 3 \text{ Scribner Decimal C} \end{aligned} \tag{8.13}$$

So the net scale of this log is the gross scale (17 Scribner Decimal C) minus the scaling deduction (3 Scribner Decimal C), which equals 14 Scribner Decimal C board feet.

4. Diameter-deduction Method. The diameter-deduction method is used primarily for sap defects and surface checking. With this method, the scaling diameter is reduced by twice the average depth of the surface defect. Care must be taken to ensure that the surface defects are not just due to drying at the ends of the log, but are actual defects extending throughout the length of the log. An 18-inch log with an average 2-inch depth of sap rot extending for the entire log circumference is scaled as a 14-inch log of the same length.

5. Sweep-deduction Method. Sweep deductions are used when a log has a gradual and consistent deviation from a straight centerline. One inch of sweep is allowed without a scaling deduction in logs of less than 12 feet in length; 2 inches of sweep are allowed without a deduction for logs of 12 feet or more in length. The amount of sweep is measured as the deflection of the centerline of the log. Sweep is expressed as a proportion of the scaling diameter:

$$\begin{aligned} \text{Board foot deduction} &= \left[\frac{S-1}{D}\right] \cdot V_G \text{ if } L < 12 \\ &= \left[\frac{S-2}{D}\right] \cdot V_G \text{ if } L \geq 12 \end{aligned} \tag{8.14}$$

where S is the maximum deviation from the straight centerline (inches), D is the scaling diameter, V_G is the gross volume, and L is the log length in feet. If a 16-foot long, 16-inch diameter sawlog has a gross Scribner Decimal C volume of 17 (170 board feet) and has 6 inches of sweep, the scaling deduction is:

$$\begin{aligned} \text{Board foot deduction} &= \left[\frac{6-2}{16}\right] \cdot 17 \\ &= 4 \text{ Scribner Decimal C} \end{aligned} \tag{8.15}$$

This log, therefore, has a net volume of 17 − 4 = 13 Scribner Decimal C (130 board feet).

6. Crook-deduction Method. A crook is a sharp deflection from a straight centerline that extends over less than one-half of the log length. The scaling deduction resulting from crook is calculated as:

$$\text{Board foot deduction} = \left[\frac{L_C \cdot C}{L \cdot D}\right] \cdot V_G \qquad [8.16]$$

where L_C is the length of the crook (feet), C is the maximum deflection of the centerline of the log within the crook (inches), L is the log length (feet), D is the scaling diameter, and V_G is the gross volume. If a 16-foot, 16-inch log has a gross Scribner Decimal C scale of 17, and has a 4-foot crook with a maximum deviation of 8 inches, the scaling defect is calculated as:

$$\begin{aligned}\text{Board foot deduction} &= \left[\frac{4 \cdot 8}{16 \cdot 16}\right] \cdot 17 \\ &= 2 \text{ Scribner Decimal C} \qquad [8.17]\end{aligned}$$

This log, therefore, has a net scale of 17 − 2 = 15 Scribner Decimal C (150 board feet).

7. Crotch-deduction Method. A crotch occurs when one end of a log contains a fork or an extremely heavy branch. The deduction for a crotch defect is achieved simply by reducing the scaling length of the log to the next even-foot log length below the crotch. A 16-foot log with a 3-foot crotch at the upper end is scaled as a 12-foot log with a scaling diameter of the diameter inside bark measured below the crotch at the 12-foot length mark. In all actuality, inasmuch as the log length will be reduced to a point below the crotch, it is better to butt-off the crotch in the woods or at the landing prior to transporting the log to the mill or woodyard.

There are some general considerations to keep in mind when scaling logs. First, the sound portion of the log must be able to yield lumber of minimum length and dimension that is acceptable by local practice. In the Lake States, the minimum acceptable lumber length is 4 feet and the minimum acceptable width is 4 inches. If the sound portion of a log cannot yield lumber of at least these minimum dimensions, the affected portion of the log is included in the defect calculations. Second, when measuring defects near the surface at the large end of the log, it is important to remember that problems outside the scaling cylinder (defined by the small-end diameter, inside bark) will be lost to slabs and processing and should not be counted as defects against gross volume. Third, it is sometimes difficult to determine how far end defects extend into a log. A scaling stick may be used to probe holes into the ends of logs. Clues to the extent of a defect may be obtained by examining the extent of possible swelling on the log surface, and knots extending to the surface may provide clues to the extent of rot along the length of the log.

Many sawlogs contain more than one scaling defect. In this case, the deduction for each defect is estimated independently, then summed to reach the total deduction. Consider a log 16 feet in length with an 18-inch small end diameter. If this log has sap rot extending an average depth of 1 inch across its entire length and circumference, it is scaled as a 16-inch log, giving a Scribner Decimal C scale of 17 (170 board feet). If, in addition, it has a spiral seam extending one-quarter of the circumference of the log for its entire length, there is an additional deduction of 25% of its volume, leading to a net scale of 13 Scribner Decimal C (130 board feet). Care must be taken to ensure that portions of the log affected by more than one defect are deducted only once.

This may seem like a lot of effort in order to scale sawlogs. In reality, experienced log scalers are quite efficient and accurate in estimating net volume. In addition, the value of sawlogs, especially high-quality sawlogs, make the effort of scaling cost-effective, especially in competitive markets.

Because of the inexact nature of both log scaling and milling practices, the board feet of actual lumber produced rarely, if ever, equals the net scaled volume of raw sawlogs entering a mill. If a mill is sawing a greater volume of lumber than net scale volume entering the mill, the mill is said to be in *overrun*. If a mill is producing fewer board feet of lumber than net scaled input, the mill is said to be in *underrun*. Overrun or underrun is usually expressed as a percentage of net scale volume entering the mill. For example, if a mill purchases 10 MMBF of net scale raw sawlogs and produces 11 MMBF of sawn lumber, the mill is said to have a 10% overrun. Efficient mills that tend to be in overrun can be more competitive in purchasing raw sawlogs in the open market.

Log scaling is really just the first step in log grading. *Log grading* systems classify logs into one of several quality classes and are designed to (1) serve as a basis for the settlement of price between the buyer and the seller; (2) provide an indication of the quality of product that may be manufactured from the log; and (3) provide an additional check of mill efficiency against accepted utilization standards. Grading rules identify defects that result in a loss of quality, although these defects may not result in a loss of volume unless they are also scaling defects. Grading rules define the minimum quality sawlog that will meet a particular log grade. In doing so, they establish minimum utilization standards (minimum length, minimum scaling diameter, and maximum amount of scaling defect).

In grading sawlogs, a log is mentally divided into four quarters of circumference, called *faces*. The log grade is based on the three best faces; defects in the poorest-quality face are ignored in all but veneer-quality logs, where all four faces are considered. Grading rules differ by species according to utilization standards. Grading rules promoted for hardwood and softwood sawlogs by the Northern Hardwood and Pine Manufacturer's Association (1976) are given in Tables 8.5 and 8.6. Most log grading systems classify sawlogs within several grade classes: Prime Veneer, Veneer, Grade 1, Grade 2, Grade 3, and

TABLE 8.5 Grading Rules for Hardwood Sawlogs

	Log Grade		
Grade Factors	*1*	*2*	*3*
Minimum Diameter	12 in.	10 in.	10 in.
Minimum Length	8 ft	8 ft	8 ft
Maximum Deduction			
Total Cull	40%	50%	50%
Sweep and Crook	15%	30%	50%
Clear Cuttings (3 best faces)			
Minimum Length in Face			
12–15 in. Diameter	7 ft	3 ft	3 ft
16–19 in. Diameter	5 ft	3 ft	3 ft
20+ in. Diameter	3 ft	3 ft	3 ft
Maximum Number on Face			
8–11 ft	2	2	Unlimited
12+ ft	2	2	Unlimited
Minimum Yield in Face (length)	5/6	4/6	3/6
Discoloration (maximum proportion of diameter)			
<16 in. Diameter	1/2	1/2	Unlimited
16+ in. Diameter	1/2	3/5	Unlimited

Source: Adapted from Northern Hardwood and Pine Manufacturer's Association (1976).

Exceptions:
Grade 2 10-inch diameter must be Grade 1 in surface quality.
Grade 2 11-inch diameter limited to 2 three foot cuttings.
Grade 2 8- to 9-ft lengths limited to 12-inch diameter; 3/4 yield in not more than 2 3 ft cuttings.
Sweep and crook allowance reduced to 1/3 in logs with more than 1/4 diameter in sound end defects.
Sixty percent cull deduction allowed in Grade 2 if otherwise of Grade 1 quality.
Sixty percent cull deduction permitted in Grade 3 if otherwise of Grade 2 quality.

Nonmerchantable. Local market conditions sometimes alter grading standards in practice, particularly for high-value products such as Prime Veneer, which may be sold in the export market. As noted by Lyon (1986), value can differ by a factor of up to 6 in successive grades; in the mid-1990s, Prime Veneer sugar maple sawlogs for the export market were selling for as much as 200 to 300 times the price of normal Grade 1 sawlogs. Accurate grading, therefore, can have a tremendous impact on the economic return to landowners, the business success of processing operations, and the nature of future management practices utilized by landowners.

TABLE 8.6 Grading Rules for Softwood Sawlogs

Grading Factor	Log Grade 1	2	3	4
Minimum Diameter (in.)	14[a]	8	6	6
Minimum Length (ft)	10[b]	10	8	8
Maximum Number of Weevil Injuries	None	None	2[c]	No limit
Maximum Scaling Deduction (%)				
Total	50	50	50	66 2/3
Sweep or Crook	20	30	40	66 2/3
Minimum Face Requirements	2 full, or 4 50%[d]	None	None	None
Maximum Size of Knots on 3 Best Faces				
Red Knots	0	3 in., 1/6 diam.	5 in., 1/3 diam.	No limit
Dead Knots	0	1.5 in., 1/12 diam.	2.5 in., 1/6 diam.	No limit
Degrade (Apply the largest applicable.)				
Conks, Pine Borer Damage on Surface, Punk Knots				
Present on 1 face	1	1	1	–
Present on 2 faces	2	2	–	–
Present on 3 faces	3	–	–	
Log End Defects (Consider log as having a total of 8 quarters—4 per log end.)				
Present in 2 quarters	1	1	1	–
Present in 3 or 4 quarters	2	2	–	–
Present in 5 or more quarters	3	–	–	–

Source: Adapted from Northern Hardwood and Pine Manufacturer's Association (1976).

[a] Twelve- and 13-inch logs with four full-length good faces are acceptable.

[b] Eight-foot logs with four full-length good faces are acceptable.

[c] Eight-foot Number 3 logs limited to one weevil injury.

[d] Minimum 50% length good face must be at least 8 feet.

Grading is usually done at the same time a log is scaled to determine net volume. The species, scaling diameter, length, and soundness establish the maximum grade of a sawlog. The three grading faces are then evaluated for the length of clear cutting on each. The length of clear cutting on a face is defined as the length of the face that is free of grading defects; the minimum length of a clear cutting is defined by the grading system, but is usually at least 2 feet. As indicated in Tables 8.5 and 8.6, the minimum length of acceptable clear cutting for a given grade depends on the log diameter, length, and species. *Grading defects* are imperfections that result in lowering the quality of product that may be recovered from a sawlog. Grading defects for hardwoods include bulges or bumps on the log surface (which may indicate overgrown rot), burls, mineral stain if excessive, the presence of conks or cankers (which may indicate rot), knots, splits, or seams, if they penetrate more than an inch or so below the surface, bird pecks, and so on, as indicated in the grading rules.

Grading rules are periodically reexamined as milling technology and market conditions change. The Timber Producers Association of Michigan and Wisconsin (1988) slightly revised the standards given in Tables 8.5 and 8.6, for example. There are also alternative grading systems. Rast, Sonderman, and Gammon (1973) presented grading rules for hardwood sawlogs, for example. The underlying principles of these various systems are quite similar, but it is possible for some logs to be assigned different grades by different systems. The most important consideration for landowners and land managers is not which system is being used, but that buyers and sellers understand and agree on the system and that the prices are consistent with the grading standards being used.

In summary, the process of scaling and grading forest products is of vital interest to landowners, managers, contractors, and processors. It is important for landowners and managers to be at least minimally familiar with the standard procedures for buying and selling raw forest products in order to properly evaluate management options and to protect themselves from unscrupulous buyers. Even if financial return is not a primary management goal, familiarity with scaling and grading rules can lead to opportunities for increased economic return during management operations. Identification of commercially valuable trees during vegetation management planning can lead to financial returns from the management operation, regardless of the primary purpose of the activity. Moreover, judicious selection of trees for removal can also increase the potential for future economic returns.

The basic process for determining the value of sawlogs through scaling and grading can be broken into three steps:

1. Identify the species of the log, the scaling diameter, the log length, and the gross scale.
2. Determine the volume of scaling defects and estimate net scale.
3. Determine the maximum log grade, based on the log dimensions and soundness; adjust log grade according to conditions of the three best grading faces.

INDIVIDUAL TREE MEASUREMENTS

Now that we have discussed the process of determining the commercial value of harvested forest commodities, we can use this knowledge for inventories and assessments of standing trees. During an inventory of the forest overstory, measurements are made on individual sample trees in selected sample units. The various types of forest overstory inventories are discussed later in this chapter, and the statistical aspects of inventories are discussed in Chapter 15. But first, the types of measurements made on individual trees must be addressed.

Measurements are made on individual sample trees in the field and aggregated to arrive at a characterization of the forest overstory at the stand level. Many of the individual tree characteristics discussed here will be collected in almost every inventory effort, regardless of the underlying management objectives. Other characteristics may not be routinely collected, but will prove useful in some situations.

Species

It may seem obvious, but it is very important to characterize forest overstory structure and composition by species. Different species provide different habitat requirements for wildlife, they regenerate and grow under different physical conditions, they are utilized for different products, and, for the same product, they may have vastly different values.

Diameter

Tree dimensions are important characteristics that are highly related to biomass content; nutrient, light, and moisture requirements; and commodity values. Diameter is measured at a standard reference point, usually referred to as *breast height,* defined as the height 1.37 m (4.5 feet) from the ground line (Figure 8.2). This is a convenient measurement location for most people; for most trees, it is above significant butt swell near the ground line and within the portion of the tree of greatest commercial value. Diameter measurements are made perpendicular to the main stem, so if the tree is leaning, the diameter measurement must be adjusted accordingly (Figure 8.2). If the tree forks below breast height, it should be recorded as two different trees, each measured above the fork and above any stem swell associated with the fork. If the tree forks above breast height, it is measured as a single stem at a point as near as possible to breast height that will avoid any extraordinary stem swell.

Diameters may be recorded to the nearest 0.1 cm or 0.1 inch, but in most operational inventories diameters are recorded by diameter classes. A diameter class is simply an interval for convenient characterization of stand structure. In some parts of the United States, for instance, it is common practice to record tree diameters by 2-inch (5 cm) diameter classes. In this case, a 6-inch tree is any tree with a diameter at breast height (d.b.h.) of between 5.0 and 6.9 inches, inclusive. In other parts of the United States, 1-inch diameter

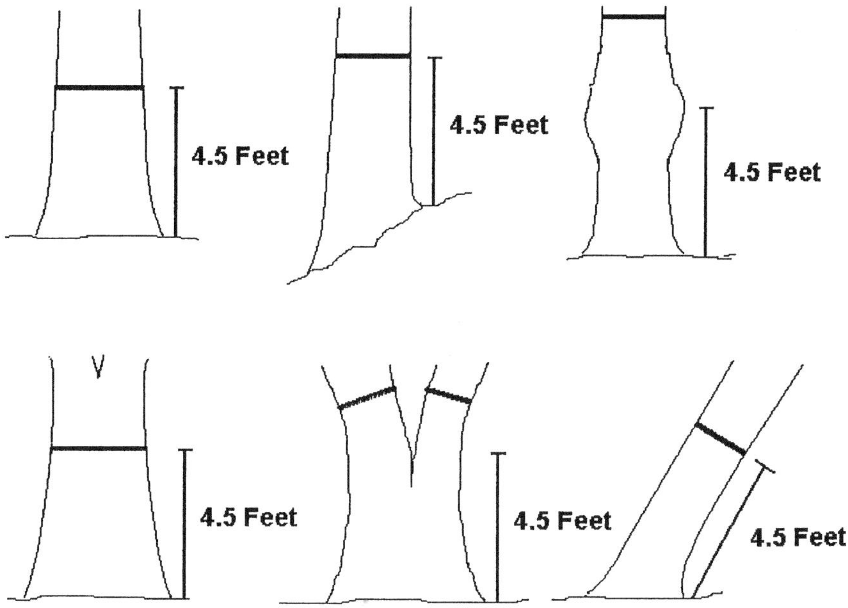

FIGURE 8.2 Diameter measurement points for several common situations.

classes are more common. In using 1-inch diameter classes, the 6-inch class includes all trees between 5.5 and 6.4 inches d.b.h., inclusive.

There are a number of instruments used for measuring diameter at breast height. Experienced personnel often use ocular estimates of diameter class, usually coupled with periodic check measurements using instruments. Other common d.b.h. measurement devices include diameter tapes (Figure 8.3) and calipers (Figure 8.4).

The diameter tape, as illustrated in Figure 8.3, is placed around the circumference of the tree stem. Assuming a circular cross section, the tape is graduated, utilizing the relationship between the circumference of a circle and its diameter (circumference $= \pi \cdot$ diameter). This allows the observer to read the diameter of the tree stem directly from the tape.

Calipers, on the other hand, measure diameter directly and do not rely on the assumption of a circular cross section. As many caliper measurements can be made as necessary to characterize an irregularly shaped tree stem. Usually, two caliper measurements are made at right angles along the major and minor axes of the stem profile (Figure 8.4). The two values are averaged to obtain the tree's recorded diameter.

Basal Area

The cross-sectional area of a tree at breast height is referred to as its *basal area*. Cross-sectional area is usually linearly related to tree volume or biomass

FIGURE 8.3 Use of a diameter tape to measure tree outside bark diameter.

FIGURE 8.4 Use of calipers to measure tree outside bark diameter.

and is indicative of the resources required to support and maintain the living tree. Basal area is not usually measured directly, but is calculated from diameter at breast height by assuming a circular cross section. The basal area, therefore, is equivalent to the area of a circle of the same diameter as the tree stem:

$$\mathrm{B} = \pi \cdot \left(\frac{D}{2}\right)^2 \quad [8.18]$$

where *B* is tree basal area and *D* is tree d.b.h. Basal area is usually expressed in m^2, and diameter is usually measured in cm. To incorporate this conversion of units:

$$\begin{aligned} B &= \frac{\pi \cdot (D/2)^2}{10{,}000} \\ &= 0.00007854\ D^2\ \mathrm{m}^2 \end{aligned} \quad [8.19]$$

A tree with a d.b.h. of 10.2 cm, therefore, has a basal area of .00817 m^2. In using English units, basal area is usually expressed in square feet, and diameter is measured in inches, leading to:

$$\begin{aligned} B &= \frac{\pi \cdot (D/2)^2}{144} \\ &= 0.005454\ D^2\ \mathrm{ft}^2 \end{aligned} \quad [8.20]$$

A tree with a 10.2-inch d.b.h., therefore, has a basal area of 0.567 ft^2.

Height

Tree height can either be measured as *total height,* from the ground line to the top of the tallest point on the tree, or as *merchantable height,* the height from the stump height to a merchantable diameter limit. The *stump height* is the height above the ground at which the tree will be severed during harvesting. This is usually assumed to be 0.3 m or 1 foot in central and eastern North America. The *merchantable diameter limit* is the minimum upper stem diameter limit for a particular product. For sawlogs, the merchantable diameter limit is usually 20 or 25 cm (8 or 10 inches). Merchantable height, therefore, is the height from the top of the stump to the minimum upper stem diameter limit for the particular product and market. Pulpwood minimum diameter limits are typically 5 or 10 cm (2 or 4 inches).

Total height is usually measured in meters or feet, whereas merchantable height is often measured in product units (number of sawlogs or number of pulpsticks). Merchantable height measurements are always rounded down-

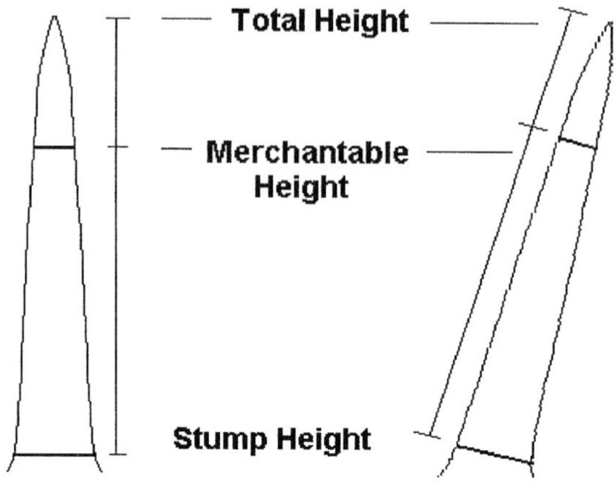

FIGURE 8.5 Height measurements commonly made on standing trees.

ward. In other words, a tree must contain two full sawlogs before it is recorded as a two-sawlog tree.

The goal of height measurement, as related to commodity production, is to obtain an estimate of the length of the tree stem (Figure 8.5). For some purposes, such as the estimation of vertical foliar structure, vertical height to the live crown, is the desired measurement. Care must be taken to ensure that the appropriate quantity is measured to address the management question. In addition, it is important when measuring leaning trees to ensure that there are no biases introduced into the measurements resulting from the line-of-sight. A clear line-of-sight, perpendicular to the lean of the tree, is necessary to obtain an unbiased measurement of height.

There are many different instruments and methods for measuring tree heights. Most are based on the principle of equivalent triangles, as illustrated in Figure 8.6. In applying this principle, the observer is at a known distance

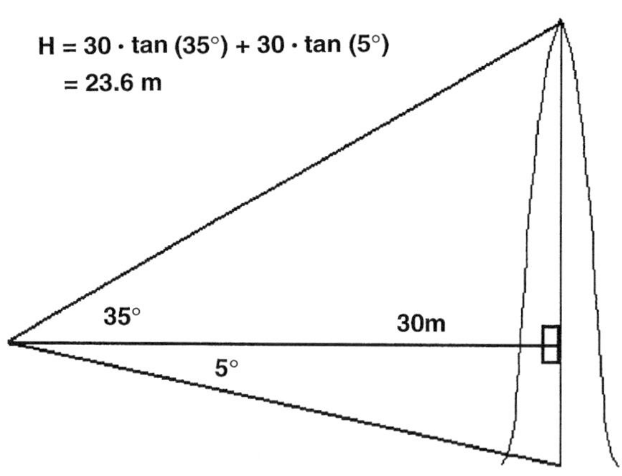

FIGURE 8.6 General method for measuring tree heights based on the principle of similar triangles.

from the tree and the angle to the top of the tree and the angle to the bottom of the tree are measured. The height of the tree is calculated as the sum of the tangents of the two angles times the distance from the tree. In other words, for an observer 30 m from a tree who observes a 35° angle to the tree top and a 5° angle to the tree base, the total height of the tree is:

$$\begin{aligned} H &= 30 \cdot \tan(35) + 30 \cdot \tan(5) \\ &= 21.0 + 2.6 \\ &= 23.6 \text{ m} \end{aligned} \tag{8.21}$$

Many instruments have specially designed scales for reading height directly from certain fixed distances, such as 10, 20, or 30 m, from a tree. In English units, common distances are 100 feet and one chain (66 feet). The 100-foot scale is often labeled the *percent scale,* whereas the 1 chain scale is often labeled the *topographic scale.*

Biomass

Biomass is the mass of an organism. In common use, in referring to plants the term *biomass* usually refers to aboveground biomass. It is not possible to observe directly the biomass of living trees without destructive sampling, so biomass is estimated from the tree dimensions of diameter and height. The relationship between tree dimensions and biomass varies by species, but is often represented as an allometric relationship. Host et al. (1989) developed a computer program that contains equations for estimating the biomass of a large number of forest tree species found in the northern temperate region. Similar programs and relationships are available for almost all major forest tree species of North America. Many biomass estimation systems allow for the estimation of biomass of different tree components, such as wood, bark, and foliage.

Volume

Like biomass, individual tree volume is rarely measured directly, but is estimated based on the tree dimensions of diameter and height. Volume can be expressed in any appropriate units, including weight, cubic meters, cubic feet, cords, or board feet.

Volumes of standing trees are estimated using a *volume table,* which relates tree dimension to volume. Gevorkiantz and Olsen (1955) presented composite volume tables, which are applicable to many species of the Great Lakes region (Tables 8.7 and 8.8). Table 8.7 provides estimates of the number of cords per tree based on the tree's 1-inch diameter class and merchantable height as measured by the number of full eight-foot sticks up to the merchantable

TABLE 8.7 Number of Rough Cords for Standing Trees in the Lake States by 1-Inch Diameter Class and Number of 8-Foot Pulp Sticks to a Variable Top Diameter Limit*

Diameter (inches)	**Number of 8-Foot Bolts**							
	1	*2*	*3*	*4*	*5*	*6*	*7*	*8*
4	**0.007**	**0.011**	–	–	–	–	–	–
5	0.011	0.019	**0.022**	–	–	–	–	–
6	0.017	0.028	0.040	**0.047**	–	–	–	–
7	0.023	0.038	0.053	0.068	**0.076**	–	–	–
8	0.031	0.050	0.068	0.087	0.106	**0.116**	–	–
9	0.040	0.065	0.088	0.109	0.130	0.153	**0.170**	–
10	0.049	0.082	0.111	0.133	0.160	0.188	**0.211**	–
11	0.060	0.100	0.137	0.165	0.190	0.221	**0.250**	**0.270**
12	0.070	0.121	0.165	0.198	0.225	0.260	0.300	**0.330**
13	0.082	0.143	0.197	0.236	0.268	0.305	0.350	0.420
14	0.095	0.167	0.228	0.273	0.311	0.353	0.400	0.470
15	0.107	0.193	0.262	0.318	0.364	0.410	0.460	0.520
16	0.122	0.220	0.300	0.367	0.420	0.470	0.530	0.590
17	0.138	0.250	0.340	0.420	0.480	0.540	0.590	0.660
18	0.155	0.282	0.382	0.470	0.550	0.600	0.650	0.730
19	0.173	0.318	0.430	0.530	0.610	0.680	0.730	0.810
20	0.194	0.353	0.480	0.590	0.680	0.760	0.810	0.890
21	0.217	0.395	0.540	0.660	0.760	0.840	0.900	0.980
22	0.240	0.440	0.600	0.730	0.840	0.930	1.000	1.070
23	0.262	0.480	0.660	0.800	0.920	1.030	1.100	1.170
24	0.288	0.520	0.720	0.880	1.000	1.120	1.210	1.280
25	0.312	0.580	0.780	0.960	1.100	1.230	1.330	1.380
26	0.340	0.620	0.840	1.040	1.190	1.330	1.440	1.510
27	0.363	0.670	0.910	1.120	1.290	1.450	1.560	1.630
28	0.388	0.720	0.970	1.200	1.380	1.550	1.670	1.760
29	0.410	0.760	1.030	1.290	1.490	1.660	1.800	1.900
30	0.430	0.800	1.100	1.370	1.590	1.770	1.930	2.040

Source: Gevorkiantz and Olsen (1955).

*Bold figures in upper portion of table are to a minimum top diameter (inside bark) of 3.0 or more, but less than 4.0 inches. Other top diameters are variable but not less than 4.0 inches.

TABLE 8.8 Gross Board Feet per Tree, Scribner Rule, for Standing Trees in the Lake States by 1-Inch Diameter Class and Number of 16-Foot Sawlogs*

Diameter (inches)	Number of 16-Foot Sawlogs								
	0.5	*1*	*1.5*	*2*	*2.5*	*3*	*3.5*	*4*	*5*
8	10	16	24	31	–	–	–	–	–
9	13	23	31	39	46	–	–	–	–
10	17	30	40	49	57	62	–	–	–
11	22	38	51	62	71	78	–	–	–
12	28	48	66	78	89	100	108	–	–
13	34	59	81	96	112	126	138	145	–
14	40	70	96	116	141	160	170	178	–
15	47	81	113	137	166	188	204	220	–
16	54	93	129	158	191	224	248	263	–
17	63	106	148	182	218	257	285	308	340
18	72	122	168	207	248	292	325	355	395
19	81	137	190	234	280	328	368	405	455
20	90	156	212	262	317	366	415	450	520
21	100	173	238	293	351	405	460	505	585
22	111	194	262	328	392	450	510	560	660
23	123	215	290	360	435	500	560	620	730
24	137	236	319	400	470	550	620	690	800
25	149	258	348	440	520	600	680	760	880
26	165	281	381	480	565	650	740	820	950
27	179	305	415	520	620	710	800	890	1,030
28	195	331	450	560	670	760	860	960	1,120
29	210	356	485	600	720	830	930	1,030	1,200
30	227	383	520	650	770	890	1,000	1,110	1,290
31	245	410	560	700	830	950	1,080	1,200	1,380
32	260	440	600	740	890	1,020	1,150	1,280	1,470
33	279	470	640	790	950	1,080	1,230	1,370	1,560
34	294	500	680	840	1,010	1,160	1,300	1,460	1,670
35	312	530	720	900	1,080	1,230	1,390	1,560	1,790
36	330	565	770	960	1,140	1,310	1,480	1,650	1,900
37	349	600	820	1,020	1,210	1,390	1,570	1,750	2,010
38	365	630	860	1,070	1,270	1,470	1,660	1,840	2,120
39	384	660	900	1,130	1,330	1,550	1,750	1,940	2,240
40	405	700	950	1,180	1,400	1,630	1,850	2,050	2,350

Source: Gevorkiantz and Olsen (1955).

*Bold figures in the upper portion of the table show volume to a top diameter of 6.0 inches or more, but less than 8.0 inches, and hence are applicable only to softwoods.

diameter limit. In measuring the number of pulpsticks in standing trees, an eight-foot stick actually requires 100 inches of stem length in order to accommodate a 4-inch trim allowance during processing. A 10-inch d.b.h. tree with four eight-foot pulpsticks contains 0.133 cords. Table 8.8 provides estimates of the number of board feet per tree (Scribner log rule) based on the 1-inch diameter class and the number of 16-foot (actually 16-foot, 4-inch, including trim allowance) sawlogs to the specified minimum top diameter limit. Table 8.8 assumes that the Scribner Decimal C log rule (Table 8.2) will be used for scaling individual logs. A 16-inch diameter tree with two full 16-foot sawlogs is expected to contain 158 board feet, gross scale.

Grade

Raw materials produced from forests are often separated by grade, which relates to the value of the product that can be processed from the material. This is particularly true of deciduous species in which high-quality pieces may be utilized for very high-value specialty products. The volume of such material may be very low; Erickson, Reed, and Mroz (1990) report a situation in northern hardwoods where high-quality veneer material constituted less than 1 % of the total volume harvested, but contributed more than one-half of the total revenue from the harvest. Obviously, the financial appraisal of such material requires the ability to grade the material that could be recovered from standing trees. The grade of such high-value material depends on much more than just size; Stott (1961) developed a grading system for northern hardwood sawtimber that basically assigns a tree the grade of the highest-value butt log. He extended this system to include expected volumes in material of different grades based on tree diameter and merchantable height. Table 8.9 contains grading rules for standing sugar maple that correspond to the log grading rules of the Northern Hardwood and Pine Manufacturer's Association (1976) given in Table 8.5.

Soundness

Not all material in standing trees is of commercial quality; material that is of such low quality as to have no commercial value is termed *cull* material. The *soundness* of a piece or a tree is usually expressed as a percentage of the total or gross volume that has commercial value. In a tree with a soundness of 90 percent, therefore, 10 percent of the gross volume is cull and 90 percent is net, or sound, volume. The estimation of soundness in standing trees can be accomplished through the use of tables that describe the relative amount of total volume in different portions of the tree bole (Table 8.10). These tables depend on the volume distribution and are species specific, or at least specific to groups of species with similar growth patterns. The procedure is

TABLE 8.9 Procedure for Determining the Butt-Log Tree Grade Corresponding to the Northern Hardwood and Pine Manufacturer's Association (1976) Log Grading Rules for Standing Sugar Maple in the Lake States

Step	*Description*
1	Measure diameter at breast height to the nearest 0.1 inch.
2	Use the soundness percentage distribution (Table 8.10) to determine whether the tree is cull. If not cull, proceed to Step 3.
3	Determine the lower limit of the grading section (i.e., stump height).
4	The grading section is the 16 feet above stump height; locate the best 10-foot portion of the grading section.
5	Estimate diameter inside bark at the top of the 10-foot section to be graded.
6	Use sweep, crook, and total cull percentage to determine the maximum grade in the 10-foot section.
7	If the tree has the minimum diameter inside bark to meet Grade 1 or 2 standards, evaluate the faces using the log grading standards (Table 8.5).
8	Evaluate whether shifting faces will improve grade; if not, record the final tree grade.

to determine any tree sections that are cull, estimate the percent of volume in those sections using Table 8.10, and subtract this cull percent from 100 to obtain an estimate of soundness for the tree.

Crown Position

Crown position refers to the position of a tree within the canopy. This is usually an ocular estimate, placing each tree into one of four crown position classes. *Dominant* trees are those trees in or above the main canopy that receive full sunlight on the top and at least one side of the crown. *Codominant* trees are those trees in the main canopy that receive direct sunlight on the top of the crown. *Intermediate* trees are those that are in the main canopy, but do not receive direct sunlight because they are overtopped by dominant and/or codominant trees. *Suppressed* trees are trees that are below the main canopy and do not receive direct sunlight.

Live Crown Ratio

Live crown ratio is the percentage of total tree height occupied by living foliage. This is relatively easy to estimate in conifers, because the base of the live crown is relatively easy to identify; the point of measurement for the base of the live crown is usually defined arbitrarily, such as the point where the

TABLE 8.10 Percentage of the Total Merchantable Board Foot Volume in Different Parts of the Tree Bole: Northern Hardwood Sawtimber in Northern Michigan

Height Above Ground (feet)	Number of 8-Foot Merchantable Sections in Tree							
	1	2	3	4	5	6	7	8
60–64	–	–	–	–	–	–	–	2
56–60	–	–	–	–	–	–	–	3
52–56	–	–	–	–	–	–	3	4
48–52	–	–	–	–	–	–	3	4
44–48	–	–	–	–	–	5	4	4
40–44	–	–	–	–	–	5	5	5
36–40	–	–	–	–	7	6	6	6
32–36	–	–	–	–	7	7	7	6
28–32	–	–	–	9	8	7	7	6
24–28	–	–	–	10	9	8	7	7
20–24	–	–	13	11	9	9	8	8
16–20	–	–	14	11	10	9	9	8
12–16	–	21	15	12	11	9	9	8
8–12	–	23	17	14	12	11	10	9
4–8	39	26	19	16	13	12	11	10
0–4	61	30	22	17	14	12	11	10

Source: Michigan Technological University, Ford Forestry Center. Used with permission.

third living branch above the ground enters the main bole. In deciduous species, it is more difficult to identify the base of the live crown in the field. The more complex branching patterns found in many deciduous species make the point where the third living branch enters the main bole difficult to identify. With practice, however, it is possible to estimate live crown ratio in deciduous species with almost the same precision as other height measurements. Live crown ratio is calculated as:

$$\text{Live crown ratio} = \frac{\text{total height} - \text{height to live crown}}{\text{total height}} \cdot 100 \quad [8.22]$$

Crown Condition

Measurement of crown condition is made to obtain information regarding tree vigor and health. There is no single measure that integrates all factors of crown condition. Rather, there are several variables that, taken together, provide an indication of tree vigor and health. Crown condition is a major variable in the U.S. Forest Health Monitoring program, and this program plans to publish summaries of regional information on a regular basis (see Brooks et al. (1992) for an example). Millers et al. (1991) describe field methods designed to evaluate crown condition that are being used in an international project monitoring the health of sugar maple. Four basic variables are often utilized in describing crown condition: crown dieback, crown transparency, crown discoloration, and needle retention.

Measurement of crown dieback examines the process of stem and branch mortality. *Crown dieback* is defined as branch mortality that begins at the terminal end of the branch and extends toward the main bole. Dieback occurs on both conifers and hardwoods and is different from normal self-pruning, which occurs naturally when branches in the lower portion of the tree bole are shed. Dieback occurs in the upper crown above the region of self-pruning. Some extent of crown dieback is normal, but excessive crown dieback is a possible indicator of serious health problems. In 1990, for example, the Forest Health Monitoring Program in New England found that 92% of individual trees showed none or very light crown dieback (0 to 10%), 3% of individuals showed very heavy dieback (>50%), and the remainder were somewhere in between (Brooks et al., 1992). Crown dieback is ocularly estimated, and trained field crews can obtain repeated estimates of dieback within 10% of each other more than 90% of the time. By establishing regional norms for such variables, it is easier to place observations of crown dieback at specific sites in context.

Crown transparency is a measure of crown density and is obtained as an ocular estimate of the amount of light penetration through the crown. Transparency can be measured for both conifers and hardwoods, although normal transparency rates differ by species. Transparency does not measure the extent of the crown, but the light penetration through the main portion of the crown. Like crown dieback, some level of transparency is normal, and the presence of regional, species-specific norms allow observations of particular stands to be placed in context. Trained field crews can obtain repeated estimates within 10% more than 85% of the time (Brooks et al., 1992).

Crown discoloration is the percentage of foliage exhibiting abnormal color; it is ocularly observed on both conifers and hardwoods. This is an easy variable to measure, with trained field crews able to obtain repeated estimates within 10% more than 99% of the time (Brooks et al., 1992).

Needle retention, a measure specific to conifers, is the number of years of needle growth retained within the live crown. Different species have different potential needle lifetimes; balsam fir and red spruce are able to hold eight or more years of needle classes, whereas red pine rarely holds more than three years of needles. Premature shedding of needles may be an indication of health and vigor problems. This condition is observed ocularly, and careful field observation can yield a very precise measurement.

Tree Location

If trees in a sample area are to be remeasured in the future, the mapping of tree locations can often assist in reestablishing the sample area, especially if several years pass between measurements. Some models used to project stands into the future require maps of tree locations for the projection. Moreover, there is increasing interest in statistical analyses of spatial pattern and the spatial dynamics of tree establishment, growth, and mortality (Ripley, 1983; Cressie, 1993). All mapping methods work best when there is little brush or other visual obstacles; they also work best on small areas that do not require the movement of equipment. In mapping trees, tree location is usually measured to the estimated center of the main bole at either the ground line or breast height.

Several methods have been proposed for mapping tree locations. The simplest to execute in the field is to observe the bearing and distance from a known point, such as the center of the sample unit, to the centers of individual trees. Relatively small errors in recording distance or, especially, bearing can result in relatively large errors in the calculated coordinates of tree location.

Rohlf and Archie (1978) proposed a method based on geometric triangulation, in which the distance and bearing of a tree is measured from three known locations and a least squares method is used to adjust coordinates to a single value. Kenkel (1988) modified this method and increased its accuracy. Hall (1991) provided a critical review of the method and the associated computational algorithms. Hall noted that the propagation of errors was the most serious problem with this method, but that "with great redundancy of effort" it could be used to make accurate maps.

A third method, which is relatively fast, was proposed by Reed, Liechty, and Burton (1989). They proposed establishing a Cartesian coordinate system in the field, using measurement tapes and right-angle prisms to directly record tree coordinates. The disadvantages of this method are that it requires some time to set up the coordinate system and that it works best with a field crew of three to four people. Its advantages are that the actual mapping is fast and Cartesian coordinates of tree locations are observed directly in the field, which makes error checking much easier during the mapping process.

Other Tree Characteristics

Other tree characteristics may be of interest and inventoried during assessment exercises. A primary example is the recording of trees either being used or suitable for use as den or cavity nesting sites. It is impossible to anticipate every possible information need for every resource assessment, but additional characteristics of individual trees may be measured as appropriate for a given situation and reported for the stand as described in the following section.

INVENTORY PLANNING

A *resource inventory* is the process of determining area, amount, and condition of specified resources or ecological processes within a specified geographic area; *monitoring* is the process of determining rates of change over time. The methods appropriate for use in conducting a field inventory vary according to the objectives of the inventory. The previous sections in this chapter discuss the types of measurements that might be made on individual trees during a forest inventory; methods for collecting and summarizing this information at the stand and landscape levels are discussed in the following section.

The information needs that generate resource assessment exercises in forested landscapes may differ from those that motivate traditional, stand-level forest inventory efforts. In many ways, the major difference is in the spatial scale and the range of ecological processes that will be considered in developing and evaluating management options. Landscape-level information needs may include detailed information required for generating specific operational options for smaller areas or compartments within the landscape of interest. Integrated resource assessment efforts supporting the development of a management plan for a small operational area may require intensive information for the operational area and less intensive information for some landscape characteristics over a larger area. Both Husch, Miller, and Beers (1983) and Burkhart, Barrett, and Lund (1984) provide outlines of model inventory plans for intensive inventories of single stands. These outlines are similar in content, but differ somewhat in organization. The outline given by Burkhart, Barrett, and Lund (1984) is modified in Chapter 1 with the idea of application to a larger scale and a broader range of ecological processes. At any scale, thought must be given to several areas prior to beginning field work. Clearly specified inventory objectives, analyses of management information needs and the field methods that will most efficiently provide this information, clear statements of operational constraints, and the procedures for data processing and archiving must be in place before entering the field, regardless of whether the inventory effort is for a very specific purpose on a small area or to provide a wider range of information at the landscape level.

Davis (1966) states that:

> forest inventory control [refers] to the gathering of information needed to serve the purposes of management. It does not mean collecting data for its own sake or being enamored by the wonders of modern statistical techniques. Purposes must come first and procedures second.

The timelessness of this statement cannot be too heavily emphasized. It is especially tempting, with the availability of sophisticated data-gathering techniques and the use of geographic information systems for displaying and analyzing resource information, for managers to become enmeshed in technology and lose sight of the underlying justification for the existence of such technology, which is to support management decision making. There should never be any question as to how a variable measured in the field will be used or how it will be summarized. Likewise, there should never be a question as to the utility of a map or other summary information generated during the processing of field data. A thorough analysis of information needs, comprising specification of both existing and additional needed information to support management decision making, should be included in a resource assessment plan (see Table 1.1). During this process, identification of needed information will be facilitated, and the contribution of any particular type of data collection effort to the overall management planning process will be clear.

Existing information will rarely be sufficient. It usually will be necessary to collect field information regarding some resources or processes, either on small operational areas or within larger landscapes. The *population* is the area or group of individuals for which information is required. The description of the population may be simple, such as the legal description of the operational area to be addressed in the resource assessment, or it may be more complex as the situation warrants. The definition of the population is dependent on the goals and objectives of the resource assessment process. There may also be different populations for different landscape characteristics that must be addressed in the assessment. This is the result of the differences in spatial and temporal scales appropriate for addressing different ecological processes (see Chapter 1). The inventory plan specifies the *sample units,* which define the individual elements to be measured at each sample location, and the *sample design,* which is the method of selecting the sample units to be observed from the population of interest (see Chapter 15). The *sampling frame* is a list or description of the sample units that make up the population, from which the sample design identifies sample units for observation. It is critical to confirm that the sampling frame is representative of the population of interest. Typically, it is not necessary to begin anew in the development of a plan for every data collection effort; the formulation of organizational policy on precision requirements, analytical procedures, reporting formats, and archival procedures may provide a starting point, although there is always a need to inspect areas to be sampled and determine whether any standard methods should be modified to fit a particular situation.

FOREST OVERSTORY INVENTORY METHODS

The development of forest overstory inventory methods originated centuries ago; Cook (1717), for example, discusses the use of volume tables and the estimation of volume and value of timber. There has been a great deal of theoretical development underlying the design of sample units to be used in the field and a great deal of both empirical and theoretical work examining sampling issues. The current state of this knowledge is summarized by Schreuder, Gregoire, and Wood (1993), Philip (1994), Avery and Burkhart

(1994), and Shiver and Borders (1996). Field inventory methods for use in assessing the forest overstory are discussed here, but more background information and the underlying theory may be found in the aforementioned sources.

Sampling Units

The design of individual sample units for use in estimating forest overstory structure and condition has received a great deal of attention. Oderwald (1981b) provided a unifying framework that illustrates the similarities and differences between the major alternative forest overstory sampling methods; this work may be used to develop new methods for application to particular circumstances.

Oderwald's (1981b) development of his unifying conceptual framework is based on the visualization of a circular plot surrounding each overstory tree in the forest. The location of a sampling unit is selected based on the sampling design used in the survey (see Chapter 15). At each sampling location, all overstory trees whose circular plots encompass the sampling location are considered to be elements of the sample unit and are measured or observed. Figure 8.7 illustrates this concept for two instances: (1) when the circular plots surrounding the individual trees are of identical size (Figure 8.7a) and (2) when the circular plots surrounding the individual trees are of differing sizes (Figure 8.7b).

A tree's *sampling characteristic* (C) is the characteristic of the tree that determines the likelihood of its being sampled. The sampling characteristic can be a tree's diameter, basal area, height, crown width, or any other variable of interest. Each tree's plot size is proportional to C:

$$P = \alpha C \qquad [8.23]$$

The proportionality constant, α, is identical for all trees. A tree that has a large value of its sampling characteristic, C, will have a large plot size and a higher probability of being included in the sample.

The *characteristic expansion factor* (CF) is the ratio of the sampling characteristic to the plot size (P):

$$\begin{aligned} CF &= \frac{C}{P} \\ &= \frac{C}{\alpha C} \\ &= \frac{1}{\alpha} \end{aligned} \qquad [8.24]$$

The characteristic expansion factor indicates how many units of the sampling characteristic (C) are represented by each tree included in the inventory.

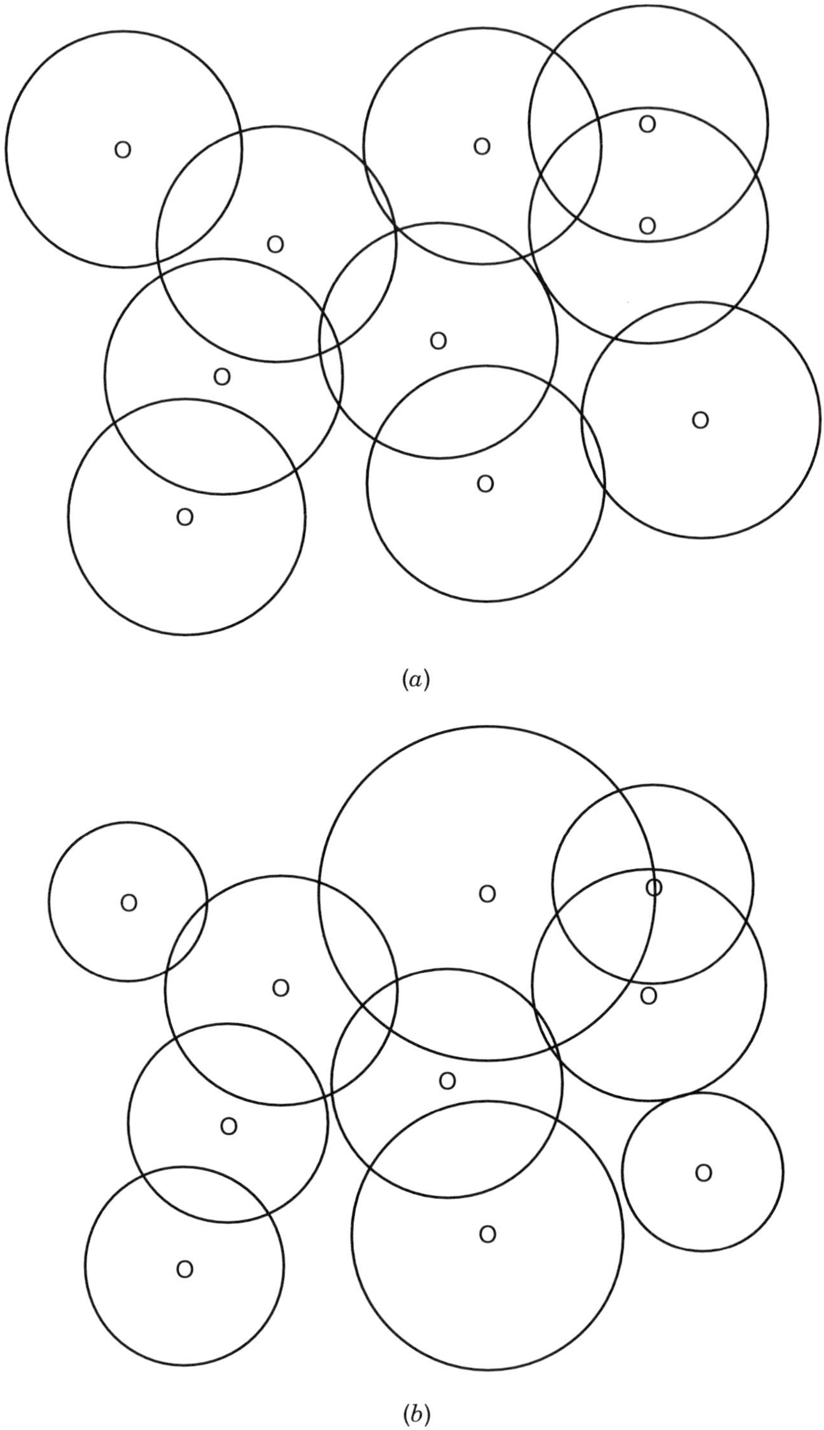

FIGURE 8.7 Oderwald's (1981) imaginary circular plots of (*a*) equal size and (*b*) unequal size surrounding overstory trees.

The *estimation characteristic* (X) is the tree characteristic for which an estimated value per unit area is desired. In developing estimates of volume per hectare, for example, the estimation characteristic is tree volume. The estimation characteristic (X) may or may not be the same as the sampling characteristic (C).

The sampling characteristic (C), the estimation characteristic (X), and the characteristic expansion factor ($1/\alpha$) are chosen prior to sampling. The plot sizes of individual trees are derived from these specified values, as in Equation [8.23]. As noted earlier, the sample locations are selected based on the sampling design, then the individual trees to be measured at a sampling location are determined based on their individual plot sizes (P). Both the sampling characteristic (C) and the estimation characteristic (X) are measured for each tree included in the sample. The estimated value of X per unit area is calculated by taking the ratio of the estimation characteristic to the sampling characteristic, resulting in the units of the estimation characteristic per unit of the sampling characteristic, for all trees at all sampling locations. These values are multiplied by the characteristic expansion factor, resulting in the units of the sampling characteristic per unit area represented by each sampled tree. These values are summed to obtain the per-unit area estimate of X for each sampling unit:

$$X_j = CF \sum_{i=1}^{n_j} \frac{X_{ij}}{C_{ij}} \qquad [8.25]$$

where X_{ij} and C_{ij} denote the values of the estimation characteristic and the sampling characteristic, respectively, for the ith tree in the jth sampling unit, n_j is the number of trees sampled in the jth sampling unit, and CF is the characteristic expansion factor. This is a mean-of-ratios estimator, in which the likelihood of sampling is proportional to the value of the auxiliary variable (the sampling characteristic), resulting in unbiased and highly efficient estimates (see Chapter 15). Note that if X and C are the same, then the estimate of X per unit area is obtained by multiplying the characteristic expansion factor by the number of trees included in the sampling unit. The estimate of the mean value of the estimation characteristic, denoted by $\overline{x}$, for the population is obtained by summing the values for the individual sampling units and dividing by the number of sampling units (see Chapter 15):

$$\overline{x} = \frac{\sum_{j=1}^{t} X_j}{t} \qquad [8.26]$$

where the X_j values are obtained for each sampling unit using Equation [8.25] and t is the number of sampling units established in the population.

So how does all this relate to sampling in the field? One of the oldest methods of inventorying forest resources involved the use of *fixed area plots,* in which sample locations were established and all trees within a certain distance of the center were measured. In the aforementioned framework, the sampling characteristic of fixed area plot sampling is simply tree presence and the value of the sampling characteristic is set to 1. This implies that the imaginary plots surrounding the individual trees are of equal size ($1/\alpha$, Equation [8.24]). The characteristic expansion factor for fixed area plot sampling, then, is the inverse of the plot size. For 0.1 ha plots, for example, per hectare values are obtained by multiplying the plot-level totals by 10 ($10 = 1/0.1$).

The decision to use fixed area plots is really a choice of the sampling characteristic (C) defined earlier. To implement fixed area plot sampling, the next decision is the choice of plot size, which is equivalent to the choice of the characteristic expansion factor. The plot size is usually specified in areal units, such as 0.1 hectare. A number of empirical studies have investigated the efficiency of different plot sizes in estimating particular population characteristics, such as commercial volume per acre. As noted by Philip (1994), large plots tend to minimize the variance between plots, but they are more costly to install and to measure. Simulation studies utilizing homogeneous populations indicate that the variance decreases in a negative exponential fashion with increasing plot size. Freese (1962) gave the following identity relating two different fixed area plot sizes to the resulting coefficients of variation (Chapter 15) of plot level characteristics, such as basal area or volume:

$$CV_2^2 = CV_1^2 \cdot \sqrt{\frac{P_1}{P_2}} \qquad [8.27]$$

where P_1 and P_2 are two plot sizes and CV_1 and CV_2 are the resulting coefficients of variation. If a pilot survey conducted using a plot size of 0.10 ha resulted in a coefficient of variation of basal area per hectare of 30%, a sample taken in the same population using plot sizes of 0.02 hectare would be expected to result in a coefficient of variation of about 45%:

$$\begin{aligned} CV_2^2 &= (30^2) \cdot \sqrt{\frac{0.10}{0.02}} \\ CV_2 &= 45\% \end{aligned} \qquad [8.28]$$

In reality, however, ecosystems are not homogeneous. Lessard, Reed, and Monkevich (1994) established 0.1- and 0.2-acre circular plots with the same center points in three forest types: northern hardwoods, planted red pine, and aspen-paper birch. Their results for individual stands sometimes diverged substantially from the result predicted by Equation [8.27]. In general, Equation [8.27] tended to overestimate the coefficient of variation of their 0.2-acre plots. This may be due to the relatively small initial plot size and the fact that they did not rerandomize plot locations. In any case, the relationship given earlier should be used only as a guideline to obtain approximate results applicable for a relatively narrow range of plot sizes. At some point, the spatial scale of the process being examined may be reached and increased heterogeneity may be found within larger plots. As a rule of thumb, therefore, Equation [8.27] may be used to approximate the likely precision resulting from plot sizes that differ by less than an order of magnitude or so, but probably should not be assumed to be valid for much larger differences in plot sizes.

It is easiest to install fixed area plots if they are circular, with a specified radius resulting in the desired plot area. A number of empirical studies have investigated the effect of plot shape on the resulting precision. Kulow (1966), for instance, examined circular, square, and rectangular plots in the estimation of overstory commercial volume and found very little difference in the resulting estimates of precision. Circular plots minimize the perimeter-to-area ratio, which decreases the likelihood of field crews having to make decisions about whether to include borderline trees in the sample. This is usually desirable for inventories attempting to estimate commercial volume or basal area per hectare, or to develop tables giving stand structure. For other processes, however, it may be desirable to increase the perimeter-to-area ratio of fixed area sampling units by using square or rectangular plots. Square quadrats may be logistically easier to install at some spatial scales; for understory vegetation plots, for example, it is easy to carry a square 1 m × 1 m sampling frame to the field.

The use of circular plots in forest overstory inventories allows field crews to determine whether trees are in the plot by simply measuring the distance from the plot center to the center of individual trees; if the distance is less than the radius of the circle of the desired plot area, the tree is measured; otherwise, the tree is not measured. The distance to many trees may be determined ocularly, but occasionally it will be necessary to measure the distance to borderline trees to confirm that they are in the sample plot. The area of a circle is equal to $\pi \cdot r^2$ where r is the radius and the area is expressed in square units of distance, so the plot radius to use during field work is given by:

$$r = \sqrt{\frac{\text{plot size}}{\pi}} \qquad [8.29]$$

Note that the areal measurements must be converted from hectares to square meters in order to calculate distance in meters. A 0.1 ha plot has an area of 1,000 m^2 and a radius of 17.84 m; all trees within 17.84 m of the plot center would be considered to be in the sample unit and should be measured. In the field, following establishment of the plot center, the plot radius is usually marked off in cardinal compass directions. Beginning at an obvious point, such as the nearest tree to the plot center or the nearest tree to due north, trees are measured and recorded sequentially in a uniform direction (clockwise). Table 8.11 includes a listing of tree measurements collected on a 0.10 ha plot in a stand of mixed conifers growing on a poorly productive wetland site.

TABLE 8.11 Forest Overstory Data Collected from One 0.10-Hectare Fixed Area Plot Sample Location in a Lowland Mixed Conifer Stand

Species	*Diameter (cm)*	*Volume* (m^3)
Black Spruce	30	0.399
Northern White-Cedar	35	0.550
Northern White-Cedar	30	0.333
Red Maple	30	0.333
Black Spruce	25	0.268
Black Spruce	15	0.056
Northern White-Cedar	25	0.224
Red Maple	35	0.460
Black Spruce	25	0.224
Northern White-Cedar	15	0.056
Northern White-Cedar	20	0.101
Northern White-Cedar	20	0.101
Northern White-Cedar	20	0.101
Red Maple	20	0.101
Red Maple	25	0.099
Bigtooth Aspen	20	0.101
Bigtooth Aspen	30	0.333
Black Spruce	20	0.101

When the inventory objective is to estimate volume per unit area, the estimation characteristic (X) is individual tree volume (V) and the volume for each sampling unit (V_j) is calculated using Equation [8.25]:

$$V_j = CF \sum_{i=1}^{n_j} \frac{V_{ij}}{C_{ij}} \qquad [8.30]$$

where V_{ij} is the volume of the ith tree in the jth sampling unit and all other notation is as defined previously. Since the sampling characteristic (C) is one for all trees in fixed area plot sampling:

$$V_j = CF \sum_{i=1}^{n_j} V_{ij} \qquad [8.31]$$

For an application like that in Table 8.11, the volume of the individual trees would be obtained from a volume table (see Tables 8.7 and 8.8 for examples) after diameter and height have been measured in the field. For fixed area plot sampling, the volumes of the trees on the plot are summed and then multiplied by the characteristic expansion factor to arrive at the estimated volume per hectare for the plot. In the example, the plot size is 0.10 ha, so the characteristic expansion factor for the plot is 10, and the sum of the volumes of trees measured on the plot is 3.941 m^3; the resulting estimated volume is 39.41 m^3 per hectare.

Similarly, an estimate of basal area per unit area, or of any other tree characteristic for that matter, is obtained by summing the individual tree values and multiplying by the characteristic expansion factor. The basal area of individual trees is calculated using Equation [8.19] for metric units and Equation [8.20] for English units. For the data in Table 8.11, the sum of the basal areas of the trees in the plot is 0.895 m^2 and the estimated stand basal area is 8.95 m^2/ha.

To estimate density, or the number of trees per unit area, in fixed area plot sampling, thc estimation characteristic is set equal to 1. This implies that the estimation characteristic and the sampling characteristic are equal in Equation [8.25]:

$$\begin{aligned} X_j &= CF \sum_{i=1}^{n_j} \frac{X_{ij}}{C_{ij}} \\ &= CF\, n_j \end{aligned} \qquad [8.32]$$

For the example in Table 8.11, there are 18 measured trees in the plot, which results in an estimated density of 180 trees per hectare.

It may sometimes be necessary to calculate volume, basal area, or density estimates for portions of the total population. For the example in Table 8.11, the number of trees per acre of a particular species, or of a given size class of a particular species, may be needed. The same calculations performed earlier for the entire sample unit are appropriate for subpopulations, but only the sampled individuals that are members of the appropriate subpopulation are included in the calculations. To estimate the number of northern white-cedar trees per hectare in this example, there are 7 northern white-cedars on the plot, which results in an estimate of 70 northern white-cedars per hectare.

To obtain population estimates of volume, basal area, or density, calculations similar to those presented earlier would have to be performed for all sample units established in the stand. These observations would then be used to calculate estimates of population means or totals, standard errors of those estimates, confidence intervals, and other statistical summary information. The calculation methods for these population-level summary statistics are reviewed in Chapter 15.

Another common type of sample unit used in forest overstory inventory is the *variable radius plot;* this procedure is also referred to as *Bitterlich sampling, point sampling, plotless sampling,* or by a number of other names. Developed by Bitterlich (1947), this method identifies trees to be measured at a sample unit with a probability proportional to their size. In the most common application, the method uses tree basal area as the sampling characteristic, but methods have also been developed to use height as the sampling characteristic. When the sampling characteristic is basal area, the characteristic expansion factor (Equation [8.24]) is expressed in units of basal area per hectare; each tree measured represents some number of units of basal area per hectare rather than a fixed number of trees per hectare. The characteristic expansion factor in variable radius plot sampling is often referred to as the *basal area factor,* or BAF.

In variable radius plot sampling, trees are selected for measurement based on their size, with larger individual trees being more likely to be measured. This is equivalent to larger trees having larger imaginary plot circles surrounding them (Figure 8.7b). The maximum distance from the center location to which a tree of a given size will be measured is referred to as the critical distance and is equivalent to the radius of the imaginary plot surrounding the tree (Figure 8.7b). Trees of different sizes will have different critical distances. The radius of the imaginary plot for each tree can be calculated using the relationship in Equation [8.24], with the plot size expressed in terms of the radius (area $= \pi \cdot r^2$):

$$\begin{aligned} CF &= \frac{C}{P} \\ &= \frac{C}{(\pi \cdot r^2)/10{,}000} \end{aligned} \qquad [8.33]$$

when using metric units and converting the plot size in hectares to square meters. This implies that the radius can be calculated as:

$$r = \sqrt{\frac{10{,}000 \cdot C}{\pi \cdot CF}} \qquad [8.34]$$

When a basal area factor of 2 m^2/ha is used, this means that each tree measured represents 2 m^2/ha of basal area, and the critical distance for a tree of 40 cm is calculated as follows:

$$\begin{aligned} r &= \sqrt{\frac{10{,}000 \cdot 0.00007854 \cdot (40^2)}{\pi \cdot 2}} \\ &= 14.14 \text{ m} \end{aligned} \qquad [8.35]$$

This means that if the center of a 40 cm diameter tree is less than 14.14 m from the plot center location, the tree should be measured; if the tree is more than 14.14 m from the center location, the tree should not be measured. The critical distance for a 30 cm tree using the same basal area factor (2 m^2/ha) is 10.61 m. Note that, except for rounding, the ratio of distance-to-diameter is constant for a given basal area factor. For the basal area factor used here (2 m^2/ha), the ratio between critical distance and diameter is about 35.35. This ratio (R_d) can be calculated directly from Equation [8.33] (note that the radius must be converted from meters to centimeters to be expressed in the same units as the diameter):

$$\begin{aligned} CF &= \frac{10{,}000 \cdot 0.00007854\ D^2}{\pi \cdot (r_{cm}/100)^2} \\ &= \frac{7{,}854\ D^2}{\pi\ r_{cm}^2} \\ &= 2{,}500 \left(\frac{D}{r_{cm}}\right)^2 \\ &= 2{,}500 \left(\frac{1}{R_d}\right)^2 \end{aligned} \qquad [8.36]$$

therefore:

$$R_d = \sqrt{\frac{2{,}500}{CF}} \qquad [8.37]$$

which allows the calculation of R_d for any basal area factor or, conversely, the calculation of the basal area factor as a function of R_d.

In using English units, the basal area and the basal area factor are expressed in square feet per acre, and the only difference in Equation [8.34] is the conversion from acres to square feet:

$$r = \sqrt{\frac{43{,}560 \cdot C}{\pi \cdot CF}} \qquad [8.38]$$

Likewise, in using English units, Equation [8.37] is equivalent to:

$$R_d = \sqrt{\frac{10{,}890}{CF}} \qquad [8.39]$$

So, for a basal area factor of 10 square feet per acre, the ratio between critical distance and diameter is 33. Note that this R_d is expressed in inches of distance per inch of diameter. The ratio can be converted to feet of distance per inch of diameter by dividing by 12. For a basal area factor of 10 square feet per acre, the ratio between critical distance in feet and diameter in inches is 2.75; the critical distance for a 10-inch diameter tree is 27.5 feet.

A decision to use variable radius plot sampling in the field portion of an inventory is really a decision to use individual tree basal area as the sampling characteristic. The choice of a basal area factor is equivalent to the selection of the characteristic expansion factor in Oderwald's (1981b) formulation. Unlike fixed area plot sampling (Equation [8.27]), there is no clear relationship between the expected precision using different basal area factors. Sukwong,

Frayer, and Mogren (1971) used a relationship developed by Smith (1938) that relates the coefficient of variation to the number of trees in a sample unit:

$$CV = \beta_0 N^{\beta_1} \quad [8.40]$$

where CV is the coefficient of variation, N is the number of trees in the sample units, and the coefficients β_0 and β_1 differ by forest type. Using this, the relationship between the coefficients of variation resulting from the use of different basal area factors is given approximately by:

$$\frac{CV_1}{CV_2} = \left(\frac{BAF_1}{BAF_2}\right)^{\beta_1} \quad [8.41]$$

As with fixed area plot sizes, the most efficient basal area factor depends on the estimation characteristic and the nature of the target ecosystem. Wiant, Yandle, and Andreas (1984), for instance, gave an example from Appalachian hardwood forests in which the efficiency of a range of basal area factors was examined. Smaller basal area factors lead to more trees being measured than do larger basal area factors. In a forest with 30 m^2/ha of basal area, the use of a basal area factor of 2 m^2/ha will, on average, result in 15 trees being measured at each sampling location, whereas the use of a basal area factor of 1 m^2/ha will result in an average of 30 trees being measured at each sampling location. In forest types with scattered large trees, it is sometimes easy to miss a large tree that is a greater distance from the center location, which then results in biased final estimates. Iles and Fall (1988), in fact, identified this problem as a major source of bias in variable radius plot sampling. Some authors have recommended that the basal area factor be selected so that the expected number of trees per sample unit falls between four and eight (Dilworth and Bell, 1984; Beers and Miller, 1964). It is also important to select a basal area factor so that the critical distance for the expected largest tree in the overstory will result in the tree's being sufficiently visible to field personnel, given the brush and visibility conditions in the ecosystem of interest. From a practical standpoint, it is difficult for field crews to change basal area factors from one stand to another, because the critical distance for a given tree size will change as the basal area factor changes. For this reason, most organizations conducting a great number of field inventories of forest overstory structure have identified a compromise basal area factor and use it in most, if not all, forest types they normally encounter.

The major steps to implement variable radius plot sampling in the field are similar to those required to implement fixed area plots. After establishing the sample location, an easily identifiable tree is located (usually the nearest tree to the center point or to a cardinal compass bearing); then the trees are measured sequentially in a clockwise fashion until returning to the starting point. The major difference, of course, is the method of identifying which trees to measure.

There are several common tools used in identifying the sample trees to measure. The most basic tool is the *angle gauge*. An angle gauge is a simple device developed from the relationships defined in Equations [8.37] and [8.39]. The angle gauge is based on the principle of similar triangles, and it consists of a sight of known width affixed to the end of a stick of known length (Figure 8.8). The end of the stick is held to the viewer's eye at the center location; if a tree's width extends beyond the sight, the tree should be measured. If the sight extends beyond the tree's width, the tree is too far away from the center location to be measured. The distance to any tree that appears borderline should be measured. If the distance to the center of the tree is greater than R_d multiplied by the tree's diameter, then the tree is too far from the center location and it should not be measured.

A specialized optical device called a *relaskop* was developed by W. Bitterlich. This is a multiple-purpose device that contains optical scales inscribed with several basal area factors, the choice of which depends on the user. It also contains scales for measuring tree heights and upper stem diameter, which are useful in identifying merchantable height. The flexibility of this instrument is as great an advantage as is its size, since it can be held easily in the hand. In the authors' experience, though, it is sometimes difficult to see the different scales under dense canopy or poor lighting conditions. If a tree appears to be a borderline tree, its diameter and distance from the center location should be measured to determine whether it should be included in the measurements made for the sample unit.

FIGURE 8.8 Angle gauge for use in variable radius plot sampling.

Another device for determining whether individual trees should be measured during variable radius plot inventories is the *wedge prism*. The optics underlying the use of this device are described in detail by Husch, Miller, and Beers (1983). To use this instrument, the prism is held over the center location. The observer looks through the prism at each tree that may be a candidate for measurement. If, in the prism, the tree bole appears completely offset, the tree is not measured (Figure 8.9). If the tree is not completely offset, it is included in the measurements made at that location. Sometimes wedge prisms that are purchased as being calibrated for one basal area factor

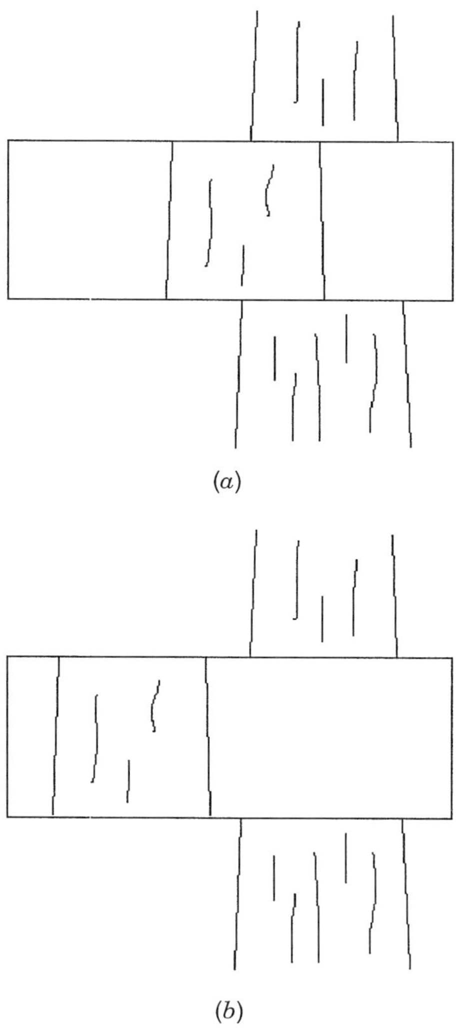

(*a*)

(*b*)

FIGURE 8.9 Use of a wedge prism to determine whether a tree is (*a*) within the sample unit or (*b*) outside the sample unit.

are not truly shaped correctly for the nominal factor. To determine the true basal area factor for a prism, an observer should chose a tree-like object of known diameter, such as a utility pole, and back away from that object until it appears borderline. The distance to the object from that point, coupled with its diameter, can be used in Equation [8.34] or [8.38] to solve for CF, the true basal area factor for the prism in question.

Equation [8.25] is used to develop the methods for obtaining estimates of volume, basal area, and number of trees per hectare in variable radius plot sampling. For volume:

$$V_j = CF \sum_{i=1}^{n_j} \frac{V_{ij}}{B_{ij}} \qquad [8.42]$$

where V_{ij} is the volume of a sampled tree, B_{ij} is its basal area, n_j is the number of trees sampled at the jth center location, CF is the characteristic expansion factor (basal area factor), and V_j is the estimate of volume per hectare (or volume per acre) for the sample unit. Table 8.12 contains data from two variable radius plots located in a stand similar to that described by the fixed area plot sample data in Table 8.11. English units are used in this example to illustrate the use of volume tables (Tables 8.7 and 8.8). The basal area factor used during the field procedures was 10 square feet per acre. The estimated number of cords per acre at the first sample unit is given by:

TABLE 8.12 Forest Overstory Data Collected from Two Basal Area Factor 10 (ft^2/ac) Variable Radius Plot Sample Locations in a Lowland Mixed Conifer Stand

Species	*Diameter (inches)*	*Height (number of 8′ pulp sticks)*	*Volume (cords)*
Sample Unit 1			
Black Spruce	12	4	0.198
Northern White-Cedar	14	4	0.273
Northern White-Cedar	12	3	0.165
Red Maple	12	3	0.165
Sample Unit 2			
Black Spruce	10	4	0.133
Black Spruce	6	2	0.028
Northern White-Cedar	10	3	0.111
Red Maple	14	3	0.228
Black Spruce	10	3	0.111

$$V_1 = 10 \cdot \left[\frac{0.198}{0.005454\ (12^2)} + \ldots + \frac{0.165}{0.005454\ (12^2)}\right]$$
$$= 10 \cdot [0.25 + 0.26 + 0.21 + 0.21]$$
$$= 9.3 \text{ cords per acre} \qquad [8.43]$$

The estimation of basal area is easy for variable radius plot sampling because the estimation characteristic and the sampling characteristic are the same:

$$B_j = CF \sum_{i=1}^{n_j} \frac{B_{ij}}{B_{ij}}$$
$$= CF \cdot n_j \qquad [8.44]$$

For the example in Table 8.12, then, the basal area per acre at sample unit 1 is 40 square feet per acre and for sample unit 2 is 50 square feet per acre. Note that these would have to be combined with the estimates from the other sample units in the population, as in Equation [8.26], to estimate the mean basal area per acre for the population.

The number of trees per acre is estimated in variable radius plot sampling as it is in fixed plot sampling, by substitution of the value "1" for the estimation characteristic for all trees included in the sample unit:

$$N_j = CF \sum_{i=1}^{n_j} \frac{1}{B_{ij}} \qquad [8.45]$$

For the first sample unit in Table 8.12, the estimated number of trees per acre for the sampling unit is:

$$N_1 = 10 \cdot \left[\frac{1}{0.005454\ (12^2)} + \ldots + \frac{1}{0.005454\ (12^2)}\right]$$
$$= 47.6 \text{ trees per acre} \qquad [8.46]$$

As in fixed area plot sampling, estimates of any estimation characteristic may be obtained for subpopulations by simply summing over only the members of the subpopulation in Equations [8.42], [8.44], and [8.45]. Thus, the estimate of basal area per acre of northern white-cedar trees from sample unit 1 in Table 8.12 is 20 square feet per acre. Likewise, the number of 12-inch red maple trees per acre estimated for sample unit 1 is 12.7 trees per acre.

The examples given here for both fixed area plot and variable radius plot demonstrate the estimation of volume, basal area, and number of trees per unit area for the population and subpopulations of interest. These are probably the most common estimation characteristics in forest overstory inventories, but the framework developed by Oderwald (1981b) is valid for any estimation characteristic of interest. The number of trees per unit area in a population that provide suitable habitat for a bird or animal nesting site may be estimated by recording an indicator variable for each tree sampled; the indicator variable takes the value of "1" if the tree provides suitable nesting habitat and the value of "0" if the tree does not provide suitable nesting habitat. In the example in Table 8.12, for instance, if the 12-inch black spruce provided suitable nesting habitat for a species, the estimated number of suitable nesting sites per acre for the species, based on the single sample unit, would be calculated as:

$$\begin{aligned} \text{Nest}_j &= 10\left[\frac{1}{(0.005454\ 12^2)} + 0 + 0 + 0\right] \\ &= 12.7 \text{ suitable nesting sites per acre} \end{aligned} \tag{8.47}$$

A number of studies have compared the efficiency of fixed area plot and variable radius plot sampling of forest overstory characteristics (e.g., O'Regan and Palley (1965), Kulow (1966), Oderwald (1981a)). A general conclusion seems to be that fixed area plot sampling is often more efficient in the estimation of trees per unit area, whereas variable radius plot sampling is often more efficient in the estimation of basal area or volume. This is not surprising, inasmuch as the sampling characteristic for fixed area plot sampling represents density and the sampling characteristic for variable radius plot sampling is basal area. Oderwald (1981a) noted that the nature of the forest, particularly the pattern of overstory tree location, can affect the comparisons. He found that variable radius plot sampling was more efficient for estimating basal area in most stands, with the exception of stands with regular spacing, such as plantations, where fixed area plots seemed to be more efficient.

The flexibility of the general sampling approach represented by Equation [8.25] has not been widely utilized by resource managers. This approach allows the development of sampling schemes designed specifically for the estimation of a wide range of ecosystem characteristics. Such sampling schemes may be developed by the appropriate choice of sampling and estimation characteristics. As an example, consider a situation in which a manager wanted to estimate the commercial quantity of small poles in a forest. The poles had to have a diameter within a given range and a minimum length. If the number of poles in a tree was selected as both the sampling characteristic and the estimation characteristic, then a count of trees within the critical distance of the center location, multiplied by the characteristic expansion factor, would provide an estimate of the number of poles per unit area. If a characteristic expansion factor of 40 poles per hectare was selected, Equation [8.34] could then be used to calculate the critical distance for trees containing one, two, or more poles. For example, a tree containing one pole would have a critical distance of:

$$\begin{aligned} r &= \sqrt{\frac{10{,}000 \cdot 1}{\pi \cdot 40}} \\ &= 8.9 \text{ m} \end{aligned} \tag{8.48}$$

whereas the critical distance for a tree containing two poles is given by:

$$\begin{aligned} r &= \sqrt{\frac{10{,}000 \cdot 2}{\pi \cdot 40}} \\ &= 12.6 \text{ m} \end{aligned} \quad [8.49]$$

Field sampling would then be implemented with the number of trees within their critical distance of the center location being recorded; if five trees were recorded as being within the sample unit, the estimate of the number of poles per hectare would be 5 times 40, or 200 poles per hectare. Custom-designed sample units could be developed for measurements to be made in English units, using Equation [8.38] for calculating the critical distances.

Sampling Design

As noted previously, the sampling design is the method of selecting sample units from the population for measurement during a sampling exercise. The basic sampling designs utilized in natural resource sampling include simple random sampling, systematic sampling, stratified random sampling, cluster sampling, and the use of ratio and regression estimation. The statistical aspects of these basic designs are reviewed in Chapter 15. The choice of a sampling design depends on the goals and objectives of the inventory effort and the type of information required.

The most commonly applied sampling design for the estimation of forest overstory characteristics is probably systematic sampling, or systematic sampling within stratified random sampling. Simple random sampling is used occasionally, but it is difficult to implement in the field because distance between sampling locations can only be approximated during most field exercises and sample location is rarely known with a high degree of certainty, although this is changing with the improvement and adoption of satellite positioning systems. This problem is complicated by the fact that it is sometimes difficult to plan efficient means of travel between randomly selected locations during field work and the fact that the random location of sample units may leave relatively large portions of the area of interest without sample units. Systematic sampling is usually preferred both from a logistic standpoint and intuitively, because it guarantees that sample units will be distributed throughout the geographic area of interest. If the area being inventoried is truly homogeneous, this latter point is difficult to defend from a statistical standpoint, but ecosystems are rarely homogeneous and distributing sample units spatially throughout the population of interest has a great deal of practical appeal. Cluster sampling is used when the goal is to estimate characteristics of individual trees rather than to describe the overstory structure as a whole, although both simple random sampling and systematic sampling have been used in estimating individual tree characteristics as well. Ratio or regression estimation is typically used in conjunction with auxiliary variables derived from aerial photographs or satellite imagery and is very efficient in the estimation of stand-level characteristics such as volume and biomass per unit area. Ratio or regression estimation may also be incorporated within either randomly or systematically located sample units for the estimation of such tree characteristics as tree grade, which require more time for evaluation in the field. As noted previously, the estimation procedure given in Equation [8.25] is really a mean-of-ratios procedure in which the probability of an element's being included in the sample is proportional to the denominator.

A number of studies have compared the results of simple random sampling and systematic sampling in ecological settings (e.g., Clapham, 1932; Hasel, 1938; Finney, 1948; 1950; Bourdeau, 1953; Mesavage and Grosenbaugh, 1956; Kulow, 1966). Most of these studies have found very little difference between simple random sampling and systematic sampling in terms of the precision of the resulting estimates or in the cost of conducting the survey. As noted earlier, some authors have favored systematic sampling because the sample units are geographically distributed throughout the population of interest. Kulow (1966), on the other hand, favored simple random sampling because "simple random sampling plots tended to be grouped together so travel time is less." He also noted, however, that the precision achieved in simple random sampling in his study tended to be somewhat better than that of systematic sampling because the sampling locations tended to be clumped together in somewhat more uniform stand conditions. This result seems to imply a heterogeneous population, which is the reason that most of the other authors tended to favor systematic sampling.

Several authors, including Finney (1950), noted that the variance estimates from systematic samples are inappropriate if the underlying population is periodic in nature. As noted in Chapter 15, periodicities are common in ecological settings. Avery and Burkhart (1994) indicated that if estimates of precision are important, then it is probably better to use simple random sampling in forest overstory inventories. They also noted that systematic sampling is generally appropriate for estimating means or totals in forest inventory exercises, largely owing to the efficiency of implementing the design in the field. If the major interest is in the estimate of the population mean or total, systematic sampling may be used and the resulting estimates of precision should be considered as approximations.

Finally, if there are subpopulations, or regions of more uniform conditions within the overall population, then stratified random sampling generally does as well or better than either simple random sampling or systematic sampling. In order for stratified sampling to provide a significant increase in precision, the strata should be chosen to minimize within strata variability while maximizing between strata variation. In other words, if there are clearly subpopulations that are relatively homogeneous, then stratification will improve the precision of the estimate of the overall population mean.

SUMMARIZING FOREST OVERSTORY INFORMATION

Like field methods, the type of summary of forest structure or condition included in a resource assessment depends on the purposes of the effort. These purposes should be clearly thought out prior to beginning any field work, and a good planning document should contain outlines of any tables that are expected to be included in the final report, including titles, column headings, variables to be summarized, and so on (see Chapter 1). There are some common summaries of forest overstory structure that are appropriate for many management purposes and reported as part of many resource assessment exercises; other methods of developing summary information are less common and may even be unique to certain landscapes. There is, therefore, no standard set of tables or methods to summarize field information that will satisfy all management information needs.

Information on forest overstory structure is needed to address resource management questions, ranging from wildlife management (Chapter 9) to the appraisal of commercial value. A *stand table,* which gives the number of trees per unit area by size class and species, is a common summary tool. Table 8.13 gives a stand table for a stand similar in composition to the stand from which the data in Tables 8.11 and 8.12 were obtained. Either fixed area plots or variable radius plots may be used in the field procedures to obtain the information needed to construct a stand table. The numbers of trees per unit area for each size class and species combination are calculated separately as subpopulations of the overall population. The values from each subpopulation may then be summed to obtain an estimate of the overall population values. For example, in Table 8.12, there are two ten-inch diameter black spruce in the two sample units, which translates to an average of 18.3 ten-inch-diameter black spruce trees per acre.

TABLE 8.13 Stand Table for a Lowland Mixed Conifer Stand

	Species				
Diameter Class (cm)	*Red Maple*	*Bigtooth Aspen*	*Northern White-Cedar*	*Black Spruce*	*Total*
15	6.3	0.0	19.1	6.3	31.7
20	0.0	3.6	10.7	7.2	21.5
25	0.0	2.3	6.9	2.3	11.5
30	1.6	0.0	1.6	1.6	4.8
35	1.2	0.0	1.2	1.2	3.6
40	0.9	0.0	0.0	0.0	0.9
Total	10.0	5.9	39.5	18.6	74.0

Another frequently compiled summary of forest overstory information is a *stock table,* which is similar to a stand table except that the volumes per unit area are entered rather than the number of trees; a *basal area table* contains the basal areas per unit area rather than volume or number of trees. Another common form of presenting stand structure information is a graph of the *diameter distribution,* which is a description of the number of trees per unit area in each size class (Figure 8.10). The stand-level diameter distribution is calculated by summing over all species for each size class, but the diameter distribution for each species is also important in developing silvicultural options (Smith, 1986).

In a landscape-level resource assessment effort, stand tables describing forest overstory structure will usually be needed for every major vegetation unit within the landscape for which management options are to be developed and evaluated. There may be areas with different ownership or other inaccessible areas within the landscape where it is impossible or inappropriate to conduct the inventory necessary to develop stand tables. In these cases, aerial photography may at least identify the type of vegetation and provide some idea of the similarity in structure to other areas that are accessible.

The spatial arrangement of different vegetation types may or may not be important to consider in any given assessment effort. There are instances in which the existence of connecting corridors of similar vegetation are desired,

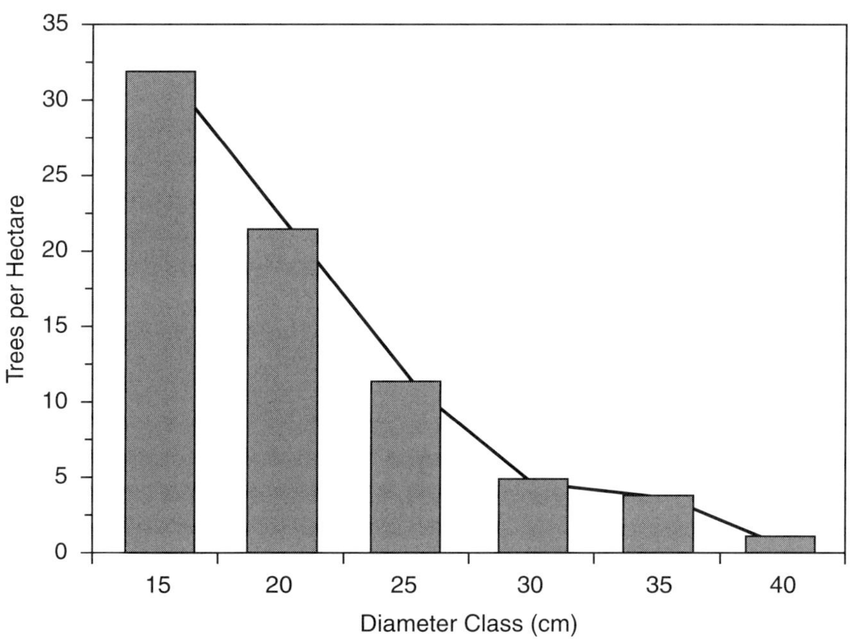

FIGURE 8.10 Frequency distribution of diameters of all species for the stand in Table 8.13.

and in such situations the landscape may be examined with the use of aerial photography or satellite imagery without detailed, on-the-ground inventories of forest overstory structure. Because larger-scale, landscape-level information may place constraints on the development of management options for specific operational areas, it is often necessary to include broader-scale information for the entire landscape together with finer-scale information on forest overstory structure for individual management compartments within the landscape. In addition, it may be necessary to provide some combined information at the landscape level. An example of this type of situation might occur in the evaluation of habitat for a particular faunal species, in which the area providing suitable nesting sites must be maintained at or above some minimal critical level in order to ensure sufficient opportunity for the establishment or the maintenance of a viable breeding population.

SUMMARY

Information concerning the forest overstory is important in assessing wildlife habitat, commercial value, some aspects of water quality, and many other resource issues. An understanding of the commercial potential of forested landscapes and the pressure to produce some revenue with which to meet the costs of ownership and management is important for all managers of forested landscapes, regardless of the primary management goals and objectives. Even if the production of income is not a primary goal of management, opportunities for actions such as the manipulation of wildlife habitat often depend, to some extent, on the commercial value of wood products. On the other hand, the pressure and opportunity to produce revenue may be great, and the need to maintain the integrity and function of ecological processes operating within the landscape must not be forgotten. Procedures for assessing the structure and condition of the forest overstory are vital tools to provide support for the development and evaluation of management options.

APPLICATIONS

8.1 Calculate the board feet in the following lumber:

2″ × 6″, 12 feet in length
1″ × 8″, 6 feet in length
4″ × 4″, 10 feet in length
2″ × 4″, 8 feet in length

8.2 Calculate the basal area, cord, and board foot volume (using Tables 8.7 and 8.8) of the following trees:

Diameter (Inches)	Number of 8-Foot Pulpsticks	Number of 16-Foot Sawlogs
16	–	1 1/2
8	3	–
24	–	2 1/2
10	5	–
18	–	2

8.3 Determine the grades of the following northern hardwood sawlogs:

Species (Diameter, Length)	Sweep Deduction (%)	Total Deduction (%)	Clear Cuttings per Face		
			1	2	3
Sugar Maple (20″, 14′)	4	20	7′	3′	9′
Basswood (18″, 16′)	0	40	8′	9′	12′
Sugar Maple (15″, 12′)	20	30	3′	8′	5′
Red Maple (11″, 8′)	0	0	4′	4′	5′
Red Oak (16″, 16′)	10	30	6′	4′	3′

8.4 Use a volume table (either Tables 8.7, 8.8, or another table appropriate for your geographic area) to estimate the volume per acre on a 0.20-acre fixed area plot. Estimate basal area and the number of trees per acre.

8.5 Use variable radius plot sampling and install a minimum of ten sample units in a forested stand. Construct a stand table for the stand.

8.6 Using Equation [8.34] or [8.38], calculate the basal area factor of your thumb.

8.7 Design a sampling scheme for estimating the number of mistletoe infestations per hectare. Use the number of mistletoe infestations per tree as both the sampling and the estimation characteristic. Select a characteristic expansion factor that would be appropriate in a stand where the crown visibility is such that mistletoe in trees more than 25 m away is difficult to observe. Give the critical distances for trees with 1, 2, 3, and 4 mistletoe infestations.

Chapter

9

Wildlife Resources

Traditionally, most of the effort in managing wildlife populations has focused on vertebrate game species. As a consequence, most of the assessment methods and examples in the literature are adapted to the particular characteristics of game species, which tend to be larger species that occupy a variety of habitats. It is becoming increasingly important for land managers to expand their concept of wildlife resources to include nongame vertebrate species and even nonvertebrate species. Interest in nonvertebrate species may center on threatened or endangered species, such as the American burying beetle, but in the most general sense may include a much wider range of taxa than traditionally considered in land management. Outside of particular situations in which the enhancement of a single species is a primary management goal, it is relatively unusual for landscape-level resource assessment efforts to concentrate on the habitat or population levels of a single species.

The assessment of wildlife resources in forested landscapes usually focuses on habitat, but may also consider population size and community structure or diversity. Hunting and trapping, of course, directly impact population numbers and community structure, but most land management activities directly impact habitat and only indirectly impact population size or community composition. In most situations, it is not necessary to quantitatively describe the components contributing to changes in population size (birth rate, death rate, immigration, and emigration), but instead it may be important to assess the possible indirect effects of planned management activity on population dynamics or community structure. For this reason, methods described here focus on the assessment of habitat and population size. The description of community structure and diversity, and the impact of proposed management practices on these properties, will also be relevant to many management situations; methods for assessing the diversity or structure of wildlife populations are similar to those used in vegetative communities (Chapter 7). The methods presented are not adequate if the situation calls for specific, quantitative assessment of the various components of population dynamics for a particular species; procedures for estimating population birth, death, immigration, and emigration rates are described by Seber (1982). Detailed discussions of the interactions among wildlife populations, habitat, and management activities are presented by Hunter (1990) and Patton (1992).

HABITAT ASSESSMENT

For any species, the suitability of a given habitat depends on the availability of adequate food, water, and cover. It is important to remember that the availability of suitable habitat is a necessary, but not sufficient, condition for a species to be present in an area. In addition, it is usually not appropriate to concentrate assessment efforts on the habitat requirements of a single species, although enhancement of a single species, such as an endangered species or a highly desired game species, may be the primary goal of management in some situations. Most resource assessment efforts focus on the assessment of habitat of selected species, often referred to as *indicator species,* which are representative of groups of species, or *guilds,* that have similar habitat requirements.

Some assessment efforts rely only on elementary descriptions of habitat characteristics. Knowledge of the general types of habitat in a landscape provides a great deal of information concerning the likelihood that various wildlife species will be present. Benyus, Beuch, and Nelson (1992), for example, provide a list of wildlife species known to utilize different general types of habitat, such as upland deciduous or lowland coniferous forest, in the upper Great Lakes region (Table 9.1). DeGraaf et al. (1992) provide information concerning habitat utilization by different wildlife species for the New England region. In other cases, much more detailed information may be required to support management planning efforts.

In some cases, there may be a great deal of basic interpretive information available from descriptions of soil type (Chapter 4) or vegetative association (Chapter 7) present in an area. The type of habitat typically occurring on different soil mapping units is usually available from soil survey interpretive information. Table 9.2, for example, contains wildlife habitat interpretive information given by Schwenner (1991) for several soil mapping units in Houghton County, Michigan. This type of information can help in assessing the habitat potential under different management alternatives.

Characteristics of landscape structure may be combined with descriptions of broad vegetative type, such as upland deciduous forest, to provide more

TABLE 9.1 Utilization of Different Plant Communities by Selected Bird Species in the Upper Great Lakes Region

	Community		
Species	*Shrub Swamp*	*Mature Upland Deciduous*	*Closed-Canopy Lowland Conifer*
American Crow	Yes	Yes	Yes
American Redstart	No	Yes	No
American Robin	Yes	Yes	No
American Widgeon	Yes	No	No
Barred Owl	No	Yes	No
Black-backed Woodpecker	No	No	Yes
Black-capped Chickadee	Yes	Yes	Yes
Blue Jay	No	Yes	Yes
Blue-gray Gnatcatcher	Yes	Yes	No
Boreal Owl	Yes	No	Yes

Source: Adapted from Benyus, Buech, and Nelson (1992).

TABLE 9.2 Wildlife Habitat Interpretive Information Given by Schwenner (1991) for Selected Soil Mapping Units in Houghton County, Michigan

	Map Symbol			
Habitat Factor	*10B*	*12*	*17A*	*32B*
Soil Name	*Munising*	*Gay*	*Croswell*	*Alcona*
Potential for Habitat Elements:				
Grain and Seed Crops	Good	Poor	Poor	Good
Grasses and Legumes	Good	Poor	Poor	Good
Wild Herbaceous Plants	Good	Fair	Fair	Good
Hardwood Trees	Fair	Fair	Fair	Good
Coniferous Plants	Fair	Fair	Fair	Good
Wetland Plants	Poor	Good	Poor	Poor
Shallow Water Areas	Poor	Good	Very Poor	Very Poor
Potential as Habitat For:				
Openland Wildlife	Good	Fair	Poor	Good
Woodland Wildlife	Fair	Fair	Fair	Good
Wetland Wildlife	Poor	Good	Very Poor	Very Poor

detailed information concerning wildlife habitat. DeGraaf et al. (1992) list four different habitat opportunity classes that are based on landscape structure:

- Habitat Opportunity Class I—Landscapes that are at least 90% forested
- Habitat Opportunity Class II—Landscapes at least 90% forested with at least 5% of the area in water and wetland nonforest types
- Habitat Opportunity Class III—Landscapes at least 70% forested with less than 5% of the area in water and wetland nonforest types
- Habitat Opportunity Class IV—Landscapes at least 70% forested with more than 5% of the area in water and wetland nonforest types

DeGraaf et al. (1992) provide a dichotomous key of use in developing management prescriptions for each of these habitat opportunity classes. They also provide general information regarding the mix of different types of elements to maintain in a New England landscape. Information concerning the structure of elements within a specific landscape may be used to develop management alternatives designed to meet the management goals for that landscape. Geographic information systems (Chapter 13) provide the means to estimate the length of edge (Chapter 7) between two habitat types, for example, which may provide a very precise estimate of the area of suitable habitat for some species. Game species, in particular, tend to inhabit such areas. Other species tend to be found away from edges, in the forest interior. If a relevant definition of what constitutes the forest interior can be developed for a particular species, then geographic information systems can provide estimates of the area of interior forest habitat by broad vegetation class, and the connections or distance between different habitat areas, within a landscape. Geographic information systems can be used to examine landscape structure and associated habitat characteristics under different management alternatives.

For some species, it is not just the spatial characteristics of the landscape that are important; the temporal aspects may also be critical. This is obviously true for migratory species, which must be able to locate suitable food, water, and cover along their migration routes at the time of migration. Nonmigratory species may also utilize certain habitats at critical times of the year. In parts of the northern United States and southern Canada, for instance, ephemeral ponds or vernal pools support a wide range of insect and vegetative species in the spring following snowmelt. These areas may provide critical habitat for many species during the period following winter conditions, but prior to the full development of summer vegetation.

There is a great deal of difference in the amount and type of information available concerning the habitat requirements of different wildlife species. For some species, particularly game species, there may be a great deal of information available, whereas for others, it is difficult to find even a few sources describing their habitat requirements. Martin, Zim, and Nelson (1951) provide a general source of information on the utilization of plants for food by a wide range of North American species. Although a great deal of information has obviously been developed since this source was published, it still provides basic information on the food requirements of many species that is difficult to find in another single source. Of particular value is a cross-listing that identifies the wildlife species that utilize particular plant species and the relative importance of such plants in their diets. Martin, Zim, and Nelson (1951), for example, list 20 avian and 12 mammalian species known to utilize dandelion for food (Table 9.3). If there is information available

TABLE 9.3 Avian and Mammalian Species Known to Utilize Dandelion (*Taraxacum* spp.) for Food

	Plant Component	*Species*
Upland Game Birds	Foliage, Seed Heads	Blue Grouse Ruffed Grouse Sage Grouse Sharp-tailed Grouse Hungarian Partridge Ring-necked Pheasant Greater Prairie Chicken Bobwhite Quail Valley Quail Wild Turkey
Songbirds	Seeds	Brewer Blackbird Yellow-headed Blackbird Goldfinch Pine Siskin Brewer Sparrow Chipping Sparrow English Sparrow Sharp-tailed Sparrow Vesper Sparrow Green-tailed Towhee
Fur and Game Mammals	Foliage, Plants	Varying Hare Western Porcupine Cottontail Rabbit
Small Mammals	Foliage, Seeds, Plants	Least Chipmunk Northwest Chipmunk Pocket Gopher Western Pocket Gopher Richardson Ground Squirrel Pine Mouse Gunnison Prairie Dog
Hoofed Browsers	Plants	Mule Deer White-tailed Deer

Source: Martin, Zim, and Nelson (1951).

concerning the vegetative associations present in a landscape, then a list of plant species occurring in the different vegetative associations can usually be developed. Even if there is no other readily available interpretive information accompanying the vegetative association descriptions, information such as that in Table 9.3 may be coupled with the list of plant species to develop some idea of the food resources available within a landscape for a range of wildlife species.

Each species is unique in the combination of habitat requirements necessary for its survival. Broad classes of vegetation type or landscape structure can certainly indicate the suitability of habitat for some species, and this may be sufficient in many resource assessment efforts. As mentioned previously, there are times when more quantitative information is required concerning the availability of habitat requirements within a landscape for particular species. This information should integrate all three components of habitat (food, water, and cover). The U.S. Fish and Wildlife Service has developed a series of indices, referred to as *habitat suitability indices,* which are available to describe the degree to which habitat requirements of selected North American species are met within a given area (U.S. Fish and Wildlife Service, 1980, 1981). These indices provide a numerical score, ranging from 0 to 1, which describes the relative suitability of a given area as habitat for the selected species. A high value does not guarantee that the species in question will actually be present in an area, but it does provide information that can be used in the assessment process to describe current habitat suitability and possible changes in habitat suitability as a consequence of the management alternatives being considered. In most situations, it will be desirable to perform such analyses for several indicator species of different guilds in order to obtain an overall evaluation of the possible consequences of management alternatives on the range of wildlife species present in a landscape. Almost every management alternative will lead to improved habitat for some species or guilds, and less suitable habitat for others.

For example, Allen (1983) provides a habitat suitability index for fisher. Habitat suitability index models are developed following a review of the literature available for a given species and are designed to focus on the most restrictive habitat requirements for the species, such as winter–early spring habitat for fisher. A critical habitat component is selected, such as winter cover for fisher, and variables associated with this component are identified and their relationships with the appropriate levels of the habitat component are quantified. For the fisher model, the habitat variables are percentage of canopy closure (X_1), average overstory tree diameter (X_2), tree canopy diversity (X_3), and percentage of the overstory in deciduous species (X_4). Thomasma (1988), who evaluated this index, estimated average overstory diameter based on forest inventory records and the remaining variables from aerial photographs. A suitability index value is obtained for each of these habitat variables for a particular location (Figure 9.1). The suitability index

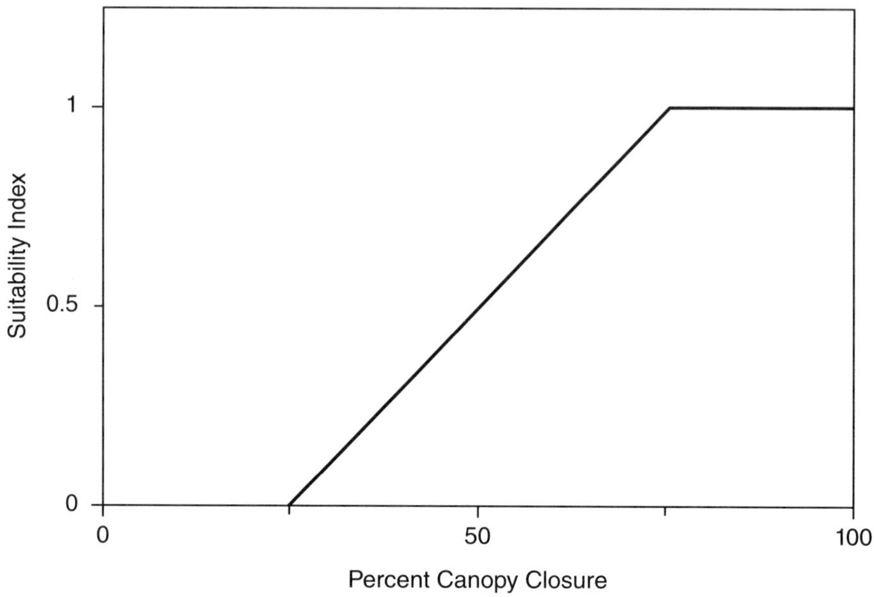

(*a*)

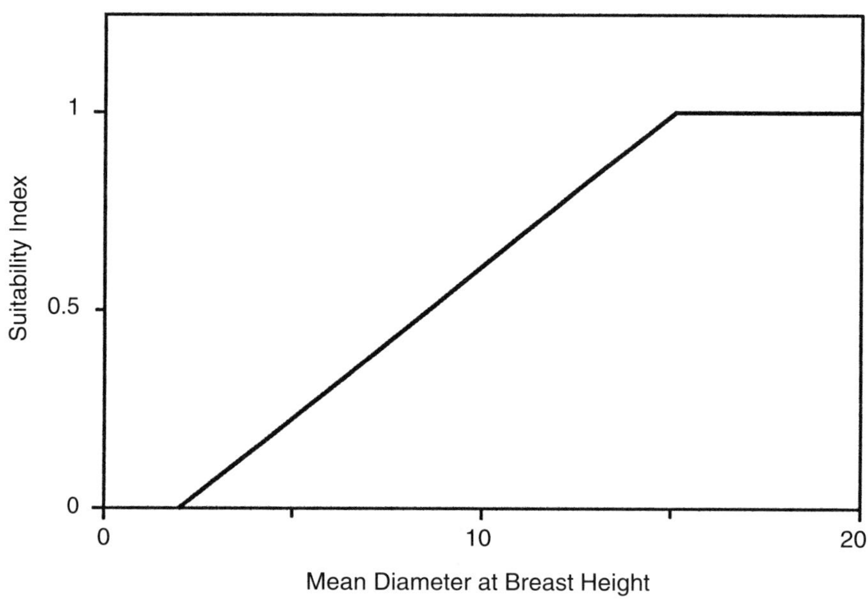

(*b*)

FIGURE 9.1 Suitability index values for calculating the habitat suitability index for fisher (*Martes pennanti*, adapted from Allen (1983) and Thomasma (1988)): (*a*) percentage of canopy closure, (*b*) average overstory tree diameter at breast height, (*c*) tree canopy diversity class, and (*d*) percentage of the overstory in deciduous species.

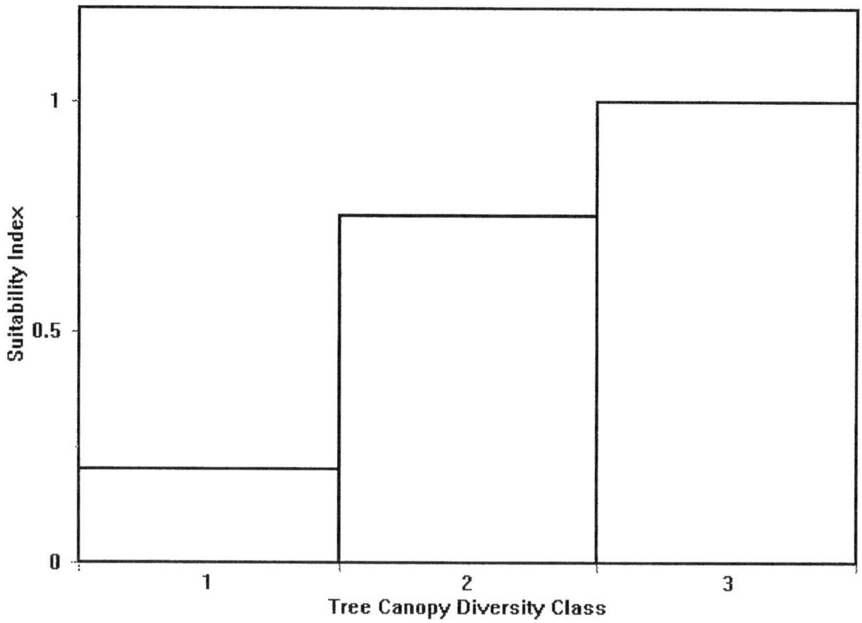

(c)

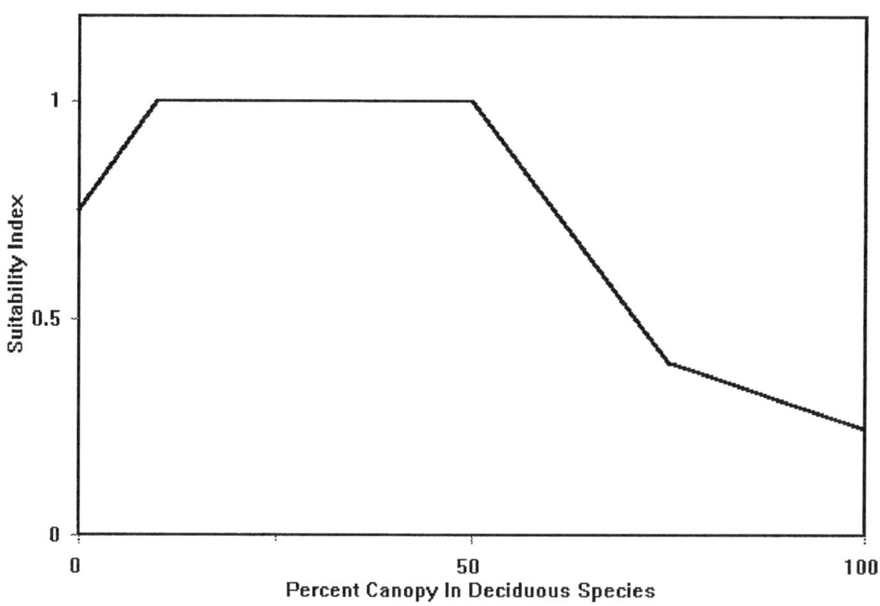

(d)

FIGURE 9.1 *(Continued)*

values are combined to obtain the habitat suitability index for a given location:

$$\text{Fisher } HSI = (X_1 \cdot X_2 \cdot X_3)^{1/3} \cdot X_4 \qquad [9.1]$$

where Fisher HSI is the habitat suitability index for fisher, which ranges from 0 to 1, and the X_i variables are the suitability indices for the respective habitat variables from Figure 9.1. For an area with 90% canopy closure, an average forest overstory diameter of 10 inches, two layers of canopy foliage, and containing 50% coniferous species, the resulting fisher habitat suitability index, using the suitability indices from Figure 9.1, is:

$$\begin{aligned}\text{Fisher } HSI &= (1.0 \cdot 0.5 \cdot 0.75)^{1/3} \cdot 1.0 \\ &= 0.72 \qquad [9.2]\end{aligned}$$

The value of this index will change over time as forest overstory trees increase in size and as stand composition changes in response to succession and management activity. As noted previously, estimation of a number of these indices for a suite of species occupying a range of habitats provides an indication of the overall impacts of management alternatives on wildlife populations present within a landscape.

In a number of cases, vegetation within a landscape may be utilized for grazing of domesticated livestock. Grazing may be the dominant land use in some areas, but it may also be a secondary land use in an area primarily managed for other purposes. A specialized type of habitat analysis is needed to assess the grazing resources within a landscape. There are a number of vegetative assessment techniques that are specifically tailored for use in *rangelands,* which are lands supporting vegetation utilized for grazing of domesticated livestock or herbivorous wildlife species. Instead of a habitat suitability index that ranges from 0 to 1, range resources are usually expressed in terms of the numbers of animals that the vegetation can support. A specialized measurement unit, the *animal unit month,* is used to quantify range resources. An animal unit month is defined as the equivalent of 1,000 pounds of animal live weight that the vegetation within an area can support. Approximately 900 to 1,050 pounds of forage are required to support one animal unit month (Avery and Burkhart, 1994).

The assessment of the number of animal unit months that may be supported within a landscape requires estimation of the biomass of forage species. Chapter 15 describes general sampling methods that may be used for these purposes, and the procedures described for estimating biomass of overstory trees in Chapter 7 can be used to estimate biomass of forage species with only minor modification. Usually, small fixed area plots are used and forage vegetation is removed, or "clipped," dried, and weighed to obtain estimates of the dry weight of forage species per unit area. This is relatively expensive, and a double sampling technique is often utilized to reduce inventory costs. With the use of this method, the dry weight of forage species is ocularly estimated for a large number of plots, and on a small number of plots, it is both estimated and measured. The relationship between estimated and measured weights is used to adjust the estimated forage weight per unit area, using ratio or regression estimation (Chapter 15).

It is important that only those plant species actually utilized by animals for grazing or browsing be included in the estimates of available forage biomass. In addition, only portions of some plants, such as foliage and small twigs, may be utilized by herbivores. Because different herbivorous species may prefer and utilize different plant species and even different portions of the same plants, it is important to consider the preferences of the herbivorous species present within the landscape during the development of field sampling procedures for estimating forage availability. Animal unit months were originally defined for cattle, but the Range Term Glossary Committee (1964) gave a number of *animal unit conversion factors* that may be used to convert animal unit months to values relevant to other species. The forage required by a white-tailed deer, for instance, is approximately equivalent to the forage required to support 0.14 animal unit months.

Once the appropriate plant species to sample and the portions of each plant species to include are determined, field sampling can be undertaken to estimate the forage biomass per unit area. To determine grazing capacity, the total forage biomass per unit area is summed over all species and divided by the forage weight required to support an animal unit month:

$$\text{Grazing capacity} = \frac{\sum_{i=1}^{k} W_i}{\text{Animal unit requirement}} \qquad [9.3]$$

The grazing capacity estimated in Equation [9.3] is the estimated number of animal unit months per unit area that the forage resources in a given area can support, W_i is the estimated biomass per unit area of the ith forage species, and the animal use requirement is the estimated weight of forage required to support one animal unit month. Grazing capacity can be adjusted to estimate the number of individuals of a given species that an area can support by estimating the appropriate plant biomass values for the herbivorous species of interest and the animal unit conversion factors noted earlier; dividing grazing capacity by 0.14, for example, provides an estimate of the number of white-tailed deer that could be supported per unit area.

Forage resources are not static. They vary annually in response to weather, and there are seasonal differences in the amount of forage available within a year. These factors must be considered during field sampling and the assessment process. In addition, herbivory itself, especially in extreme amounts, can alter the plant community composition in an area. This, in turn, can lead to different levels of forage quality and availability in the future. McLaren and Peterson (1994) provide an example in which moose herbivory of balsam fir altered succession and forest overstory composition over long time periods on Isle Royale, Michigan. During the assessment process, therefore, if animal unit months or forage availability is being used as an indicator of wildlife habitat or domestic animal grazing potential, it is not only important to include the direct effects of management alternatives on forage availability, but subsequent vegetative responses to herbivory must also be considered.

ESTIMATION OF POPULATION SIZE

The problem of estimating the number of individuals in a particular population is complicated by several factors: (1) wildlife species do not utilize all portions of an area uniformly, but tend to prefer some areas and avoid others; (2) wildlife species are mobile and can sometimes travel great distances during the observation period; and (3) individuals may be attracted or repelled by the observation process, making it difficult to obtain unbiased population estimates. In addition, the processes of reproduction, mortality, immigration, and emigration add and subtract individuals from all populations. *Open* populations are subject to changes in population size during the observation period; populations are considered to be *closed* populations if, for some interval of closure, they are not subject to changes in population size. Most populations in natural settings are actually open except in relatively unusual cases or for relatively short time intervals. Reproduction is seasonal for many species, but mortality can occur at any time. The degree of emigration and immigration can vary tremendously, even for the same species, in different areas. If a population is open to even a small degree, this must be considered in the process of estimating population numbers in order to avoid biasing the final results.

There are several commonly used methods to estimate the size of wildlife populations. Some of these methods are analogous to procedures used in forest overstory estimation (Chapter 8). Quadrat or transect sampling, for example, is equivalent to fixed area plot sampling in that uniformly sized plots are established in the field. In wildlife sampling, the population may be directly observed by counting individuals, or indirectly observed by counting tracks or fecal pellets. In many ways, line transect sampling is similar to variable radius plot sampling, as discussed in Chapter 8 for forest overstory estimation, but there are also important differences. In line transect sampling for wildlife populations, the likelihood of an individual's being detected by the observer, even if it is present, is considered in the estimation process. Finally, perhaps the most familiar methods for assessing population numbers are capture-recapture methods. In these procedures, a number of individuals are captured and permanently marked, then released, and the ratio of marked-to-unmarked individuals in a subsequent capture effort is used to develop an estimate of population size. Each of these three methods, quadrat and transect sampling, line transect sampling, and capture-recapture sampling are discussed in more detail in the following paragraphs. Each has its advantages and disadvantages for different situations and for species with different characteristic patterns of behavior and habitat preference.

Quadrat and Transect Sampling

The terms *quadrat sampling* and *transect sampling,* in the context of estimating wildlife population sizes, are used to refer to the use of fixed area plots to estimate, either directly or indirectly, the number of individuals per unit area. *Quadrat sampling* is the more common term and is used to refer to plots of all shapes and sizes; the term *transect sampling* is sometimes used to refer to elongated rectangular plots. In sampling wildlife resources, sizes of quadrats may be much larger than the sizes of the fixed area plots normally used for sampling the vegetative community or forest overstory. If the population size is to be estimated directly (i.e., if the number of individuals in an area are being observed and counted), the size of the quadrat must be large enough to be expected to contain some individuals. This implies that the quadrat must be larger than, and maybe several times larger than, the expected home range of an individual or mating pair. As noted by Burnham, Anderson, and Lake (1980), direct estimation of population size using quadrats is best applied to relatively immobile populations or to organisms with small home ranges. Quadrat sampling is usually not appropriate for large animals such as black bear or gray wolf that may range over thousands of hectares to obtain resources. Quadrat sampling may be used to estimate the number of colonies or nesting sites rather than the number of individuals. All this implies that a single quadrat size that is appropriate for one species will not be appropriate for a wide range of species of different guilds.

Individual quadrats are sample units (Chapter 15) and, as discussed in Chapter 8, the inverse of the plot size is the expansion factor used to obtain estimates of *population density,* or the number of individuals per unit area. Population density is an important variable in describing wildlife populations because many population processes are density-dependent. In a landscape assessment, it may be equally important, or even more important, to obtain an estimate of the total population size in a given area.

Any sampling design discussed in Chapter 15 may be adapted to estimate wildlife population size or density. The discussion and examples that follow assume that simple random sampling is used, but other sampling designs may be used with appropriate modification. If the population density is estimated directly, then quadrat sampling is just an application of simple random sampling. The overall number of sampling units in an area (the sampling population, not to be confused with the wildlife population being observed) is obtained by dividing the total area by the size of the individual quadrats. In a 1,000 ha area, for example, there are 4,000 possible 0.25 ha quadrats ($N = 1{,}000/0.25 = 4{,}000$). The average number of individuals per ha is estimated as follows:

$$\bar{d} = \frac{\sum_{i=1}^{n} d_i}{\alpha \cdot n} \qquad [9.4]$$

where $\bar{d}$ is the population density, or the number of individuals per unit area, d_i is the number of individuals observed on the ith quadrat, n is the number of sampled quadrats, and α is the quadrat size in hectares or acres, as appropriate. The standard deviation of this estimate is equivalent to the standard error of the mean discussed in Chapter 15. Assuming that the sampling population is large, the standard deviation of the mean number of individuals per unit area is given by:

$$s_{\bar{d}} = \frac{s}{\alpha \cdot \sqrt{n}} \qquad [9.5]$$

where s, the standard deviation of the number of individuals observed on the quadrats, is estimated as follows:

$$s = \sqrt{\frac{\sum_{i=1}^{n} (d_i - \alpha\bar{d})^2}{n - 1}} \qquad [9.6]$$

or, equivalently:

$$s = \sqrt{\frac{\sum_{i=1}^{n} d_i^2 - \frac{\left(\sum_{i=1}^{n} d_i\right)^2}{n}}{n - 1}} \qquad [9.7]$$

If the area of observed quadrats is more than 5% of the total area, the finite population correction factor (Chapter 15) should be used to adjust the estimate of the standard deviation of the mean density from Equation [9.5]:

$$s_{\bar{d}} = \frac{s}{\alpha \cdot \sqrt{n}} \cdot \sqrt{\frac{N - n}{N}} \qquad [9.8]$$

Table 9.4 contains example data from a quadrat-based sample of small mammal populations in a riparian area. The total riparian area in the landscape of interest is 47 ha; fifteen 5 m × 5 m quadrats were randomly established throughout the riparian area. The quadrat size (α) is 25 m^2 or 0.0025 ha. The estimated number of individuals per hectare is calculated using Equation [9.4]:

TABLE 9.4 Example Data Collected During Quadrat Sampling of Small Mammal Populations in a Riparian Area

Quadrat	*Number of Small Mammals Observed*
1	1
2	0
3	4
4	1
5	0
6	2
7	1
8	5
9	0
10	0
11	8
12	3
13	2
14	2
15	5
$\sum_{i=1}^{15} d_i$	34

$$\bar{d} = \frac{34}{(0.0025) \cdot (15)}$$

$$= 906.7 \approx 907 \text{ individuals per hectare} \qquad [9.9]$$

Because less than 5% of the total area was sampled, the standard deviation of the estimated population density is calculated using Equation [9.5]:

$$s_{\bar{d}} = \frac{2.34}{(0.0025) \cdot (\sqrt{15})}$$

$$= 241.67 \approx 242 \text{ individuals per hectare} \qquad [9.10]$$

The standard deviation can be used to calculate confidence limits for the estimated mean as described in Chapter 15. An estimate of the total number of individuals in the area can be found by multiplying the total area (47 ha) by the estimated number of individuals per hectare.

$$D = \bar{d} \cdot \text{Area} \qquad [9.11]$$

In the example, the total number of small mammals in the riparian area is estimated as:

$$\begin{aligned} D &= (906.7) \cdot (47) \\ &= 42{,}614.9 \approx 42{,}615 \text{ individuals} \end{aligned} \qquad [9.12]$$

The standard deviation of the estimated total number of individuals in a population is simply the standard deviation of the mean population density multiplied by the size of the area:

$$s_{\mathrm{D}} = s_{\bar{d}} \cdot \text{Area} \qquad [9.13]$$

In the example, the standard deviation of the estimated total number of small mammals in the riparian area is:

$$\begin{aligned} s_D &= (241.67) \cdot (47) \\ &= 11{,}358.49 \approx 11{,}358 \text{ individuals} \end{aligned} \qquad [9.14]$$

In quadrat sampling, it is assumed that individuals within a quadrat will all be counted once. In practice, it is sometimes difficult to ensure that no individuals are missed, especially because many species try to avoid detection by the observer. It is also sometimes difficult to be sure that individuals are only counted once, especially because both they and the observer may be moving. These two problems are less serious on small quadrats, but even then there may be a difficulty in some situations.

Another implicit assumption of quadrat sampling is that individuals are not moving in and out of the quadrats during the observation period. In fact, there is a further implicit assumption that the overall population is closed. This is reasonable for cases such as the example, in which individuals do not tend to travel great distances, travel corridors are reasonably well defined, and individuals tend to enter or leave the sample quadrats using only a minimum number of paths. In many situations, however, populations are open and the use of quadrat sampling, at best, provides a one-point-in-time approximation of the number of individuals within the area of interest.

Another phenomenon in quadrat sampling is the frequent number of quadrats where no individuals are observed; in the example in Table 9.4, almost one-third of the quadrats contained no observed individuals. Pennington (1983) suggested that zero observations be treated separately, and Seber (1986) suggested the use of data transformations to improve the statistical analysis. There is no hard-and-fast rule, but if the numbers of zero observations are not excessive, then special analytical methods are not usually required. Generally, quadrat sizes should be chosen to minimize the likelihood of an excessive number of zero observations.

Indirect observation of population size using quadrat sampling involves the use of surrogate variables such as tracks or fecal pellets to indicate the presence of individuals in an area. Track counts or fecal pellet counts are performed on each quadrat, and the average values are estimated as in Equation [9.4]. A conversion factor is necessary to convert the value of the indicator variable into the number of individual animals to estimate population size or density. Pellet counts or track counts may also be used to provide an index of population size. This does not require the assumption of a conversion factor to calculate number of individuals. Nonvisual observation may be used in avian populations, in which the number of singing individuals are recorded rather than the number of sighted individuals. Adjustment factors may also be necessary to estimate total population size for some species in which only one sex is likely to be identified.

Line Transect Sampling

Many organisms attempt to avoid detection by a human observer. Line transect sampling, which is related to line intercept sampling (Chapter 4), is a technique for estimating wildlife population size or density that does not require the assumption that every animal within the transect boundaries will be detected. Transects may be temporary, or they may be permanently marked in order to facilitate comparison of population density estimates over time. Temporary transects are usually established at the time of observation. The observer walks a predetermined distance along a preselected compass bearing and records sightings of individual animals as they occur. The method assumes that individuals are not counted more than once and that sightings of different individuals are independent of each other.

In line transect sampling, observations are made along a line of travel. In some applications the number of sighted individuals is recorded. In others, each time an individual is sighted, the angle from the line of travel and the distance to the individual from the observer are recorded (Figure 9.2) in order to estimate the distance of the sighted individual from the centerline of the transect. The perpendicular distance from the transect centerline is estimated using a trigonometric relationship:

$$y_i = r_i \sin \theta_i \qquad [9.15]$$

where y_i is the calculated perpendicular distance from the line of travel to the sighted individual, r_i is the distance from the observer to the individual when it is sighted, and θ_i is the angle from the line of travel at which the individual was sighted. It is assumed that r_i and θ_i are observed without error. If it is possible to directly observe the perpendicular distance from the line of travel, then that is preferable and the angle and initial sighting distance need not be recorded. This would be the case, for example, if line transect sampling was being used to census immobile organisms or objects, such as ant colonies or squirrel nests.

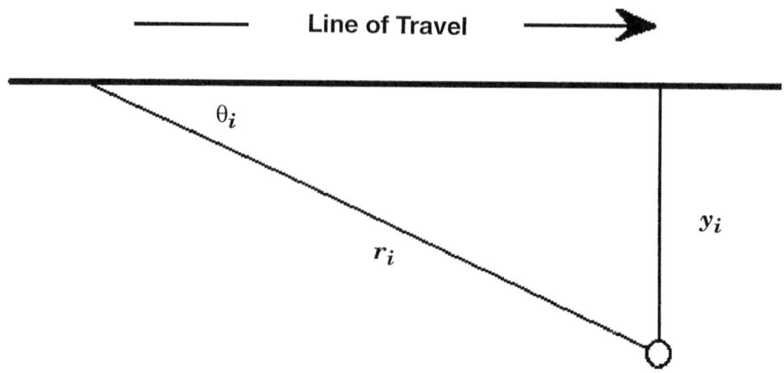

FIGURE 9.2 Schematic diagram of line transect sampling showing the relationships among the line of travel, sighting angle, sighting distance, and perpendicular distance to the individual.

A *detection function* is used to describe the likelihood that an individual is detected as a function of its distance from the line of travel (Figure 9.3). This function could be a recognized statistical density function, such as the exponential function, or it could be described nonparametrically using a step function or some other specification of the likelihood of detection at certain fixed distances. It is important to note that the overall detection function may be constant for all organisms, but in some instances there may be a mixture of different detection functions for different subpopulations. It may be more

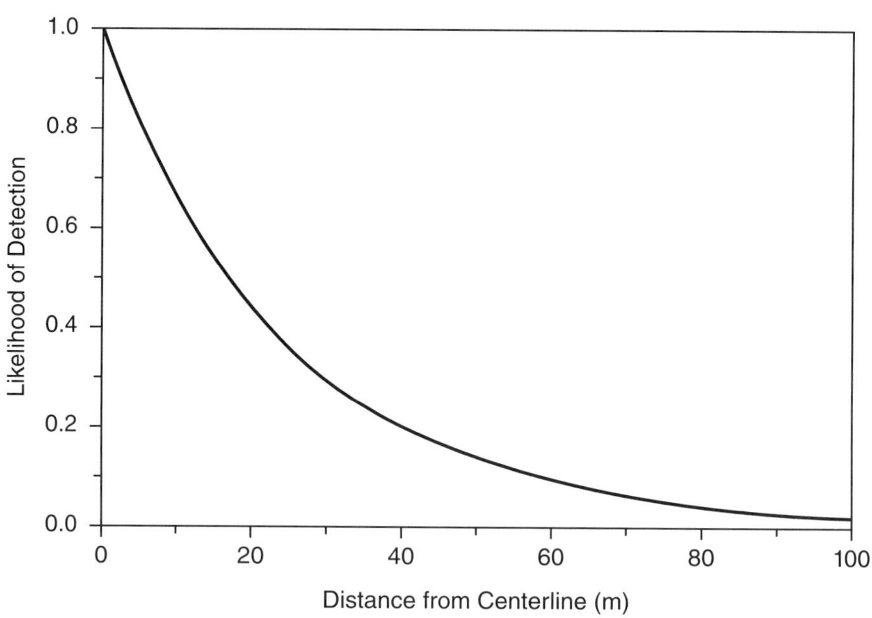

FIGURE 9.3 Illustration of an exponential detection function for use in line transect sampling.

difficult to see very young individuals than to see to mature individuals, for example. Mixtures of detection functions also may be used if it is more difficult to detect animals under some habitat conditions versus others, such as in lowland coniferous forests as compared with upland deciduous forests. The general form of the line transect sampling estimator for population density is given by:

$$\overline{d} = \frac{n}{2\ L\ \hat{a}} \qquad [9.16]$$

where $\overline{d}$ is the estimated density per unit area, n is the number of individuals sighted along the transect, and L is the length of the transect. The detection function is integrated over the effective width of the transect to obtain an estimate of $\hat{a}$:

$$\hat{a} = \int_0^w g(x)\ dx \qquad [9.17]$$

The detection function is defined, either parametrically or nonparametrically, by $g(x)$, and w is one-half of the effective width of the transect. The transect width may be infinite for some applications; this implies that sightings will be recorded regardless of how far the individual is from the centerline. In other cases, the distance may be capped at some distance w, and sightings at greater distances are not recorded.

Gates, Marshall, and Olson (1968) found that use of the exponential function (Figure 9.3) as the detection function led to the following estimate of the population density:

$$\overline{d} = \frac{n}{2\ L\ \hat{C}} \qquad [9.18]$$

where $\hat{C}$ is

$$\hat{C} = \frac{\sum_{i=1}^{n} y_i}{n - 1} \qquad [9.19]$$

Table 9.5 contains data collected along one 100 m transect. The variables recorded in the field are the distance and angle to the individual when first sighted. The perpendicular distance from the transect centerline to each sighted individual is calculated using Equation [9.15]. For the first individual:

$$\begin{aligned} y_i &= 10 \sin (30°) \\ &= 5.0 \text{ m} \end{aligned} \qquad [9.20]$$

TABLE 9.5 Example Data Collected During Line Transect Sampling Along a 100 m Transect

Individual	Observed Variables θ (°)	r (m)	Calculated Perpendicular Distance (m)
1	30	10	5.0
2	47	15	11.0
3	38	12	7.4
4	52	21	16.5
5	21	16	5.7
6	42	18	12.0
7	16	27	7.4
8	33	15	8.2
$\sum_{i=1}^{8} y_i$			73.2
$\bar{y}$			9.15

The total calculated perpendicular distance for all of the observed individuals in the example is 73.2 m. This leads to an estimated value of $\hat{C}$ of:

$$\hat{C} = \frac{73.2}{8 - 1} = 10.46 \text{ m} \qquad [9.21]$$

Combining the results of Equations [9.20] and [9.21] in Equation [9.18] results in the following estimate of the number of individuals per square meter:

$$\bar{d} = \frac{8}{2\,(100)\,(10.46)} = 0.00382 \text{ individuals per square meter} \qquad [9.22]$$

which is equivalent to 38.2 individuals per hectare.

Observations from multiple individual transects are required to obtain an estimate of the variance of the number of individuals per unit area. When there are observations from multiple transects, the overall estimate of the mean population density is a weighted mean obtained by weighting the estimates of the population densities from the individual transects by the length of the respective transects:

$$\bar{D} = \frac{\sum_{i=1}^{m} L_i \bar{d}_i}{\sum_{i=1}^{m} L_i} \qquad [9.23]$$

where m is the number of observed transects. The variance of this estimate is given by:

$$s_{\overline{D}}^2 = \frac{\sum_{i=1}^{m} L_i (\overline{d}_i - \overline{D})^2}{(m - 1) \sum_{i-1}^{m} L_i} \qquad [9.24]$$

If the data in Table 9.5 were from one of four 100 m transects established in an area, and if the four estimates of population density were 38.2, 50.7, 25.5, and 40.1 individuals per hectare, then the overall estimate of population density is given, with some algebraic simplification because the transects are all the same length, by:

$$\begin{aligned} \overline{D} &= \frac{100\ (38.2 + 50.7 + 25.5 + 40.1)}{400} \\ &= 38.6 \text{ individuals per hectare} \end{aligned} \qquad [9.25]$$

The variance of this estimate, with appropriate simplification owing to equal transect lengths, is given by:

$$\begin{aligned} s_{\overline{D}}^2 &= \frac{100\ [(38.2 - 38.6)^2 + \ldots + (40.1 - 38.6)^2]}{400\ (4 - 1)} \\ &= \frac{100\ [0.16 + 146.41 + 171.61 + 2.25]}{1{,}200} \\ &= 26.7 \text{ (individuals per hectare)}^2 \end{aligned} \qquad [9.26]$$

This estimate of the variance of the estimated population density may be used to calculate confidence intervals for the mean number of individuals per hectare, as described in Chapter 15.

There are any number of detection functions that may be appropriate in different situations; Burnham, Anderson, and Lake (1980) provide examples and discussion of numerous different detection functions for use in line transect sampling. It is important to utilize a detection function that is realistic for a given situation. A different function, for example, will be required if the studied species tends to hide and then flushes when the observer gets within a certain distance, versus a species that tends to move to avoid detection.

Line transect sampling can be used in aerial census efforts with little or no modification. In this case, the flight line is considered to be the transect centerline, and either a simple count or the perpendicular distance of each sighted individual from the centerline is the recorded variable. Distances are necessarily estimates, as they are in most applications of line transect sampling, and it is important to practice estimating distances prior to the actual field sampling effort in order to avoid introducing bias into the population estimates.

One modification of line transect sampling involves the observer's traveling along a predetermined transect; but instead of making continuous observations, the observer stops at preselected points, waits for a short period of time to allow any effects of observer movement to dissipate, and then records sightings or calls heard during a fixed observation interval. This method is particularly useful in censusing avian populations.

Capture-Recapture Sampling

One of the most widely known methods to estimate the size of wildlife populations is the capture-recapture method. In this procedure, a sample of animals is caught, tagged, and released back into the population. Subsequently, after the marked individuals have had time to disperse throughout the population, a second sample is caught and the ratio of previously tagged-to-untagged individuals in the second sample is used to estimate the total number of individuals in the population.

Several assumptions must be made in order to obtain estimates of population size using capture-recapture methods: (1) tagged individuals are distributed randomly throughout the population at the time of their second capture; (2) the initial capture does not affect the likelihood of a second capture (i.e., individuals do not become "trap-happy" or "trap-wise" following the first capture); (3) no tags are lost between capture efforts; and (4) all tags are noticed and recorded at the second capture. Shiver and Borders (1996) note that it is particularly difficult to ensure that the likelihood of second capture is not affected by the initial capture. As they indicate, Otis et al. (1978) have presented methods for use when it is likely that animal behavior regarding the second capture effort has been altered by the initial capture. In addition to the assumptions required for statistical analyses, it should be noted that efforts to implement capture-recapture procedures for vertebrate species may be governed by the Laboratory Animal Welfare Act of 1966 as subsequently amended. Special training in animal handling and/or a regulatory permitting process may be required before implementation of these methods in resource assessment efforts. Organizations that handle animals regularly must have an Institutional Animal Care and Use Committee that is responsible for ensuring humane treatment of animals.

Capture-recapture methods can be divided into *direct* and *inverse* procedures, which differ in the means of implementing the second capture effort. In direct procedures, a fixed number of individuals are captured at the second time and the number of tagged individuals is recorded; in inverse procedures, the second capture effort continues until a predetermined number of tagged individuals are observed. The following estimator, known as the *Petersen Estimator* (Petersen, 1896) or the *Lincoln Index* (Lincoln, 1930), is used for direct sampling in closed populations. It is biased, but it is a maximum likelihood estimator under some distributional assumptions (Schreuder, Gregoire, and Wood 1993).

$$\hat{N} = \frac{n_1 n_2}{m} \qquad [9.27]$$

where $\hat{N}$ is the estimate of the total number of individuals in the population, n_1 is the number of individuals initially captured and tagged, n_2 is the total number of individuals captured during the second capture effort, and m is the number of tagged individuals captured during the second capture effort. If 50 animals are initially captured, marked, and released back into a population, and if a second effort also captures 50 animals, 17 of which are observed as having marks from the initial capture, the estimate of population size is:

$$\begin{aligned}\hat{N} &= \frac{50 \cdot 50}{17} \\ &= 147.1 \approx 147 \text{ individuals}\end{aligned} \qquad [9.28]$$

Shiver and Borders (1996) note that this estimator is biased and give a second estimator, the *Chapman Estimator,* which is less biased, for use in direct sampling of closed populations:

$$\hat{N} = \frac{(n_1 + 1)(n_2 + 1)}{(m + 1)} - 1 \qquad [9.29]$$

For the example noted earlier, the Chapman Estimator of population size is:

$$\begin{aligned}\hat{N} &= \frac{(50 + 1)(50 + 1)}{(17 + 1)} - 1 \\ &= 143.5 \approx 144 \text{ individuals}\end{aligned} \qquad [9.30]$$

As a rule of thumb, in using either of these two estimators, there should be 10 or more marked animals captured during the second capture effort in order to obtain reasonable estimates of population size. Seber (1982) gave the following approximate expression of the variance of the Chapman Estimator:

$$s^2_{\hat{N}} = \frac{(n_1 + 1)(n_2 + 1)(n_1 - m)(n_2 - m)}{(m + 1)^2 (m + 2)} \qquad [9.31]$$

For the example:

$$s_{\hat{N}}^2 = \frac{(50 + 1)\,(50 + 1)\,(50 - 17)\,(50 - 17)}{(17 + 1)^2\,(17 + 2)}$$
$$= \frac{51 \cdot 51 \cdot 33 \cdot 33}{18^2\;19}$$
$$= 460.1 \text{ individuals}^2 \qquad [9.32]$$

and

$$s_{\hat{N}} = 21.5 \text{ individuals} \qquad [9.32a]$$

In inverse sampling, the second capture effort is continued until a predetermined number of tagged individuals are observed. In this case, if the second capture effort is conducted without replacement, the total number of individuals in the population is estimated by:

$$\hat{N} = \frac{(n_1 + 1)\,n_2}{m} - 1 \qquad [9.33]$$

Again, the difference between inverse and direct sampling is that in inverse sampling, m is predetermined for the second capture effort, whereas in direct sampling, n_2 is predetermined for the second capture effort. As an example, assume that 50 individuals are initially captured and released. After an appropriate time interval to allow for dispersion of tagged animals throughout the population, a second capture effort is undertaken and continued until 20 tagged animals are recaptured. If this requires the capture of 62 animals during the second capture effort, the population size is estimated to be:

$$\hat{N} = \frac{(50 + 1)\,62}{20} - 1$$
$$= 157.1 \approx 157 \text{ individuals} \qquad [9.34]$$

In inverse sampling, the approximate variance of $\hat{N}$ is given by:

$$s_{\hat{N}}^2 = \frac{(n_1 - m + 1)\,(\hat{N} + 1)\,(\hat{N} - n_1)}{m\,(n_1 + 2)} \qquad [9.35]$$

Continuing the example:

$$s_{\hat{N}}^2 = \frac{(50 - 20 + 1)\,(157.1 + 1)\,(157.1 - 50)}{20\,(50 + 2)}$$
$$= 504.7 \text{ individuals}^2 \qquad [9.36]$$

and

$$s_{\hat{N}} = 22.5 \text{ individuals} \qquad [9.36a]$$

If the second capture effort in inverse sampling is conducted with replacement, then the population size is estimated as in direct sampling, using Equation [9.27]. The variance of the estimated population size in this case is approximated by:

$$s_{\hat{N}}^2 = \frac{n_1^2 \, n_2 \, (n_2 - m)}{m^2 \, (m + 1)} \qquad [9.37]$$

All of the capture-recapture methods described here assume that sampling is from a closed population. This assumption is probably not violated to a great extent for most capture-recapture efforts designed for use in resource assessment efforts as long as the second capture effort is implemented as soon as it is likely that the tagged animals have dispersed throughout the population. The closed-population assumption is almost always violated to some degree, however, and resource managers should be mindful of this restriction on these methods.

In some resource assessment efforts, longer-term monitoring efforts are established in which capture-recapture methods may be utilized in subsequent observation intervals to assess changes in population size. Seber (1986) and Pollock et al. (1990) discuss various methods that may be used when there are multiple recapture efforts, but the assumption that a population remains closed is increasingly violated with the passage of time. Lebreton et al. (1992) provide a survey of capture-recapture methods that may be suitable for such efforts in open populations. The basic underlying sampling model is referred to as the Jolly-Seber model (Jolly, 1965; Seber, 1965), which uses multiple recapture efforts to estimate population size and survival rate. Methods of establishing long-term multiple recapture efforts to monitor changes in wildlife populations are beyond the scope of this text, and readers are referred to Pollock et al. (1990), Seber (1986, 1992), and Lebreton et al. (1992) for guidance in the establishment of appropriate population monitoring methods.

THREATENED AND ENDANGERED SPECIES

The Endangered Species Act of 1973 provides special protection for species whose survival is at risk. A species is listed as *endangered* if it is in danger of extinction throughout all or a significant portion of its range, or *threatened* if it is likely to become endangered within the foreseeable future. The Act provides protection to both plant and animal species, including both vertebrate and invertebrate species as well as subspecies and local populations. The Endangered Species Act makes it unlawful to *take* an endangered species, which means to harass, harm, pursue, hunt, shoot, wound, kill, trap, capture and collect, or to attempt to engage in such conduct. The U.S. Fish and Wildlife Service has defined *take* to include any land use action, whether on public or private land, that adversely modifies habitat (50 CFR §§ 17.3 and

§ 17.31 (a)). It is these provisions of the Endangered Species Act, which deal with the habitat of listed species, that land managers must be most aware of during resource assessment and management planning efforts. States may also have legislation designed to protect species whose survival is at risk. These acts may provide for the listing and protection of species in addition to those that are provided federal protection.

It is somewhat paradoxical that species of high interest, certainly from a regulatory viewpoint, are the most difficult to observe. It is extremely difficult to design a survey to obtain information about rare populations. This is especially true when, as in many cases, there is no assurance that they even occur within the area of interest. In many states, there is an agency that maintains a Natural Features Inventory that can provide assistance to landowners wanting to assess the likelihood of a listed species being present within a local area. For logical reasons concerning security and protection, these agencies usually will not disclose the exact location of known populations of listed species, but they will provide the identities of listed species that may be encountered in different counties within the state. For some regions, there are very good published sources of information on the occurrence of threatened or endangered species. Evers (1992), for example, provides detailed habitat requirements for all state and federally listed species occurring in Michigan, including a county-by-county listing of where they might occur. Even less specific analyses, such as that given by Benyus, Beuch, and Nelson (1992), provide lists of species known to utilize different types of habitat within a region. All of these sources of information can be combined in planning and executing an assessment effort. Table 9.6, for example, gives state and federally listed threatened and endangered species that may be found in Houghton County, Michigan. Evers (1992) provides detailed habitat requirements for each of these species, which are of great value in identifying areas within a landscape that may provide critical habitat for these protected species.

Preparation is the key to identification of listed species and their habitats during a resource assessment process. In most cases, it is extremely unlikely that individuals of listed species will be actually sighted or observed. Most assessments are conducted on the basis of habitat and known occurrence in an area. It is important that field personnel, from surveyors to inventory crews to equipment operators, be instructed in the recognition of listed species within a given area and the proper action to take if individuals of listed species are observed.

COMMUNITY STRUCTURE

In many resource assessment efforts, it is more important to obtain information on community structure and composition than estimates of population numbers for a few species. Some form of quadrat or line transect sampling may be used to simply develop a list of species present in an area or to

TABLE 9.6 Federally and State-Listed Threatened and Endangered Species Found in Houghton County, Michigan

	Species	*Status*
Plant	Wild Chives	State Threatened
	Green Spleenwort	State Threatened
	Northern Reedgrass	State Threatened
	Walking Fern	State Threatened
	Farwell's Water-milfoil	State Threatened
	Rayless Mountain Ragwort	State Threatened
Mammal	Marten	State Threatened
Bird	Common Loon	State Threatened
	Bald Eagle	Federal Endangered
	Osprey	State Threatened
Reptile and Amphibian	Eastern Box Turtle	State Threatened
Fish	Lake Sturgeon	State Threatened

Source: Adapted from Michigan Department of Natural Resources (1994a).

note their relative abundance. Once a species list is available, measures of community richness and diversity can be calculated as described for vegetative species in Chapter 7; measures of evenness require information on the relative abundance of species in the community. Similarity indices, also described in Chapter 7, may be calculated to compare composition and structure between communities in different areas or within the same community sampled at different times, such as before and after implementation of a management activity.

SUMMARY

The assessment of wildlife resources is very important during the formulation and evaluation of management alternatives. Management goals often involve the enhancement of wildlife habitat or increasing population sizes, and there may be regulatory concerns, especially in regard to threatened or endangered species and their habitat. It is not usually necessary to estimate the number of individuals of a given species in an area except in cases in which management goals and objectives specifically provide for the improvement of habitat or increase in population sizes of individual species. During resource assessment efforts, it is usually more practical to concentrate on habitat quality and availability rather than the number of individuals present in an area. Measures of community structure and indices of structural similarity between different communities provide additional means to characterize wildlife resources, to compare management alternatives, and to evaluate success in achieving management goals.

APPLICATIONS

9.1 For a particular type of habitat in your region, such as lowland coniferous forest, develop a list of avian, amphibian, reptilian, and mammalian species that may utilize the habitat all or part of the year.

9.2 Locate a habitat suitability index model for a species found in your region. Determine the variables that must be collected to calculate the index, and observe these variables in at least three different locations. Calculate the value of the index for each location and determine which provides the most suitable habitat for the species.

9.3 The following data are numbers of white-tailed deer observed on each quadrat in a sample of 1 ha quadrats established within a 90 ha management area. Estimate the population density, its standard deviation, the total population number, and its standard deviation:

1	0	2
3	2	0
1	0	0
0	2	1
2	0	3

9.4 Line transect sampling is used to estimate the number of moose in an aerial survey. On a five-kilometer transect, the following observations were made of individual animals:

Distance (m)	*Angle from Centerline* (°)
200	30
75	50
250	25

Assume an exponential detection function and estimate the population density based on this single transect.

9.5 Five 600-foot-transects were established in a 600-acre oak-hickory forest to estimate the number of squirrel nesting sites in the area. Assuming an exponential detection function, estimate the nesting density (nests per acre), the total number of nests, and the associated standard deviations, for the following data; each value given is the perpendicular distance in feet to the nest from the centerline of each transect:

	Transect				
	1	*2*	*3*	*4*	*5*
Nest					
1	25	50	30	15	35
2	35	20		40	
3	70				

9.6 In a 6 ha area, 38 mice are captured, marked, and released over a two-day period. One week later, 40 mice are captured over a two-day period, 12 of which were marked the previous week. Estimate the number of individuals in the area, assuming a closed population.

9.7 In a 120-acre watershed, 16 squirrels are captured, tagged, and released. One month later, a second capture effort is initiated which continues until 8 tagged squirrels are recaptured; a total of 22 squirrels are captured at this time. Estimate the number of squirrels inhabiting the watershed, assuming a closed population.

9.8 For a selected county, develop a list of threatened and endangered species that are known to occur in the county and identify the types of habitat where they may be found.

PART

IV

Social Context

Chapter

10

Legal and Regulatory Environment

The legal and regulatory environment affecting public and private land management is a complex, ever-changing mix of federal, state, and local laws. *Legislation* refers to the actual laws and ordinances passed by governing bodies; *regulations* are rules to implement the intent of legislation that are developed by the designated responsible agency. Land managers must be aware of legislative and regulatory requirements applicable to their management activity. Some legislation is applicable only to particular ownerships, such as federal land, and some is applicable to all land management enterprises. A text such as this is not a reliable source of legal information; managers need to have competent, current legal advice in order to ensure compliance with applicable legislation.

The legal and regulatory environment develops as new legislation is passed by governing bodies or public referenda and as existing legislation is interpreted by the judicial system. The current legal and regulatory environment has developed over an extended period of time and has its basis in legislation and common practice developed in past centuries. The following sections of this chapter refer to the major federal legislation affecting public and private land management and summarize the major types of state and local land management legislation. This is not an exhaustive presentation; excellent and comprehensive reviews of the development of forest policy in the United States are given by Dana and Fairfax (1980), Ellefson (1992), and Cubbage, O'Laughlin, and Bullock (1993).

FEDERAL LAWS AND REGULATIONS

There are several major types of federal legislation affecting land management. Federal laws and regulations set the management goals and direct the activities of public natural resource management agencies, but they are also applicable to private land management in some instances. Federal legislation affecting private land management is usually designed to ensure protection of environmental quality or to encourage certain types of management activity. In some cases, federal legislation may not directly affect private land management, but state legislation enacted in response to federal requirements may actually be the relevant legislation that applies to private land managers. This action by individual states introduces state-by-state variability into the regulatory details with which land managers must comply.

More than 280 million hectares (690 million acres) in the United States are owned and managed by the federal government. Various agencies have responsibility for the management of these areas. The *Bureau of Land Management,* in the U.S. Department of the Interior, is responsible for the management of more land area than any other federal agency; the Bureau manages 109 million hectares (270 million acres) of forest, rangeland, and desert and administers the mineral rights (see Chapter 3) associated with more than 230 million hectares (570 million acres) of land in the United States. The *U.S. Forest Service* in the Department of Agriculture is responsible for the management of the 77 million-hectare (191 million-acre) National Forest System. Other federal land management agencies include the following:

- The *U.S. National Park Service,* in the Department of the Interior, is responsible for more than 350 areas, including national parks, lakeshores, monuments, and historic and recreation areas, totaling more than 33 million hectares (83 million acres);
- The *U.S. Fish and Wildlife Service,* also in the Department of the Interior, is responsible for more than 500 Wildlife Refuges covering more than 36 million hectares (90 million acres);
- The *Bureau of Indian Affairs,* in the Department of the Interior, is responsible for the management of more than 22 million hectares (55 million acres) of Native American lands held in trust by the federal government;
- The *Department of Defense* is responsible for almost 11 million hectares (27 million acres) of military reservations and installations;
- The *Department of Energy* manages more than 50 sites and a total of approximately one million hectares (2.5 million acres); and
- The *National Oceanic and Atmospheric Administration,* in the Department of Commerce, is responsible for national marine sanctuaries and estuarine research reserves.

The Bureau of Land Management and the Forest Service are guided by principles of multiple use. The other federal land management agencies have more directed missions. Federal agencies periodically publish summaries of the relevant federal legislation affecting their particular agencies (e.g. USDA Forest Service, 1983) and specific types of activity (e.g. U.S. Department of the Interior, 1993). This forms an extensive body of legislative activity; more than 135 federal laws apply to the U.S. Forest Service (USDA Forest Service, 1983), most of which were passed since 1960. The major legislative mandates of the U.S. Forest Service include the management of more forest land than any other agency in more than 150 National Forests, provision of cooperative assistance to private landowners, the conduct of research in eight research stations, and the management of programs providing international assistance and training. The major legislation affecting the activities of the Forest Service is summarized in Table 10.1.

The *Multiple Use Sustained Yield Act of 1960* requires the U.S. Forest Service to manage land under its jurisdiction for "outdoor recreation, range, timber, watershed, and wildlife and fish purposes." In addition to land directly under Forest Service jurisdiction, the *Forest and Rangeland Renewable Resources Planning Act of 1974* requires the Forest Service to analyze long-term resource supply and demand from all forest land in the United States, including both public and private holdings. The Forest Service is required to periodically prepare summaries of the state of these resources and plan for their sustained management. These planning requirements are further enhanced by the *National Forest Management Act of 1976,* which requires the development of a management plan for each National Forest. This act places further requirements on Forest Service land management activity, including requirements to maintain productive capacity, existing native vertebrate species, and the diversity of tree species present at the initiation of the management plan, as well as to preserve and enhance biological diversity to equal or exceed that of natural, unmanaged forests. For federal land, then, general management goals are specified by legislation; specific goals for each operational management area have to be developed by the responsible agency for the local conditions, but must be consistent with the general management goals.

As mentioned earlier, the Bureau of Land Management is responsible for the management of more land area than any other agency, including areas of forest, rangeland, and desert. The Bureau also administers mineral leases associated with federal mining rights beneath public and private land. The *Federal Land Policy and Management Act of 1976* requires the Bureau to manage land under its jurisdiction for multiple purposes and to prepare management plans documenting management intentions for specific areas. Other legislation directing activities of the Bureau of Land Management is summarized in Table 10.2.

The other federal resource management agencies have more specific missions than does either the Forest Service or the Bureau of Land Management. The National Park Service, for instance, has more than 350 units consisting of National Parks, Monuments, Battlefields, Preserves, Lakeshores, Memorials, Parkways, Historic Sites, and Trails. The National Park Service is charged

TABLE 10.1 Major Legislation and Administrative Actions Defining the U.S. Forest Service Mission and Management Goals

Legislation/Administrative Action	*Effect*
Forest Reserves Act of 1891	Allowed President to set aside public lands as forest reserves
Organic Act of 1897	Mandated that forest reserves be managed to "improve and protect the forest within the boundaries, for the purpose of favorable conditions of water flows, and to furnish a continuous supply of timber"
Transfer Act of 1905	Transferred Bureau of Forestry to USDA and created the Forest Service
Weeks Act of 1911	Enabled cooperative work with states in fire protection and purchase of land in headwaters of navigable streams
Clarke-McNary Act of 1924	Assist states with fire control, reforestation, and management
L-Regulations of 1924	Allowed "primitive areas" to be established
Knutsen-Vandenberg Act of 1930	Created fund for reforestation, mitigation, and other silvicultural activities
U-Regulations of 1939	Allowed wilderness, wild, and recreational areas to be designated on National Forests
Cooperative Forest Management Act of 1951	Allowed work with states to provide assistance to private landowners, operators, and manufacturers
Multiple Use Sustained Yield Act of 1960	Clarified mission to manage National Forests for recreation, range, timber, water, wildlife, and fish in a "combination that will best meet the needs of the American people"
Forest and Rangeland Resources Planning Act of 1974	Mandated multiresource management program to be updated every five years and an assessment of supply and demand for natural resources every 10 years
National Forest Management Act of 1976	Required land management plan for every National Forest, to be updated every 10 years; required both economic and social values to be considered in decision making

TABLE 10.2 Major Legislation and Administrative Mandates Defining the Bureau of Land Management Mission and Goals

Legislation/Administrative Action	*Effect*
General Mining Law of 1872	Provided for private mining of mineral deposits on public land for a nominal fee
Mineral Leasing Act of 1920	Provided for leasing of mineral deposits
Taylor Grazing Act of 1934	Authorized grazing districts and issuance of permits (for a fee) for private individuals to graze livestock on public land
Oregon and California Act of 1937	Mandated sustained yield management of Oregon and California lands (with timber uses being dominant)
Executive Reorganization No. 3 of 1946	Created Bureau of Land Management with some land management responsibilities for mineral leasing, grazing, timber production, and real estate
Federal Land Policy and Management Act of 1976	Authorizing statute for agency; requires long-range management plans and management under a multiple use concept similar to that of the Forest Service.

with maintaining these areas in an unimpaired state and with making these areas accessible for public use (Table 10.3).

The U.S. Fish and Wildlife Service is charged with conservation, protection, and enhancement of fish and wildlife, which includes habitat protection (Table 10.4). Migratory birds, threatened and endangered species, selected marine

TABLE 10.3 Major Legislation and Administrative Mandates Establishing the Mission and Goals of the National Park Service

Legislation/Administrative Action	*Effect*
Yellowstone Act of 1872	Designated first U.S. National Park
Antiquities Act of 1906	Authorized designation of National Monuments
National Park Service Organic Act of 1916	Established the National Park Service with the mission of preserving National Parks, Monuments, Battlefields, Lakeshores, Memorials, Parkways, Historic Sites, and Trails and making them accessible to the public in their natural states
Mining in National Parks Act of 1976	Limited applications of federal mining law in National Parks

TABLE 10.4 Major Legislation and Administrative Actions Defining the Mission and Goals of the Fish and Wildlife Service

Legislation/Administrative Action	*Effect*
Joint Resolution of February 9, 1871	Authorized Commission on Fish and Fisheries
1885 Appropriations Act	Authorized funds to study economic ornithology
Pelican Island National Wildlife Refuge Act of 1903	Established first National Wildlife Refuge
Migratory Bird Conservation Act of 1929	Authorized purchase of private lands for migratory bird refuges
Fish and Wildlife Coordination Act of 1934	Authorized agency to consult with federal dam-building agencies to support wildlife when possible
Pittman-Robertson Act of 1937	Authorized funding (by tax on ammunition and arms) to aid state agencies in restoring game populations
Dingell-Johnson Act of 1950	Authorized annual appropriation for cooperative state programs of fish enhancement
Aquatic Resources Trust Fund of 1984	Expanded equipment tax and set up Aquatic Resources Trust Fund as well as sport fish restoration and boating safety accounts

mammals, and approximately 500 National Wildlife Refuges are under the jurisdiction of the Fish and Wildlife Service. The Fish and Wildlife Service also manages more than 70 fish hatcheries, several fish technology centers, and several other research centers.

The other major federal agency with responsibility for natural resource management is the Bureau of Indian Affairs, which acts as the United States agency in administering Native American lands held in trust by the federal government (Table 10.5). The Bureau of Indian Affairs also provides assistance in resource management to tribal authorities. The *Indian Self-Determination and Educational Assistance Act of 1975* explicitly gives tribal authorities an active role in the management of natural resources under their jurisdiction. Because there were a very large number of treaties with different Native American tribes, management of Native American land and resources varies greatly throughout the United States.

Federal legislation governing land use and management by these different agencies can be quite varied; the U.S. Department of the Interior (1993) describes legislation affecting federal historic preservation activity that includes the National Park Service Organic Act of 1916, relevant sections of the Internal Revenue Code, and such legislation as the Amtrak Improvement Act of 1974. Legislation provides general management goals, and federal

TABLE 10.5 Major Legislation and Administrative Action Defining the Land Managment Mission and Goals of the Bureau of Indian Affairs

Legislation/Administrative Action	*Effect*
Office of Indian Affairs Act of 1824	Established Office of Indian Affairs in the War Department
Indian Appropriations Act of 1909	Provided first appropriations for Indian forestry
The Act of June 25, 1910	Permitted sale of timber from reservations; proceeds to benefit the owners
The Act of September 20, 1922	Authorized protection of timber on reservations from insects, disease, and fire
Indian Reorganization Act of 1934	Directed that Indian forestry units be operated and managed on a sustained-yield basis
Indian Self-Determination and Education Assistance Act of 1975	Provided for active Indian participation in resource management

regulations are issued to implement the legislation (Table 10.6). The details in the regulations are at least as important to the day-to-day activity of land managers as the broad legislative management goals. Regulations are published in the Federal Register, which is available in most large libraries and is for sale through the Government Printing Office.

Some legislation applies to all federal agencies. The *National Environmental Policy Act of 1969,* for example, requires all federal agencies to prepare *environmental impact statements* describing the environmental consequences of any proposed development project that may significantly alter the environment. Discussed in more detail in Chapter 14, environmental impact statements are to describe (1) the environmental consequences of the proposed

TABLE 10.6 Selected Federal Regulations Dealing with Historic Preservation

Reference	*Subject*
23 CFR Part 771	Environmental impact and related procedures
36 CFR Part 60	National Register of Historic Places
36 CFR Part 61	Procedures for approved state and local preservation programs
36 CFR Part 65	National Historic Landmarks program
40 CFR Parts 1500–1517	Regulations of the Council of Environmental Quality
43 CFR Part 3	Preservation of American antiquities
43 CFR Part 7	Protection of archaeological resources

Source: U.S. National Park Service (1993).

action; (2) adverse effects that cannot be avoided if the proposal is implemented; (3) alternative actions; and (4) the trade-off between short-term uses and long-term productivity. The National Environmental Policy Act also requires environmental impact statements to include plans to mitigate adverse environmental effects. Environmental impact statements are subject to public review and comment, during which time other agencies, as well as private citizens and organizations, may comment on the proposed activity.

All federal agencies must also comply with the *Endangered Species Act of 1973*. The Endangered Species Act requires the U.S. Fish and Wildlife Service to designate species in danger of extinction through all or a significant part of their range as *endangered;* species are listed as *threatened* if they are likely to become endangered in the foreseeable future. The National Marine Fisheries Service under the Department of Commerce considers the listing of anadromous fish, salmon, and most marine species. The Endangered Species Act makes it unlawful to take a listed species. The definition of *take,* from Section 3 of the Act, is to "harass, harm, pursue, hunt, shoot, wound, kill, trap, capture, and collect, or to attempt to engage in such conduct." The Act also makes it unlawful to remove or maliciously damage or destroy listed plant species on federal lands, or to knowingly destroy such species on other lands in violation of state law. The U.S. Fish and Wildlife Service (50 CFR §§17.3 and §17.31 (a)) has defined *taking* to include any management action that adversely modifies habitat so as to injure or harm even a single individual of a listed species. The Act requires federal agencies to consult with the Fish and Wildlife Service to ensure that agency actions are not likely to jeopardize listed species or their habitat.

The only provision of the Endangered Species Act that applies to private land is Section 9, which prohibits the taking of a listed species. Application of the Endangered Species Act to private land is still being resolved in the judicial system, but the Fish and Wildlife Service has maintained that private landowners may be criminally liable for significant habitat modifications under the definition of *harm,* regardless of actions to reduce harm to individuals of listed species. This may have a significant impact on private property, especially for some listed species. The red-cockaded woodpecker, for instance, inhabits the southeastern United States and requires pine forests with some old-growth attributes as habitat; 75 percent of the species' potential habitat may be on private land (Society of American Foresters, 1992). Clearly, although public land provides many opportunities for habitat recovery, participation by private landowners is also important in ensuring the survival of many species. Private landowners may participate, in cooperation with federal or state agencies, in habitat conservation programs for the protection of listed species on their property. The details of such participation are beyond the scope of this text, but it is extremely important to consider the habitat requirements of listed species during a resource assessment process (see Chapter 9).

In addition to the Endangered Species Act, any management plan developed by the U.S. Forest Service, the Bureau of Land Management, the National

Park Service, or the U.S. Fish and Wildlife Service must be in compliance with the following legislation:

- *Fish and Wildlife Coordination Act of 1958,* which requires all federal agencies to consult with the Fish and Wildlife Service regarding any development project that may affect wildlife populations or habitat.
- *Wilderness Act of 1964,* which required all federal agencies to examine land under their management to determine suitability for inclusion in the national system of wilderness areas set aside for preservation. *Wilderness* is defined in the Act as lands "untrammeled by man, where man himself is a visitor who does not remain." Congressional action is required to assign wilderness status to a specific area.
- *Wild and Scenic Rivers Act of 1968* provides for a congressionally designated system of preserved wild, scenic, or recreational rivers managed by the agency or state through which they flow.
- *Sikes Act of 1974* requires federal agencies, that are responsible for the management of wildlife habitat, to work with state agencies, that are responsible for managing wildlife populations, to coordinate plans for conservation of wildlife species.
- *Federal Water Pollution Control Act Amendment of 1972* sets broad standards for resource management activity, including control of nonpoint-source pollution, dredge and fill activity in wetlands and navigable waters, and the development of "best management practices" by individual states.
- *Clean Water Act of 1977* requires mandatory state water quality planning and increased control of dredge and fill activity in wetlands.
- *Water Quality Act of 1987* increased requirements on states for planning and implementation of best management practices, specified nonpoint-source pollution control procedures, and specifically identified forest management practices that could negatively impact water quality.

Private land management planning is exempt from some of the requirements placed on federal management planning. The Wilderness Act, for instance, has little direct impact on private land management, although there are indirect impacts in the areas surrounding designated wilderness areas in terms of employment opportunities, recreational opportunity, and commercial timber production. The water quality legislation noted earlier requires states to implement water planning activity, increase control of dredge and fill activity in wetlands, and develop best management practices that may impact private land management activities. Most of these practices require measures to protect water quality during road building, commercial harvesting, and management activity; in most states, for example, permits are usually required for stream crossings.

Federal legislation and regulation also affects the conduct of certain land management operations on private land. The *Federal Insecticide, Fungicide, and Rodenticide Act of 1947,* as amended by the *Federal Environmental Pesticide Control Act of 1972,* regulates the application of certain pesticides and extends authority to the states to register pesticides and certify applicators. General worker safety regulations are developed by the Occupational Safety and Health Administration (OSHA), including federal logging safety standards that are enforced at the state level.

Other federal legislation impacts private land management in a nonregulatory way. The *Forest Stewardship Act of 1990,* for example, promotes stewardship on private nonindustrial forest land by providing for financial, technical, educational, and other assistance in the management of timber, wildlife, water, and recreational resources. The Act also provides for a program using conservation easements to protect certain private forestlands from development. The *Renewable Resources Extension Act of 1978* fosters federal, state, and university cooperation in educational assistance projects. The *Food Security Act of 1985* provides for the Conservation Reserve program; farmers can withdraw certain farmland from agricultural production and receive financial matching funds to apply to the cost of establishing trees or grass on designated reserve areas. Finally, provisions in the tax code can significantly impact financial returns from private resource management. The *Tax Reform Act of 1986* removed the capital gains treatment for timber income and placed restrictions on the deductibility of timber management expenses by private landowners.

Resource assessment documents should list, in detail, any restrictions placed on management activity through federal legislation. If a landscape is directly impacted by federal legislation, this should be noted and the relevant agencies or regulatory bodies should be indicated. If the landscape is home to a threatened or endangered species, for example, this should be clearly noted, and any prior agreements concerning habitat protection should be described in detail. The assessment process should clearly identify all limitations on management options that must be considered during the planning process.

STATE LAWS AND REGULATIONS

All states have laws regarding the establishment and execution of contracts, the ownership of real estate, and other activities related to the operation and conduct of commercial activity. Private landowners, whether individuals or incorporated entities, are subject to all applicable laws of their state. In addition to laws governing general business activity, Ellefson and Cheng (1994) note that all states place some sort of legal limits on private forest management. Some of these limits are in the form of measures to ensure environmental quality or fire protection. Many states also have either voluntary or mandatory guidelines for the conduct of private forest management activity. Private forest management activities are directly regulated through forest practice legislation by some states. Other states have taken an incentive-based approach, providing landowners with some incentive, usually tax-related, in return for either compliance with certain voluntary management guidelines or some other consideration, such as the provision for open, nonmotorized access to the general public for certain recreational activities, such as hunting or fishing.

Environmental Protection

As noted earlier, federal water quality legislation has required the states to develop plans to protect water quality and programs to control nonpoint-source pollution. In particular, Section 319 of the Water Quality Act of 1987 required states to initiate such programs, but they were given a wide latitude in the design of their programs. Lloyd and MacLeod (1993) examined the programs in four states spanning the range from a completely voluntary, educational approach in Ohio to a highly regulated program requiring licensing, permitting of timber harvesting activity, and inspection before, during, and after harvesting in Massachusetts. Not surprisingly, compliance increased with the increasingly regulatory nature of the program. The approach is so variable from state to state, however, that managers must become familiar with the procedures within their own state to ensure that their management activities are in compliance.

Section 404 of the 1972 Federal Water Pollution Control Act Amendments concerns the discharge of material into navigable waters and wetlands. Silvicultural activity was specifically mentioned in the Act as contributing to nonpoint-source pollution. There is a provision for the exemption of silvicultural activities, and the laws of most states allow forest harvesting activities in wetlands if they are in fulfillment of a silvicultural plan. Chapter 6 discusses some recommended best management practices for the protection of water quality in wetlands and riparian areas.

The body of state and federal legislation designed to ensure environmental protection during private land management activity can be quite extensive. Table 10.7, for example, lists state laws requiring landowners to obtain permits related to the control of nonpoint-source pollution in Michigan. The details vary from state to state, but landscape managers must be aware of the applicable legislation that includes permitting or other regulatory requirements in their state.

During the resource assessment process, physical areas in the landscape where permitting may be required, and the types of activities requiring permits, should be clearly identified. Sensitive areas, such as jurisdictional wetlands (Chapter 6), should be noted, and relevant restrictions on management options should be described. If there are any streams that would require permits to establish crossings or bridges, these should be noted together with the addresses of relevant permitting agencies. Other relevant requirements to ensure

TABLE 10.7 State of Michigan Laws Requiring Permits to Control Nonpoint-source Pollution

Act	*Regulated Activities Requiring Permits*
Forest Protection and Forest Fires Act (P.A. 329) of 1969	• Most burning on or adjacent to forest land • Prescribed burning
Natural Rivers Act (P.A. 231) of 1970	• All development and land use within 400 feet of designated Natural Rivers • Roads, bridges, and culverts within Natural River areas
Inland Lakes and Streams Act (P.A. 346) of 1972	• Dredging, filling, or construction below ordinary high water mark • Installation of docks or boat launching ramps • Interference with natural flows • Installation of permanent and temporary stream crossings
Soil Erosion Sedimentation and Control Act* (P.A. 347) of 1972	• Transportation facilities and service facilities • Subdivision or commercial development • Recreational facilities, including parks and campgrounds • Utilities • Oil and gas wells • Water impoundments and waterway construction
Sand Dune Protection and Management Act (P.A. 222) of 1976	• Construction, recreation, and timber management in certain areas
Wetlands Protection Act (P.A. 203) of 1979	• Road construction, fill, and drainage ditches in wetlands, swamps, marshes, and bogs contiguous to lakes and streams and isolated wetlands > 5 acres in size • Review of bridges and culverts to ensure they (1) can withstand 100-year floods and (2) do not increase flood stages upstream • Activities resulting in occupation, filling, or grading streambeds, channels, and lands in floodplains
Water Pollution Control Act of 1972 (federal)	• Dredging and filling all streams adjacent to certain areas (essentially duplicates Inland Lakes and Streams Act); administered by U.S. Army Corps of Engineers

*Applicable to activities that change the natural cover or the topography of more than 1 acre or within 500 feet of a lake or stream

Source: Summarized from Michigan Department of Natural Resources (1994b).

environmental protection should also be noted; it is imperative that these factors be clearly understood by managers during the process of developing management alternatives and in the implementation of management decisions.

Forest Practices

States may also directly regulate forest practices beyond the minimum environmental protection requirements established by federal law. Cubbage (1991, 1995) notes that at least 16 states have passed laws controlling forest practices to some degree on private and public land (Table 10.8). These laws may take a very regulatory approach, as in the case of the Massachusetts law noted earlier, which requires the licensing of loggers, the issuance of permits for timber harvesting from private land, and state inspection before, during, and after harvesting. Other states require less inspection and licensing, but may require the filing of approved management plans for certain private forest land. Wisconsin's law, for example, requires a landowner to file a management plan with the Department of Natural Resources. Operational areas are inspected both prior to and after harvesting activity; for some forest types, stand conditions must meet minimum stocking guidelines specified by the State Silviculturist. Most plans required by such laws must include a statement of management goals and intentions as well as some form of resource assessment, although this may be quite abbreviated in many instances.

Other states have taken an incentive-based approach to influencing forest management on private lands. Private forestland may receive favorable tax treatment in return for implementation of certain management practices or other considerations, such as open access to the public for recreational activity. Michigan's Commercial Forest Act, for example, provides for reduced property taxes to landowners in return for open public access for hunting and fishing purposes.

State legislation, if there is any, governing forest practices should be clearly summarized in the resource assessment document. This summary should

TABLE 10.8 States with Some Form of Forest Practice Regulatory Program

Type of Regulatory Program	*States*
Forest Practice Laws	Alaska, California, Connecticut, Idaho, Maine, Massachusetts, New Mexico, Nevada, Oregon, Washington
Forest Practice Regulatory Program (variety of laws)	Florida, Maryland, Montana, Virginia
Regulatory Best Management Practices	West Virginia

Source: Summarized from Ellefson and Cheng (1994) and Cubbage (1995).

indicate whether the guidelines are voluntary or mandatory and include a discussion of the guidelines applicable to specific areas within the landscape of interest. If there are requirements for filing a management plan, the use of licensed operators, or provisions for inspection, these must be known and considered during the implementation of management decisions. If landowners have the option of participating in incentive programs, the benefits of such participation and the concessions required of the landowners should be clearly described in the assessment document so that participation can be considered during the planning process. Permitting requirements should be clearly described during the assessment; in most cases, addresses of appropriate regulatory agencies should be included and procedures for filing management plans, obtaining permits, and/or scheduling inspections should be described.

LOCAL LAWS AND REGULATIONS

Local units of government, including counties, townships, municipalities, and others, may enact laws regulating some aspect of forest management practices on private land. Greene and Siegel (1994) defined a forest ordinance as an ordinance, zoning law, or tree protection article that had been, or could be, used to restrict harvesting or silvicultural practice, or the hauling of forest products. Hickman, Siegal, and Martus (1991) identified approximately 500 laws passed by local units of government in the United States with provisions affecting forest management; Greene and Siegel (1994) identified 508 such laws in the eastern United States alone.

Martus, Haney, and Siegel (1995) surveyed the local units of government that were identified by Greene and Siegel (1994) as having laws governing some aspect of forest management. They found that about 80% of the 508 laws were passed since 1983 and about 50% had been passed since 1987. Given these findings, it is clear that the number of local laws regulating some aspect of forest management practices is increasing. Eastern states with the greatest number of such laws are Pennsylvania, New Jersey, Maine, New York, Georgia, and Virginia; of the 508 local laws identified by Greene and Siegel (1994), a total of 368 (72%) were laws within these six states (Martus, Haney, and Siegel, 1995). There is no single, comprehensive listing of local laws affecting forest management; land managers must stay abreast of developments in their region and be prepared to incorporate the provisions of such laws into their management activities.

Given the fact that there are thousands of local units of government, it is not surprising that there is great variety in the types of land management activities affected by local statutes. Many local laws affecting land management are enacted in response to state-level legislation; Virginia and Maryland both require local units of government near Chesapeake Bay to enact legislation to protect aspects of water quality in the Bay, for example. Other local laws are passed in response to local initiatives or are spurred by particular local issues. Suburban neighborhoods are expanding into once-rural areas, and the resulting population increases in former agricultural and forested areas is prompting increased local legislation, which is impacting private land management.

Martus, Haney, and Siegel (1995) divided local laws affecting forest management and silviculture into five classes: tree harvesting ordinances, public property protection ordinances, tree protection ordinances, environmental protection ordinances, and special feature ordinances. Timber harvesting ordinances were defined as those ordinances that explicitly regulate harvesting and silvicultural activity. They may require permits and/or management plans for such activity, or may specify standards for silvicultural activity, road construction, and so on. Public property protection ordinances restrict activities in order to prevent damage to public infrastructure, such as roads, bridges, and so on, or to ensure public safety by limiting the impact on driving conditions caused by the hauling of forest products. This type of ordinance may require permitting or inspection of the harvesting activity.

Tree protection ordinances usually impact the removal of trees during land development activity. They usually do not impact commercial forestry operations, but they may in areas where there is a high level of development activity. They may require permitting, the development of site plans, and provisions for replanting or regeneration. Martus, Haney, and Siegel (1995) did not include urban tree ordinances in this category.

Environmental protection ordinances are designed to protect such features as water quality and wildlife habitat and the associated rights of adjacent landowners. In addition to state and federal laws, local ordinances may further restrict types of harvesting activity or require specified buffer zones along streams and roadways. Permits and plans may be required. Finally, special feature ordinances are usually developed in response to local concerns and issues. They usually deal with sensitive or fragile areas and may place limits or restrictions on harvesting activity.

In the North, most local laws were found to be timber harvesting ordinances, followed by environmental protection, special feature, and tree protection ordinances. In the South, public property ordinances were most common, followed by special feature, environmental protection, and tree protection ordinances (Martus, Haney, and Siegel, 1995). These regional differences reflect local government and cultural traditions, but they also reflect differences in the types of forest management activities typically employed in the various regions. The bottom line for land managers, again, is that regulatory activity at the local government level is increasing and managers must stay abreast of developments and changes in local legislation in their region. Assessment reports should note whether there are any local laws or ordinances regarding land use, zoning, or management practices, identify local regulatory bodies to contact for information, permits, or inspections, and note any local restrictions on harvesting or hauling of commercial products that may be applicable within the landscape of interest.

SUMMARY

There is a mix of federal, state, and local legislation that affects the management of both public and private forests. Federal land management agencies have their missions and management goals specified by legislation and administrative action. In addition, federal legislation pertaining to environmental protection and wildlife resources places constraints on both public and private land management. States place limits on land management to ensure environmental quality and, in some cases, directly regulate forest management practices. Other states take an indirect approach by providing incentives to landowners who implement certain practices. There is also an increasing body of local government legislation affecting the management of private land. Resource managers must be well aware of every component of legal and regulatory limits placed on management activity in their area. Resource assessment plans should provide details of the limitations placed on management options within a landscape through federal, state, or local legislation and regulatory activity. If there are any permitting requirements, or if there are formal planning or inspection requirements, these should be clearly described. The names and addresses of relevant regulatory agencies should also be provided.

In general, public resource management agencies may have substantially more legislation guiding their activity, but private landowners are also affected and often have less formal guidance in understanding and interpreting applicable laws and regulations. If in doubt, one should not hesitate to obtain competent legal advice before implementing any possibly regulated management activity.

APPLICATIONS

10.1 Describe the mission and general management goals for a federal, state, or local land management agency. Identify the major legislation defining the agency's goals and activities.

10.2 Obtain the land use plan from a National Forest. What are the management goals for the forest? What is the mix of values provided by the forest?

10.3 Does your state require landowners to use a set of "best management practices" to ensure water quality? Are there voluntary guidelines? Describe guidelines for activities in a riparian corridor.

10.4 What agencies in your state are responsible for permits required for stream crossings? Under what conditions are permits required?

10.5 Briefly describe any state-level restrictions on the management of private forestland in your state. If a formal management plan is required, what must be included in the plan and where should it be filed?

10.6 Are there any local governments in your area that regulate land use or management activity? If so, describe the impacts of a local ordinance on private land management.

Chapter

11

Recreation and Aesthetics

Lucas et al. (1984) define outdoor recreation as "voluntary, onsite activity engaged in for pleasure that is to some degree dependent on a natural setting." This definition includes simply viewing a landscape, but does not include participation in organized sports such as football, baseball, or soccer. Recreation must be carefully managed to maintain the quality of the experience and to minimize interference with the functioning of ecological processes within a landscape. The disruption of soil and increased erosion along trails are examples of direct impacts of recreational activity on ecological processes, yet the simple presence of humans may be enough to disrupt the activity of some wildlife species, even without alteration of other habitat components. Recreational activity in forested landscapes takes place within an ecological context; it may be very difficult for managers to balance social and ecological concerns.

The many different recreational uses of a forested landscape may be in conflict with each other. Slover (1996), for instance, describes a situation in which groups interested in hiking and horseback riding came into serious conflict regarding the use of a designated wilderness area. Backpackers may desire quiet and solitude, but this is often incompatible with larger group activities or motorized vehicle use. It is impossible for landscape managers to anticipate every possible conflict or potential problem, but careful assessment procedures may greatly increase the chance of detecting potential resource-based problems before they escalate to undesirable levels. In many cases, careful assessment may also identify situations in which there may be potential for increased recreational development. Experts may then be consulted to determine whether such development is justified and likely to succeed.

In addition to specifically planning to accommodate or provide for recreational activity, landscape management must include consideration of aesthetic concerns. People are casual observers of forested landscapes from roadways or airplanes. The great English landscape designers of past centuries recognized that many people prefer mixtures of patterns in visual scenes and created extensive artificial landscapes containing a diversity of visual elements. Alteration or disruption of landscape pattern, particularly by the removal of green vegetation resulting in large, unsightly brown areas, can stimulate strong emotional responses in many people. Some areas, such as those located in bends of major highways, are visual focal points that may be very sensitive to aesthetic concerns. In many cases, only a relatively minor modification of a planned management activity may be necessary to maintain the visual appeal of a landscape, but managers must be aware of and consider these needs during the development and evaluation of management alternatives.

This chapter concentrates on the different types of recreational activity that occur within forested landscapes. Methods and concepts involved in designing or developing concentrated-use areas such as campgrounds are beyond the scope of this book, as are analyses of visual preferences and the intricacies of landscape design. There are several texts that can provide more detailed discussion of human recreational use of forested landscapes, including Brockman and Merriam (1979) and Jubenville and Twight (1993). There are also texts on more specialized topics, such as wilderness management (Hendee, Stankey, and Lucas, 1990), which provide a great deal more information on particular issues than can be included here.

CLASSIFICATION OF OUTDOOR RECREATIONAL ACTIVITIES

Recreational use of forested landscapes takes many forms, ranging from concentrated use by numerous individuals in developed areas to dispersed use by small groups or single individuals (Table 11.1). These activities are extremely varied, and the perceived qualities of the experiences are dependent on very different landscape characteristics. A solitary backpacker, for instance, requires certain conditions in order to maintain the quality of the experience; these include a low density of fellow participants, a minimum of other types of resource use, a low level of impact from previous visitors, and so on. Other recreational activities require a greater amount of infrastructure, even if it includes only signed trails or the presence of sanitary facilities, and may depend on a greater amount of social interaction to increase the quality of the experience. Clearly, outdoor recreation encompasses a very wide range of activities that require a very diverse set of conditions and management activities in order to maintain their quality.

Just as there are natural limits to the amount of biomass that can be produced in a year, or the number of individuals of a given wildlife species

TABLE 11.1 Some Forms of Recreational Activity Occurring in Forested Landscapes

Type	*Form*
Concentrated Recreation	Campgrounds Picnic areas Swimming beaches Boat launches Marinas Historic markers Scenic overlooks Waysides
Dispersed Recreation	Hiking Horseback riding Biking All-terrain vehicles Snowmobiling Hunting Fishing Birding Photography

that can survive, there are limits to the numbers of participants in different recreational activities that can be supported within a landscape. The most common unit used to measure recreational activity is the *recreation visitor day,* or one 12-hour day spent by one person on a selected activity. In some instances, 24-hour days are used to define a visitor day; a party of two people who go on a five-day backpacking trip is equivalent to ten 24-hour recreation visitor days. A given landscape may be able to support some number of recreation visitor days for a relatively primitive activity such as backpacking, but the same landscape could support a much larger number of recreation visitor days for activities requiring more developed areas, such as campgrounds. These different uses are often in conflict; the development of a campground may greatly increase the recreation visitor days within a landscape and provide a source of income for the landowner, but the quality of more solitary activities may be severely degraded or such activities may even be precluded altogether. If the provision of recreational opportunities is a management goal, managers must consider the development of an appropriate mix of activities within a landscape. If facilities are required to provide this mix, they must be developed or maintained. If some sort of limitation of use is required to maintain the quality of the experience, such as a permitting system to limit the number of back-country campers, then methods to implement the limitations must be developed and enforced. There are very few, if any, areas in North America that do not support recreational activities; these uses and their requirements must be recognized in a resource assessment.

In planning for recreational use within a landscape, recreational activities may be grouped according to the landscape characteristics they require. The U.S. Forest Service and the Bureau of Land Management have adopted the *recreational opportunity spectrum* (ROS) to assist in their land management planning efforts (Clark and Stankey, 1979). This is a system that classifies recreational activities into six classes, depending on the landscape characteristics necessary to maintain the quality of the experiences. These classes range from primitive, requiring an essentially unmodified natural environment, to modern-urban, which occurs in a substantially urbanized environment (Table 11.2.)

There are six management factors that define the recreational opportunity spectrum: access, nonrecreational resource use, on-site modification, social interaction, acceptability of visitor impacts, and acceptable regimentation. Several of these can be further divided. Access, for example, has subcategories including the degree of difficulty in reaching the area, the type of access system (roads or trails), and the means of conveyance (paved or unpaved roads, motorized off-road vehicles, horseback, and by foot). Other factors also consist of different components: onsite modification may be defined by the extent, visibility, complexity, and type of facilities required, whereas visitor impacts may be defined by their degree and prevalence within the landscape.

Managers must ensure that appropriate management factors are in place to support the recreational opportunities planned for the landscape. Activities requiring modern-urban settings with easy access, facilities with many modern comforts and conveniences, and a high degree of social interaction among visitors can occur only where the appropriate roads and facilities exist. Recreational experiences requiring primitive areas will necessarily have more difficult access, be less compatible with nonrecreational resource use, offer very low levels of social interaction with other visitors, and conflict with extensive site modification.

During the assessment and planning process, management alternatives will be defined that meet the specified management goals, but consist of various mixes of different types of recreational opportunities. The different management alternatives may identify the targeted numbers of recreational visitor days to be provided in each recreational opportunity spectrum class. They may also identify the areas within the landscape where these activities will occur and where facilities will be maintained or, if necessary, constructed. The resource assessment exercise must identify the existing uses, opportunities, and facilities within the landscape, the potential opportunities for further development or modification of existing facilities, and the consequences of taking advantage of these opportunities. Potential recreational activities must be evaluated both in terms of the ecological consequences on abiotic and biotic landscape components and in light of the associated acceptable utilization levels for nonrecreational resources.

The recreation opportunity spectrum is useful for segregating different types of outdoor recreational opportunities within large landscapes. For

TABLE 11.2 Classes Within the Recreational Opportunity Spectrum and Some Associated Landscape Characteristics

Class	*Example Activities*	*Degree of "Naturalness"*	*Size of Area*	*Access*	*Social Interactions*
Primitive	Hiking	Unmodified Natural	Very large	Very difficult	Minimal
Semiprimitive Nonmotorized	Hunting	Predominantly Natural	Large	Very difficult	Minimal
Semiprimitive Motorized	Snowmobiling	Natural-Appearing	Moderate	Difficult	Infrequent
Roaded Natural-Appearing	Auto touring	Natural-Appearing	Moderate	Moderately difficult	Occasional
Rural	Jogging	Substantially Modified	Small	Easy	Frequent
Modern-Urban	Competitive sports	Urban	Small	Easy	Frequent

Source: Adapted from Clark and Stankey (1979) and Lucas et al. (1984).

smaller areas, the ability to provide more developed recreational opportunities is largely dependent on the characteristics of the surrounding landscape, the cooperation with neighboring landowners, the local transportation infrastructure, and so forth. For smaller-scale resource assessments for small tracts or watersheds, it may be more appropriate to simplify the recreation opportunity spectrum classes. One way to do this is illustrated in Table 11.1, where the primitive and semiprimitive classes are grouped together and labeled as "dispersed" recreation and the remaining classes are grouped together as "concentrated" uses. These two designations refer to the degree of social interaction that will be experienced by participants, but they also reflect a general division between the amounts of development and infrastructure required to support the activities. Dispersed activities require a minimum amount of development, whereas concentrated activities require a much more significant set of facilities to meet user expectations and to maintain ecosystem integrity.

CONCENTRATED RECREATION

Concentrated recreation areas include campgrounds, picnic areas, and other areas (Table 11.1) where specialized recreation activity is undertaken by relatively large groups of people. These activities are typical of the three least-primitive recreation opportunity spectrum classes: roaded natural-appearing, rural, and modern-urban (Table 11.2), although some of the activities that may occur in semiprimitive areas also require some concentrated-use areas, such as boat launches or parking areas. Concentrated recreation areas usually require some sort of capital investment to develop the infrastructure necessary to support the activity, although this is not always the case. An undeveloped swimming beach may be utilized by relatively large numbers of people even if there is only minimal supporting infrastructure (Figure 11.1). In many cases, however, concentrated recreation areas require the availability of electricity, water, and sanitary facilities. There may also be staffing requirements for facility maintenance or visitor service. In return, there may be some cost to participants in order to support the infrastructure.

The assessment of impacts caused by existing concentrated recreation areas usually focuses on levels of use of the areas themselves, along with associated use of travel corridors. Assessment of current use can take several forms, depending on the nature of the activity. If a management goal is to promote recreational activity in order to generate a revenue stream from associated fees, it will probably be necessary to go beyond normal resource assessment procedures and gather information to develop marketing plans, evaluate facility development or enhancement, or to identify new visitor bases. In these cases, specialized expertise is required, which is beyond the scope of this text.

Assessment of current use is relatively easy at sites that charge fees or require visitor registration. This is especially true if there is a permanent staff

FIGURE 11.1 Concentrated use areas, such as this swimming beach, bring larger groups of participants to selected areas within the landscape.

who can monitor use and maintain records of the numbers of users over time. The amount of available space or underutilized capacity can be estimated directly from current use levels. In the assessment document, the number of visitor days and estimates of underutilized capacity may be specified. Areas in which most of the available capacity is utilized are those that may justify further study to assess the possibility of expansion or improvement of facilities. Periods of high and low use can be identified and related to concerns identified for other landscape components. If, for instance, periods of high visitor use correspond with periods of high water flow or soil saturation, then it may be necessary to examine the recreational facilities to determine whether measures are necessary to protect soil integrity in sensitive areas. This is a common situation on many hiking trails, where boardwalks are installed across seasonal wet areas to minimize soil disturbance, protect sensitive plants, and maintain aesthetic values.

Estimation of use is more difficult if there are no permanent staff or if visitors are not required to register at a facility. These situations, which may occur at places such as unsupervised swimming beaches or boat launches, will require special attention to assess current use. A sampling scheme will have to be developed to monitor use over certain periods. The total level of use may be the appropriate variable in some instances, such as in assessing impacts of recreational use on nearby wildlife habitat. But in other cases it may be the level of maximum use, rather than total use, that is a concern. This could be the case if the possibility of soil disturbance or stream sedimentation is an issue or in planning for sanitary facilities with on-site storage or treatment of wastes. Determination of use rates will probably require site visitation and the estimation of the number of visitors or other appropriate response variable, such as the number of boat launches at a given ramp. Care must be taken in identifying the times to monitor, so as to ensure that the appropriate estimates of use are obtained. If the level of maximum use is the information required for management planning, monitoring may concentrate on weekends and holidays. If total use is the quantity of interest, monitoring will have to include peak as well as nonpeak times. In both a practical and a statistical sense, it is critical to ensure that the periods being sampled are representative of the type of use that is of most concern (see Chapter 15).

It may be difficult to estimate the degree to which recreational activity may be expanded in areas that are not well monitored. In many cases, the capacity during maximum-use periods defines the degree to which recreational capacity is being utilized. Even if a free boat launch is not used at all on many days, if it is filled to capacity during weekends or holidays most visitors will consider the facilities to be inadequate. Initiation of a user fee or requiring visitation permits may be one way of restricting or discouraging overuse. Facility expansion is the best way to increase capacity, but this may not be a viable option for many landowners if the facility will not generate revenue to cover the cost of such expansion.

In addition to the activity directly associated with concentrated recreational facilities, there is an associated impact of this activity along travel corridors. The direct effects of traffic levels may require different road construction methods or increased maintenance along roads leading to campgrounds, picnic areas, and other concentrated recreational facilities. Road traffic may be directly monitored with automated devices or by personnel at selected times. Again, it may be the peak use rather than the overall level of use that may be of concern. An evaluation of road condition, water runoff and associated erosion or sedimentation, and types and conditions of stream crossings should all be included in a landscape-level resource assessment in order to identify areas that may need attention.

If a management goal is to provide for certain levels of recreational use or to generate revenue from recreational activity, the potential for increasing utilization beyond current levels should be considered during the assessment process. On private land, enhancing recreational opportunity may not be a stated management goal. If it appears that there may be potential for recreational development that could increase the income stream or meet other ownership goals, this should be noted during the assessment process. In most cases, the resource assessment will serve only to identify areas of potential development that may warrant further study. Development or investment decisions should be based on detailed economic, financial, and marketing analyses beyond those considered in most assessment efforts. These detailed analyses require special skills that most land managers do not possess, and the consequences of improper analyses can be severe for a landowner. For these reasons, a resource assessment should usually discuss only whether the basic physical conditions, such as soil and hydrology, are suitable for recreational development. In most cases, separate analyses should be conducted to explore the potential for different types of recreational activity in greater detail.

It may be difficult for many land managers to identify areas with potential to provide concentrated recreational use. Likely areas include those surrounding lakes and ponds, along streams and rivers, in unique ecosystems, and in scenic areas. The soil maps developed by the Natural Resource Conservation Service (Chapter 4) usually contain interpretive information related to recreational development, such as the potential of different soil mapping units to adequately support campgrounds, picnic areas, playgrounds, or trails (Table 11.3). Although not every area with soil conditions that can support a campground is suitable for campground development, adequate soil conditions are a prerequisite for the location and development of concentrated-use areas if other factors indicate the potential for development. Further prerequisites, besides an obvious attraction such as scenery or the potential for boating or swimming, include suitable topography, a viable transportation network, and the presence of utility service. Road development, construction or extension of power lines, and the need to establish drinking water wells should all be considered in assessing the potential for developing concentrated-use areas within a landscape.

TABLE 11.3 Restrictive Soil Features Related to Potential Recreational Development for Selected Soil Mapping Units in Houghton County, Michigan

Mapping Units (% Slope)	*Camping Areas*	*Picnic Areas*	*Playgrounds*	*Paths and Trails*
Munising				
(1–8%)	Severe	Moderate	Severe	Moderate
(8–15%)	Severe	Moderate	Severe	Moderate
(15–35%)	Severe	Severe	Severe	Severe
Pelkie				
(0–3%)	Severe	Slight	Moderate	Slight
Allouez				
(1–8%)	Moderate	Moderate	Severe	Slight
Ontonagon				
(1–6%)	Moderate	Moderate	Moderate	Slight
(6–15%)	Moderate	Moderate	Moderate	Severe
(15–35%)	Severe	Severe	Severe	Severe
Watton				
(1–8%)	Slight	Slight	Moderate	Slight

Source: Schwenner (1991).

The existence or development of concentrated recreation areas creates ongoing administrative and operational demands and liabilities. It may be necessary to hire personnel to staff and maintain a campground, for instance. Concentrated recreational activity may also create conflicts with other management activity and the achievement of other management goals. The existence of a single concentrated-use area may increase traffic along the travel corridors and have a resulting increase in the visual sensitivity of an extensive land area. The creation of concentrated recreation areas is not to be taken lightly, in that there are initial and continuing costs for their operation and they may impact the ability to achieve other management goals. On the other hand, concentrated-use areas are very visible to the general public. A simple picnic area may create a great deal of goodwill within the local community. If properly designed and maintained, concentrated-use areas can be a great contribution to the achievement of overall management goals.

DISPERSED RECREATION

Dispersed recreation activity is usually undertaken by small groups or individuals and may occur over wide areas. These activities roughly correspond to those occurring in the primitive, semiprimitive nonmotorized, and semiprimitive motorized recreation opportunity spectrum classes (Table 11.2). Examples of dispersed activities are hiking, horseback riding, hunting, fishing, and other activities as noted in Table 11.1. In aggregate, the number of participants may be far greater than the number of people utilizing concentrated-use areas within a landscape. Dispersed activity may take place along trails, rivers, or other corridors, but activities like hunting may occur over much broader areas, including virtually the entire landscape. Human presence over the entire landscape has the potential to alter wildlife behavior and lead to decreased use of an area by some species and increased use of the same area by other species. There may also be potential conflicts with other management activity, such as commercial timber production. For these reasons, it is very important for land managers to understand the types of dispersed recreational activity that are occurring within a landscape, the time of the year they are occurring, and, if possible, the general areas where they occur.

In many ways, dispersed recreational activity requires less infrastructure and is less costly to develop than many types of concentrated-use areas (Figure 11.2). On the other hand, the initial development and continued maintenance of trails for hiking, horseback riding, and other motorized and nonmotorized off-road activity certainly requires an investment of material and personnel to minimize ecological disruption and the maintenance of ecosystem function. This often entails ongoing maintenance for the protection of soil integrity and the prevention of erosion and sedimentation. There are also opportunity costs, which are essentially foregone revenue, associated with decreased non-

FIGURE 11.2 Dispersed recreational activities involve smaller groups of people spread over larger areas; trails such as this one must be maintained in order to protect soil integrity and to ensure minimal disruption of ecological processes.

recreational resource use. Such a situation will occur if, for example, commercial commodity production is excluded from trail corridors.

Dispersed recreation activity is less likely to lead to direct income for landowners than is the development of concentrated-use areas. There are some direct costs associated with such things as cross-country skiing, hiking, and riding trails: clearing, water quality protection measures, signs, and so on. There are also liability concerns. Generally, landowners may be held responsible for the prevention of potentially hazardous situations arising from dead trees, fallen limbs, and poor trail condition if there are developed trails on their property; landowners may have less liability for other sorts of activity, such as hunting or fishing. Liability laws vary to some extent among the different states. It is good practice to periodically review these issues and obtain competent legal advice concerning liability as related to recreational use of property by others than the landowner.

In some areas, direct income may be derived from leasing hunting or fishing rights, but this is not possible everywhere. Commercial forest acts (Chapter 10) in some states require, among other things, that landowners allow open public hunting access in return for favorable tax treatment. Such legislative acts may forbid leasing exclusive hunting rights on registered land, although landowners receive some tax incentives for providing open access. Most benefits derived from providing dispersed recreation activity are indirect. They may include such things as the development of good public relations or increased public awareness and support of land management activity.

Dispersed recreation activity related to wildlife (fishing, hunting, photography, birding, and so on) may have little cost to landowners other than that which is incurred by general habitat improvement activity, such as the seeding of roads or skid trails and leaving snags for nesting sites. Enhanced fishing access requires the protection of riparian areas and prevention of stream sedimentation, but these actions will be required for compliance with state and federal laws designed to protect water quality (Figure 11.3; Chapters 6 and 10). Human activity related to fishing and hunting is concentrated into legal seasons. Photography and similar visual enjoyment of wildlife occurs throughout the entire year, but, again, may be concentrated in certain time periods such as migration seasons or soon after breeding when young are present. Acknowledgment of these temporal patterns of dispersed recreational activity in the assessment process is important, because other management activities such as road construction or maintenance can often be scheduled to minimize potential conflicts among the various users of the landscape.

AESTHETICS AND THE VISUAL IMPACT OF MANAGEMENT ACTIVITY

Aesthetics, in the most general sense, implies a great range of human emotional responses to natural resources. This includes the satisfaction landowners receive from being good stewards of their property, the fundamental apprecia-

FIGURE 11.3 Enhanced fishing opportunity requires protection of the surrounding riparian area and implementation of measures to protect water quality.

tion of the value of the continued survival of an endangered species, and many other issues that are difficult to quantify for a particular small ownership or watershed. Most management guidelines concerning aesthetic issues concentrate on the visual features of the landscape and the visual consequences of land management activity. Humans are, after all, visual animals, relying much more on sight than on other senses like hearing or smell. As such, people frequently tend to notice areas of contrasting color or texture within landscapes. Vernon (1968) noted that people focus their attention on items that they perceive to be interesting or important and pay much less attention to features they do not deem to be important. Although this may seem obvious, it has important implications concerning what people will notice and how they will perceive management activity. Hodgson and Thayer (1980) noted that the visual quality of a scene was perceived to be lower if people believed they were viewing a managed landscape rather than a natural landscape. On the other hand, Magill (1994) found that people viewing both managed and unmanaged landscapes tended to like forests, mountains, peaks, hills, meadows, and color contrasts, and to dislike roads, bare areas, and clearcuts.

A large number of people casually view forested landscapes while traveling along public highways. In some areas, scenic vistas entice people to travel a road for the visual experience itself. The Blue Ridge Parkway in Virginia and North Carolina attracts many people for its scenic beauty. Such activity may be concentrated at certain times; the fall color season in the northern portions of the United States and southern Canada attracts a great many travelers during a short time span. Travel along smaller roads also brings people into visual contact with forested landscapes, and the presence or the development of concentrated recreational-use areas will increase the importance of visual appearance over large areas along travel corridors. Perception also varies by season; the presence or absence of deciduous foliage or snow cover may alter the visual significance of management activity.

Several states have developed aesthetic management criteria. The Minnesota Department of Natural Resources (1994), for example, developed a classification scheme for defining visually sensitive areas. This scheme is based on three factors:

1. The perceived degree of sensitivity of travelers or recreational users in an area to aesthetic concerns;
2. The volume and type of travel or use in an area; and
3. The speed of travel, or, conversely, the duration of the visual experience within an area.

Three levels of visual sensitivity are identified in the Minnesota classification scheme. The most visually sensitive areas are those with significant public use and where visual quality is of high concern to many of the users; examples include areas along recreational lakes and rivers, recreational trails, and areas along some highways and roads where there is a high level of scenic quality.

Moderately sensitive areas are those where visual quality is of moderate concern to most users and there is less significant public use, even if the scenic quality is very high. Finally, the least sensitive areas are those where visual quality is of little concern to most users. This last class includes some public highways, local roads where the level of use is low, and nonrecreational lakes and rivers.

The State of Wisconsin has a system of classifying aesthetic management zones that is based on essentially the same concerns as those underlying the Minnesota system (Wisconsin Department of Natural Resources, 1990). The Wisconsin system identifies four types of aesthetic management zones: Class A, in which aesthetic concerns predominate, includes parks, special-use trails, areas along recreational lakes, and so on; Class B, in which no one use predominates, includes areas along roads with light to medium use and lakes and streams that do not have significant recreational value; Class C, in which other resource management concerns predominate, includes most areas of commercial timber production; and Class D, which includes designated special-use areas, such as designated wild and scenic rivers and wilderness areas. Specific guidelines for silvicultural activity, road construction, and other management activities are developed for each class of aesthetic management zone. As in the Minnesota system, the goal of this system is to recognize the importance of aesthetic values in the landscape and to reduce the negative visual impacts of all management activities.

In addition to public agencies, many corporations have policies and procedures designed to protect aesthetic values. Schuh (1995) presented a case study of a visual sensitivity analysis conducted by a private company in the Pacific Northwest. The company utilized a visual sensitivity rating system based on four different types of viewing: motor traffic, residential, recreational, and a "unique other" category. A visual sensitivity index was estimated for each of the four categories and summed to arrive at a visual sensitivity rating for a site or operational unit. The indices for each type of viewing were based on the number of viewers, the duration of their view, and an indicator of the distinctiveness of a site within the surrounding landscape.

A third method of analyzing the visual sensitivity of an area was presented by Bergen et al. (1995). Their analytical procedure can be conducted for large areas and utilizes a digital elevation model within a geographic information system (see Chapter 13). Using the digital elevation model and computerized analytical routines, they determined the physical areas within a landscape visible from different points along a transportation corridor. The landscape was divided into small areas, each of which was assigned a view factor depending on whether it was obscured, partially visible, or fully visible from each of several viewpoints along the road. A weighted index was calculated for each of these areas in the landscape:

$$VI = \sum_{i=1}^{n} VF_i \cdot L_i \cdot 100 \qquad [11.1]$$

where VI is the weighted viewpoint index, VF_i is the view factor for the ith road segment ($VF_i = 0$ if the area is obscured, 1 if it is partially visible, and 2 if it is fully visible from the ith road segment), L_i is the length of the ith road segment, and n is the number of road segments being considered. Following calculation of the weighted viewpoint index, each area within the landscape was assigned to one of four categories: not visible, low visibility, medium visibility, and high visibility. Management options for operational units within the landscape were dependent on the proportions of the unit in each visibility category. As visibility increased, less visually disruptive options were considered in planning harvesting operations. This type of analysis concentrates on the physical features of a landscape. In such a quantitative analysis, it is sometimes easy to forget that personal characteristics and attitudes are also very important in the way people perceive landscapes (Zube, Sell, and Taylor, 1982; Sheppard, 1989). These concerns may be incorporated to some extent in the subjective assignment of visual sensitivity factors for different areas, but recognition of the significance of personal characteristics and attitudes must also be foremost in the interpretation of the results of quantitative analyses.

All three of these approaches utilize some degree of subjectivity in determining the visual sensitivity of areas within a landscape. This is probably inevitable, inasmuch as it is human perception and attitudes that are being considered. For areas that have a substantial degree of visual sensitivity, restrictions may be placed on the types of management activities to be considered in order to maintain visual quality. The Minnesota and Wisconsin guides to visual quality management identify a number of management activities that may raise concerns about visual quality, and they offer a number of specific suggestions to limit visual impact in areas with different levels of visual sensitivity. Jones (1995) also gives a set of guidelines designed to minimize the visual impact of commercial harvesting operations, but the guidelines apply equally well to any management activity that will result in vegetation removal or site disturbance. All of these sets of guidelines are basically commonsense procedures designed to:

1. Minimize conflict between different types of users of forested landscapes;
2. Minimize the apparent size of disturbed areas;
3. Minimize the amount of debris and exposed soil areas; and
4. Minimize the formation of sharp boundaries between different vegetation types.

During the assessment process, the major task is to identify visually sensitive areas so that managers are aware of these concerns and the locations where they are most critical. In addition to the application of the procedures discussed earlier, this may entail holding meetings with representatives of user groups or focus groups to identify issues and concerns within a given landscape. Management alternatives should be formulated to minimize deterioration of visual quality; for example, measures could be taken to minimize the extent of even-aged forest overstory communities in visually sensitive areas. In many cases, the fact that there are visual concerns along roads or rivers may provide opportunities to tailor management activity to simultaneously create habitat areas or travel corridors for wildlife.

SUMMARY

There are many types of recreational activity that take place in forested landscapes. These can be broadly divided into (1) activities occurring in concentrated-use areas where large groups of people may gather to participate, and (2) activities of smaller groups of people or individuals occurring over a dispersed area. Different types of recreational activity have different implications for management. Concentrated recreational-use areas may provide an opportunity to directly obtain revenue for the landowner, but usually require investment to develop an infrastructure as well as personnel and material to maintain the facilities; they also increase the landowner's liability concerns. Concentrated recreational-use areas may also be associated with travel corridors that increase the visual sensitivity of large areas within a landscape. The assessment process usually should concentrate on determining the level of use of existing areas, the extent of the surrounding landscape affected by the area, and, if consistent with management goals, the physical potential for developing new recreation areas. Dispersed recreation activities are less likely to directly provide revenue to landowners, but they also require less investment in infrastructure, personnel, and maintenance. Assessment efforts should concentrate on identifying the extent and timing of such activity in the landscape of interest, and on the identification of practices to reduce conflicts between dispersed recreational use of the landscape and other management activity. Management activity of all sorts may also impact the visual quality of a landscape. Visually sensitive areas should be identified during the assessment process, and methods to reduce the visual impact of management activity should be specified.

APPLICATIONS

11.1 Consider a large park in your state; what types of recreational activities occur in the park? What is the importance of dispersed activities as compared with concentrated-use areas? How are the management options for surrounding areas impacted by the park?

11.2 What are the laws of your state regarding access to private land for hunting and fishing? Is there a program to provide landowners with incentives to open their land for public recreational use, or can a landowner lease hunting rights to derive revenue? If so, what limits does this place on other management activities?

11.3 Identify an undeveloped area that may be suitable for development of a campground. Is the soil in the area suitable for campground development? Are there any physical factors that would have to be considered in the development decision?

11.4 What are some factors that would have to be considered in deciding whether to invest in the development of a concentrated recreational-use area such as a campground?

11.5 On a topographic map, identify visually sensitive areas in the landscape surrounding a major highway. What are the features that make some areas more visually sensitive than others?

11.6 If the management goals for an area include revenue generation through commercial timber production, what are some techniques that might be used in visually sensitive areas to minimize the visual impact of such activity? How would this differ from what might be done in similar ecosystems that have only a low degree of visual sensitivity?

Chapter

12

Prehistoric and Historic Land Use

Most of glaciated North America has been inhabited by humans for eight to ten thousand years or more, and some nonglaciated areas have been inhabited for an even longer period. By one estimate, the population of the Americas in 1492 was 53.9 million people, with an estimated 3.8 million living in North America (Denevan, 1992). The term *prehistoric* refers to the time before there were written records of human activity; *historic* refers to the time when written records were kept. A great deal of information about past population size, culture, and land use is stored in the landscape in the form of physical evidence of past human activity. Some of the physical evidence of prior land use is relatively easy to see, as in the case of an old cellar, rock foundation, or logging camp. But much of the physical evidence of prior land use is not so easily visible; traces of prehistoric human occupation or land use are rare and are often belowground, with few remaining clues on the surface. Even historic features such as earthworks or ditches may be easily mistaken for natural features.

Even though written records were kept during historic times, much of the known detail about the daily lives of people comes from the analysis of physical, rather than written, evidence. Written records are a less objective source of information than physical evidence, partly because much of the written evidence is concerned with the lives of more affluent people. Regardless of social status, more detail may be found in the analysis of the refuse of past inhabitants than by examination of their written records. Some written records, such as recorded deeds and wills, are valuable to resource assessment because they are concerned with the ownership of land and property by individuals, often providing a source of information concerning past land use that is systematically available across the landscape. There is also a growing body of information about the life of common people, but these records are not systematically available for all areas of the landscape and are not usually compiled in a central location.

Past land use can place constraints on current management options. Even low levels of human population have significant impacts on the natural resources of a landscape. Martin (1984), for instance, argues that the extinction of large Pleistocene mammals was due in large part to human hunting activity and impacts on the landscape. Extensive areas of the Midwest and Southwest were cultivated by Native Americans prior to their contact with Europeans (Whitney, 1994). There was also extensive use of fire to aid in hunting or to encourage berry production by Native American populations in prehistoric times (Whitney, 1994). Such widespread disturbance may have altered the presence or absence and the abundance of individual plant or wildlife species, as well as the current structure of vegetative and wildlife communities in a landscape. Physical impacts such as drainage or agricultural clearing may have altered soil physical and chemical properties. Although the cultural significance of knowledge about past human society is important in its own right, it is also critical to our interpretation of past land use and vegetative response to prior disturbances. This type of information can provide valuable insight to aid current management planning efforts.

There are, therefore, two major reasons to consider prehistoric and historic land use during a resource assessment exercise. The first is that the landscape is a storehouse of valuable information, providing insight into the lives of past residents and the development of modern culture. The second is that prior land use can have long-lasting impacts on soil properties, water quality, productivity, and the vegetative community, that may place limits on current management options.

FEDERAL AND STATE HISTORIC PRESERVATION PROGRAMS

National and state-level historic preservation programs have been developed to recognize the importance of past inhabitants and their activities in the development of modern culture and to provide a means for preserving culturally significant sites. These programs tend to focus on the locations of important historical events, such as battles or settlements, places connected with important persons, or places representative of once widespread cultural practices, such as pioneer homesteads. Initial participation in these programs by private landowners is usually voluntary, but there are often incentives to participate, usually in the form of favorable tax treatment.

The primary national legislation affecting preservation of prehistoric and historic resources of local, state, regional, or national significance is the *National Historic Preservation Act of 1966,* with subsequent amendments. This Act established several policies of relevance to private landowners, including the statement that the federal government shall "contribute to the preservation of nonfederally owned prehistoric and historic resources and give maximum encouragement to organizations and individuals undertaking preservation by private means" (Section 2 (4)). Sites of national significance may be designated as National Historic Landmarks or listed in the National Register of Historic Places; sites of international significance may be included on the World Heritage List in accordance with international programs. Sites of national or international significance are relatively rare, however, and are not normally encountered during most management planning efforts. Management options in areas near such sites may be limited by aesthetic concerns and the need to ensure the integrity of the historic site, but the effect is generally localized. The need to consider such sites is usually readily apparent; when they are present, they should receive considerable attention in the resource assessment and management planning processes.

The National Historic Preservation Act also provided for State Historic Preservation Programs and the appointment of State Historic Preservation Officers with the responsibility to, among other things, "direct and conduct a comprehensive statewide survey of historic properties and maintain inventories of such properties" and to "prepare and implement a comprehensive statewide preservation plan." Individual states maintain these inventories of cultural features and administer programs for the preservation of significant cultural resources. Various agencies maintain these inventories in different states.

In most states it is possible to obtain an indication of possible sites of concern within a given area by providing the responsible agency with a legal description of the area. For security reasons, and to ensure protection of sensitive sites, the legislation provides that agencies are not required to release details of site location or content to the general public. If such a site is present in an area where management activity is being considered, it is usually possible to work with agency personnel to allow management to proceed while ensuring protection of the important cultural resource. Many times, awareness of the presence of a culturally significant site can lead to its preservation with only minor modification of current management activity. Road layout, for example, might be altered to bypass an old logging camp with very little or no additional cost if the need is detected early in the process. Likewise, significant sites often occupy only a small area, and an undisturbed buffer zone may be created around such sites with relatively little impact on the conduct or the financial returns from management operations. It is neither possible nor necessary to preserve every detail of past land use, however, and judgment must be used to determine what is and what is not significant and deserving of preservation.

PREHISTORIC LAND USE

State inventories of cultural features provide an excellent starting point for the consideration of prehistoric sites during the assessment process. As noted previously, if provided with a legal description, an agency will generally indicate whether there are any known significant features within a described area. Knowledge of the existence and type of features, coupled with field reconnaissance, is often enough to guide the resource assessment and planning process to ensure preservation of the feature while allowing management activity to proceed. In most cases, the presence of cultural features will not preclude other management activity; rather minor modifications of planned management activity are usually sufficient to ensure the integrity of such sites.

Another source of information is the General Land Office Public Land Survey notes. As noted in Chapter 7, the surveyors did not just establish property corners, they also recorded a considerable amount of information regarding the surrounding landscape. Often, the presence of Native American villages, trails, and other physical evidence of human activity in the landscape was noted (Figure 12.1). These records are preserved in the surveyor's field notes and maps in state archives and are sometimes available in county courthouses or library archives. In many cases, all surface traces of these features may have been obliterated during land clearing and settlement. In some cases, however, it may be possible to find physical evidence of prehistoric activity in today's landscape.

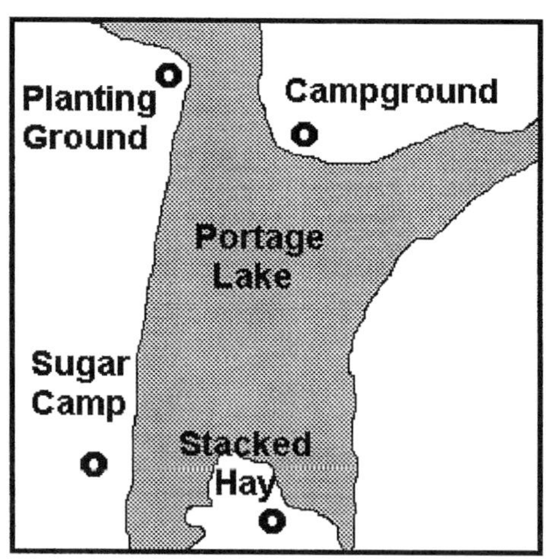

FIGURE 12.1 Notations concerning Native American land use in the original 1845 U.S. Public Land Survey of T 54 N, R 33 W in Houghton County, Michigan.

For some parts of the country, compilations of the original survey notes have been developed. Trygg (1969), for example, developed a composite map of the original General Land Office survey notes that identifies cultural features on the landscape at the time of the original survey. These features include Native American campgrounds, villages, burial grounds, cleared or cultivated areas, and trails. Activities of early European settlers are also noted, including homesites, boat landings, river crossings, mineral exploration areas, and so on. Compilations such as this provide the same information contained in the survey notes and plat maps, but the compilations contain less detail than is available on the individual survey township plats. The advantage to such compilations is the ease in visualizing regional land use patterns.

Because of past disturbance and the passage of time, it is difficult to detect physical evidence of prehistoric land use during field reconnaissance of surface features. Even substantial features, such as burial mounds, may be difficult to recognize as artificial features in the landscape. One truism to keep in mind is that humans in the past tended to enjoy and occupy the same types of places that humans today find attractive. Areas providing shelter from extreme weather conditions and easy access to terrestrial and aquatic ecosystems were probably utilized by prehistoric inhabitants. Today these areas may be popular for recreational use because they tend to be along lake margins and in scenic areas. Old lake margins, which may be far from significant bodies of water today, solid ground near the confluence of two navigable streams, and high spots in the landscape are the types of places that prehistoric peoples occupied and where physical evidence of prehistoric land use may sometimes be found.

Even if there is no prior indication of culturally significant features in a particular operational area, if evidence of ancient land use is found associated with similar features in the same general area, then the likelihood of such evidence existing on a given feature is increased. This is the case if, for example, there is an old lake margin that may extend for many miles and there are known sites elsewhere along the same feature.

Much prehistoric physical evidence is belowground and more likely to be found if there is construction activity or other activity disturbing the soil (Collins et al., 1995). If such evidence is found, due diligence should be used to attempt to minimize disturbance of the material until its significance may be assessed. Electronic sensing equipment can often identify the extent of such areas without requiring further soil disturbance. It is extremely important to make every attempt to preserve the context in which artifacts are found. The location of an artifact and associated features may provide important clues to the life and land use of past inhabitants. Much of the scientific value of an artifact may be lost if it is removed prior to careful inspection of its original context by trained personnel. Normal soil conservation efforts designed to minimize erosion are usually sufficient to protect belowground cultural resources until their significance can be determined. In many cases, the original context serves as the best storage area for artifacts until suitable techniques are developed for their proper investigation and interpretation.

During the resource assessment process, then, records from original General Land Office survey notes and surveys of known significant features maintained by State Historic Preservation Offices are two major sources of information concerning the existence of significant prehistoric sites. Even if these sources do not indicate the presence of culturally significant features in a particular management unit, if there are significant sites in the same general area that occur on landforms that extend into the operational unit, this should probably be noted during the assessment process. It is usually not necessary to undertake extensive searches for such sites within every management operational area, but care should be taken if activities like road construction are planned through an area in which significant sites are more likely to occur. The appropriate actions to be taken if physical evidence of prior land use is found, as well as the names and addresses of experts who can assess the significance of such evidence, may be noted in the assessment document; it is better to be proactive and prepared than to be forced to make decisions without adequate information. In any case, if evidence of prehistoric land use is found, the context should be preserved intact, if at all possible, until it can be examined by an expert.

HISTORIC LAND USE

Many of the best written records describing historic land use are concerned with the transfer of property, especially real estate. Such records are systematically available for all areas within a landscape, and they provide a primary source of information concerning the occupancy of an area in historic times. Descriptions of original land grants and subsequent deed transfers are usually found in county records. County boundaries changed as the population grew and European settlement proceeded; it may take some investigation to determine where records of the ownership of a particular tract are currently stored. For some areas, early records have been lost as a result of accident or disaster, such as a courthouse fire. Copies of the original records can usually be obtained if they are located, but sometimes destruction of the records makes this impossible.

Deed abstracts contain the ownership history of a parcel of land from the time of the original European owner. Abstracts are not as common as they once were, but if available they provide the names of owners and dates of ownership changes. These names and other data can provide clues for further investigation. There may be a will or probate record if the property owner died while still owning the area. Probate records and associated property inventories can provide clues to past land use as well as insights into the dailv lives of the inhabitants.

One problem with abstracts is that they are specific to individual parcels; a large landscape may require investigation of a large number of different abstracts. Over time, ownerships are subdivided and areas are combined, making it sometimes difficult to trace the ownership history of a given tract of land in its entirety. If an abstract is not available or if the expense of acquiring one is not warranted, information on past ownership may be available from old plat books. Plat books contain maps showing the property ownership boundaries and the names of landowners. Plat books provide one-point-in-time information about the landowners in an area, but they are usually available for entire counties. Plat books usually do not directly indicate land use, but include average size of ownership, proximity to towns, and other landscape characteristics that can be used as a basis for subjective evaluation of likely land use in the past. Modern plat books are often for sale in courthouses or commercial bookstores; old plat books are often found in libraries or archives.

Land that has been in individual or corporate ownership is more likely to contain physical evidence of past land use than land that has never been in private hands; almost all of the eastern United States has been in private ownership at some point in the last 200 years. Former agricultural or residential use, in particular, may impact current management options. Soil alteration through compaction or mixing of the surface horizons by grazing or plowing, possible subsequent nutrient depletion, and physical changes such as drainage ditching can all impact current productivity levels. Physical evidence, such as stone piles and fences, foundations, cellars, and so forth, all indicate former habitation (Figure 12.2). A great deal of the forest area in the eastern United

FIGURE 12.2 Abandoned foundation in a forested area; note trees growing out of the center.

States is on abandoned agricultural land; inspection of the soil can usually indicate whether an area has ever been used as an agricultural field if a plow layer is present.

Historic physical features are often difficult to identify because of revegetation and erosion that have modified their original appearance. Old drainage ditches, dams, and other earthworks are quite easily mistaken for natural features (Figure 12.3). Artificial features are commonly linear, have right-angle turns, form distinct shapes, or begin or end suddenly. Natural features tend to have more irregular shapes and to change forms gradually, as in the transition of an intermittent stream into a permanent watercourse. Earthworks rarely have such inherent cultural significance that they should be preserved throughout a landscape. They may, however, alter drainage or water movement through the landscape and have residual effects on current management options.

Clues to past land use may also be found in the vegetation. Current forest cover may be compared with witness and line tree information in the original survey notes to determine changes in forest composition following European settlement (see Chapter 7). Indications of initial logging or land clearing, and subsequent vegetative recovery, may be found in such comparisons. The past response of vegetation to disturbance is usually a good indicator of likely future response to management activities unless there have been extensive physical alterations of the site, such as severe erosion. The presence or absence of certain species in the landscape can also provide indications of past land

FIGURE 12.3 Exploratory mining trench in Houghton County, Michigan, probably from the mid- to late 1800s; the area had been burned and cleared to facilitate mineral exploration and has since returned to forest.

use. The presence of large old apple trees in an opening in the forest, for example, is a good indicator that there was once a homestead nearby (Figure 12.4).

Areas of cultural significance that should be preserved are relatively rare, but important if they are found. In most landscapes and in most resource assessment efforts, it is more important to ascertain whether there are any constraints on current management alternatives owing to past land use.

SUMMARY

The two reasons for including investigation of past land use in a resource assessment are (1) to identify areas of cultural significance that should be preserved and (2) to obtain clues concerning past land use that may place limits on current management options. Important physical evidence of past lives and culture is stored in the landscape; it is the responsibility of land managers to take prudent steps to preserve areas that are significant to the understanding of a past culture. Often this may be done with little impact on modern management activity. It is far more common, however, for land managers to use evidence of past land use to understand the impacts of past human activity on current productivity and potential management alternatives. All management activity has some type of impact on an area and on future management options. Long-lasting impacts, such as soil compaction

FIGURE 12.4 The presence of an apple tree in a forest clearing indicates the location of a former homestead and the likelihood of associated agricultural activity.

or nutrient depletion, can impact management options for many years into the future. It is these direct impacts of past land use on current management alternatives that are usually of most concern during the resource assessment process.

APPLICATIONS

12.1 Find the name and address of the agency responsible for the State Historic Preservation Program in your state; what is the procedure for obtaining information from the state inventory of historic properties?

12.2 For a selected area, use its legal description to obtain information from your state's inventory of historic properties. Are there any known cultural features of significance in the area or in the surrounding landscape?

12.3 Obtain copies of the original General Land Office survey notes and map for a given landscape. Are any features indicating prehistoric land use shown on the map or in the notes? Is there any remaining physical evidence of this activity in the landscapc?

12.4 Obtain a modern plat book for a county in your state. Are there typical patterns of tract sizes or shapes that are characteristic of agricultural areas? Are the shapes and sizes of tracts near waterways or highways different from those of tracts located away from such transportation corridors?

12.5 For your state, what percentage of the total area was forested approximately 100 years ago? What is the percent of forested area today? What land use changes and changes in economic activity have contributed to changes in forest cover over the last 100 years in your region?

12.6 What soil types in your region are more seriously impacted by past agricultural use? Are some soil mapping units more sensitive to nutrient depletion from agricultural use than others? What other impacts of past agricultural use may affect current management options?

PART V

Integration and Synthesis

Chapter

13

Geographic Information Systems

Organization and presentation of spatial information is critical to landscape management, but this need is not new. Cartography has a history extending back several thousand years, and maps have been used in the management of natural resources for centuries (Cook, 1717). What is new is the ability to quickly process and link spatial information from several sources. *Geographic information systems* are specialized computer programs for the analysis and processing of spatially explicit information. The technology has developed rapidly over the last decade and is continuing to increase in capability. It was not that long ago that a land manager's ability to consider interrelationships within a landscape was limited by the time and effort required to physically integrate the many types of information from many different sources. Today one could argue that the technology for physically linking and presenting this information has surpassed the capability of existing analytical procedures to synthesize such a tremendous amount of information into a form that resource managers can utilize.

Geographic information systems provide resource managers with the ability to custom design and rapidly produce maps of various landscape components. The most capable systems contain routines to link spatial information, such as coordinates or boundaries, with descriptive information. This linkage, and the associated processing routines, allow managers to produce tables giving the number of hectares within a landscape that have certain features, such as hydric soils, and to produce a map showing their location. Information from different sources may be combined to identify areas with combinations of features, such as coarse soils and a greater than 15% slope. Some geographic information systems have computational routines designed to produce statistical summaries describing landscape structure, such as fractal dimension or the length of edge between selected habitat types (see Chapter 7), or they can produce data files for use by free-standing analytical software to conduct such analyses. Yet perhaps the most powerful tool for resource managers is one that is only just beginning to be utilized—the linkage of spatial information with models in order to produce scenarios describing future landscape structure in the event that a selected management alternative is implemented. The graphical presentation of such analyses is an extremely efficient means of describing possible future conditions within a landscape.

This chapter concentrates on the terminology and basic information necessary for resource managers to understand in order to utilize geographic information systems. An underlying assumption is that the manager will be working with a professional geographic information system analyst to produce the maps and reports needed to support management decision making. An excellent introduction to geographic information systems, including basic descriptions of capabilities and hardware requirements, is given by Clarke (1996). More detailed information on the structure and technical aspects of geographic information systems and examples of their use are given by Aronoff (1989), Antenucci (1991), Haines-Young, Green, and Cousins (1993), and Sample (1994). More complete information about remote sensing and its use in developing classifications of landscape features is given by Lillesand and Kiefer (1994) and Campbell (1996).

TERMINOLOGY

The recent rapid development of geographic information system technology has spawned a great deal of jargon that may be difficult for nonspecialists to interpret. This is natural, but it has led to a communication gap between resource managers who utilize the results of spatial information processing and specialists concerned with the technical details of procedures to develop geographic information systems or to extract information. It is vital to link the information needs of land managers with the development of the geographic information system, yet there are many instances in which users of spatial information and developers of geographic information systems have a difficult time understanding and communicating with one another. It is important for all land managers who will be utilizing information produced via geographic information systems to understand the basic terminology involved; Congalton and Green (1992) provide an introduction to geographic information systems for resource managers, and Kessler (1992) has developed a glossary of terms that users of geographic information systems may encounter; Clarke (1996) also provides an excellent glossary. It is at least as important for the developers of geographic information systems to understand the information needs and concerns of resource managers.

Spatial information may consist of *points,* which define single locations, such as the geographic coordinates of a bald eagle nest or a water quality sampling station, *lines,* which are two-dimensional features, or *polygons,* which define the aerial extent of similar condition, such as a soil mapping unit or the watershed of a first-order stream. Information stored about points, lines, or polygons is referred to as the *attributes* of the spatial entity. An attribute may simply be a classification, such as the name of the mapping unit used to define a polygon, or the attributes may include more complete descriptive information that can be retrieved from a database, such as forest inventory data for a particular operational unit.

Variables used to define features or that are stored as attributes may be *continuous variables,* which assume values anywhere along a numeric scale, or *discrete variables,* which assume only a finite number of values. Discrete variables can be either numeric codes or nonnumeric classifications; they can be *nominal,* in which no value ranking is implied in the classification, or *ordinal,* in which value ranking is implied in the classification. An example of a nominal variable is vegetative association; an *Acer-Tsuga-Dryopteris* type is not "greater" than an *Acer-Viola-Ozmorhyza* type, although there may be associated attributes of nominal variables, such as the expected productivity of a particular forest species, which may imply a ranking along a numeric scale. An ordinal variable would be created if the habitat suitability index for a particular wildlife species was calculated and classified as high if the index was greater than 0.7, intermediate if the index was between 0.4 and 0.7, and low if the index was less than 0.4. A single point or polygon may have attributes of more than one type; attributes stored for a single soil mapping unit may include a taxonomic classification, which is nominal, a slope class, which is ordinal, and an average water holding capacity, which is continuous.

Geographic information systems differ in the way they store information. A system that stores the locations of points and the boundaries of polygons is referred to as a *vector-based* system. This is opposed to a system that is *raster-based,* in which an area is divided into very small units that are assigned to different polygons or that contain specified points. In raster-based systems, the boundaries between polygons are defined by differing classifications of neighboring units; in vector-based systems the location of points defining a boundary are stored directly. Locations of vectors or rasters are stored using some common coordinate system that may be used to link different types of information. Latitude and longitude, for example, form a familiar coordinate system for referring to spatially explicit information. Because of the curvature of the earth's surface, however, a degree of latitude or longitude is of different length at different locations on the earth's surface. There are numerous alternate coordinate systems that may be used to define spatial location. The *Universal Transverse Mercator,* or UTM, coordinate system is also commonly used to define geographic location. The UTM system avoids many of the problems associated with latitude and longitude. There are computational routines for converting from latitude and longitude to UTM coordinates.

STRUCTURE

Geographic information systems consist of many different types of information, which are stored as *layers* of information (Figure 13.1). A soils layer may define the boundaries of soil mapping units, a forest-cover-type layer may define the boundaries of forest cover types and contain forest inventory information stored as attributes of the different stands, a hydrology layer may define waterways and lakes, and so on. In some ways, a single layer may be thought of as being analogous to a paper map, but paper maps usually contain information stored in several layers of a geographic information system. Roads, political subdivisions, and waterways, for example, are on most paper maps used in resource management, but may be stored in three different layers of a geographic information system.

The selection of the types of information to include as layers in a geographic information system must be related to the information needed to support the management decision-making process. The specifics of which information to include in each layer, the number of layers, and data storage procedures should be left to technical experts, but users involved in management decision making must be involved in the process of selecting the information to include, the identification of information sources, the specification of acceptable error rates, and the design of reporting formats. The degree of utility of a geographic information system as a tool for supporting management decision making depends on the extent of cooperation between the technical personnel involved

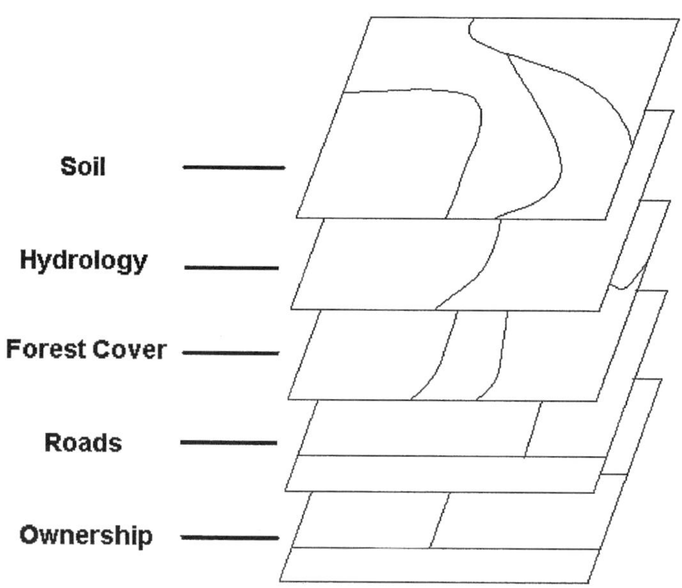

FIGURE 13.1 Structure of a basic geographic information system for use in land management.

in system development and the land managers making final use of the information.

The data entry process follows system design and the specification of the detailed information to be included. Information from existing maps can be entered into a geographic information system through a procedure called *digitizing*, which involves using electronic equipment to convert map features, such as point locations, lines, or polygon boundaries, into electronic coordinates through a process somewhat analogous to tracing. This process is tedious and involves time-consuming, labor-intensive handwork. An alternative is to electronically *scan* the maps using optical scanners, which are somewhat analogous to copying machines, and specialized software to convert the information from paper copies to digital computer files. Optical scanning works best when the source maps are of high contrast and without a lot of subtle shading or cross-hatching. Following scanning, visual inspection and manual editing is required to ensure that errors have not been introduced owing to the resolution or color separation of the optical scanner.

There are commercial sources of a great deal of natural resource information that, for a fee, provide data layers that can be directly integrated into a geographic information system. There are also an increasing number of public sources of digital natural resource information available either on optical disk or by electronic transfer over the Internet at low or no cost. The amount of available information is increasing rapidly, and any printed list is quickly out of date. Examples of available information include digital soil maps of selected areas that are available from the Natural Resources Conservation Service (formerly the Soil Conservation Service). Figure 13.2, for example, illustrates the soil associations present in Delta County, Michigan. In most cases, however, it is still necessary to digitize soil mapping information from county soil survey reports (see Chapter 4).

A *digital elevation model* is a three-dimensional representation of topographic information. The U.S. Geological Survey has developed digital elevation models of the United States that are electronically available to the general public (Figure 13.3). These are simply digitized versions of the USGS quadrangle map sheets that have been available in paper form for many years. The available map scale varies by location, but the resolution is improving with time.

Other sources of natural resource information that can usually be obtained in digital form are hydrological maps (Figure 13.4). Hydrological maps are usually available at the state level; in Michigan, the Department of Natural Resources maintains such data. Like the digital elevation data described earlier, hydrological data can often be retrieved electronically over computer networks. Other information that is not yet available for retrieval over computer networks is readily available on CD-ROM or other storage media. The U.S. Forest Service, for example, recently published a map of ecological units of the eastern United States, which is available in the format of widely used geographic information system software (Keys et al., 1995). For some areas

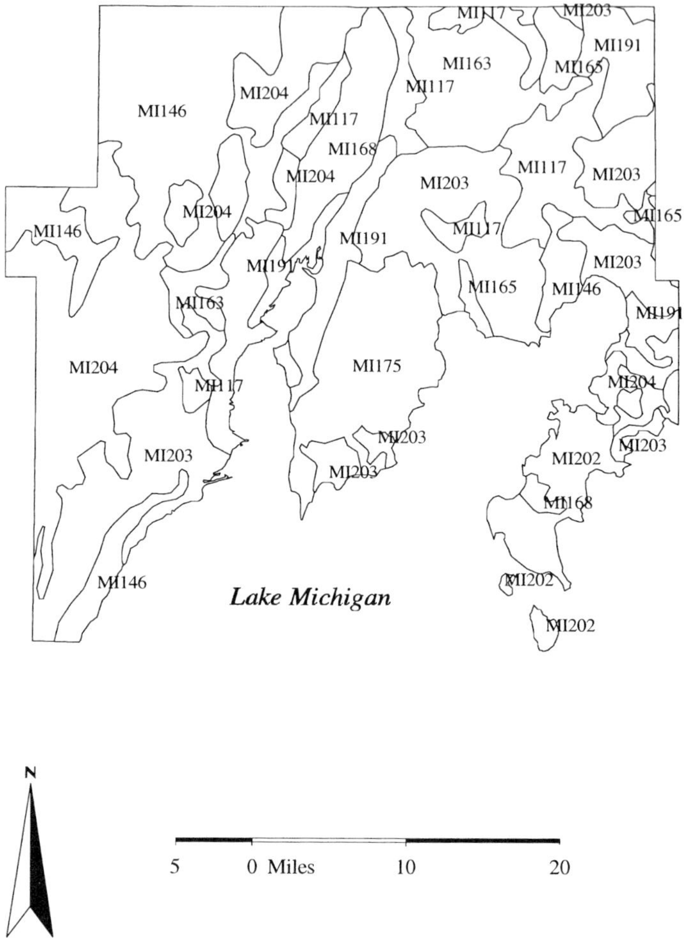

FIGURE 13.2 Digital map of soil associations for Delta County, Michigan (adapted from data available from the Natural Resources Conservation Service).

of the country, the U.S. Forest Service can provide digital maps of ecological landtypes and landtype associations (see Figure 7.3, Chapter 7).

Other vegetative and land use information is available from satellite imagery. There is now an extensive library of such imagery, primarily from the U.S. LANDSAT and French SPOT satellites, which is available for public purchase. These images extend back to the 1970s in some cases and can be valuable sources of information on land use and landscape change over the last couple of decades. The processing and interpretation of satellite imagery is not a trivial exercise and requires special expertise and techniques (e.g., Lillesand and Kiefer, 1994; Campbell, 1996).

There are many other types of information of value to land management. Many states provide digital maps of road networks and political boundaries. In addition, in many states there are public agencies or consortia that can provide other natural resource information. In Michigan, for example, the

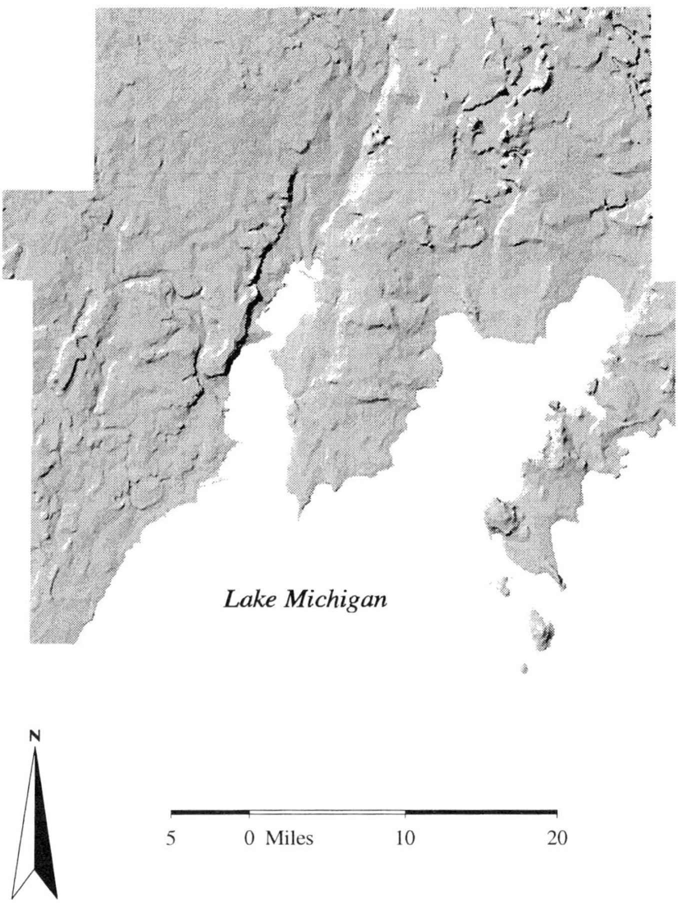

FIGURE 13.3 Digital elevation model of Delta County, Michigan (adapted from data electronically available from the U.S. Geological Survey).

state Natural Features Inventory is interpreting the original General Land Office survey notes (see Chapter 7) and developing maps of pre-European-settlement forest cover types. Printed versions of this type of map may be available at no cost, but digital versions of some of these types of data may be available only for a fee.

There is also a great deal of publicly available point data that can be obtained for low cost, or even no cost in some cases. The U.S. Forest Service, for example, has developed the Eastwide database (Hansen et al., 1992), which contains plot and tree records from the periodic Forest Inventory and Analysis (FIA) surveys of each state; there is a similar database for the western United States (Woudenberg and Farrenkopf, 1995). The spatial resolution of such data varies with the source. The plot locations in the Eastwide database are stored only to the nearest 100 seconds of latitude and longitude, for example.

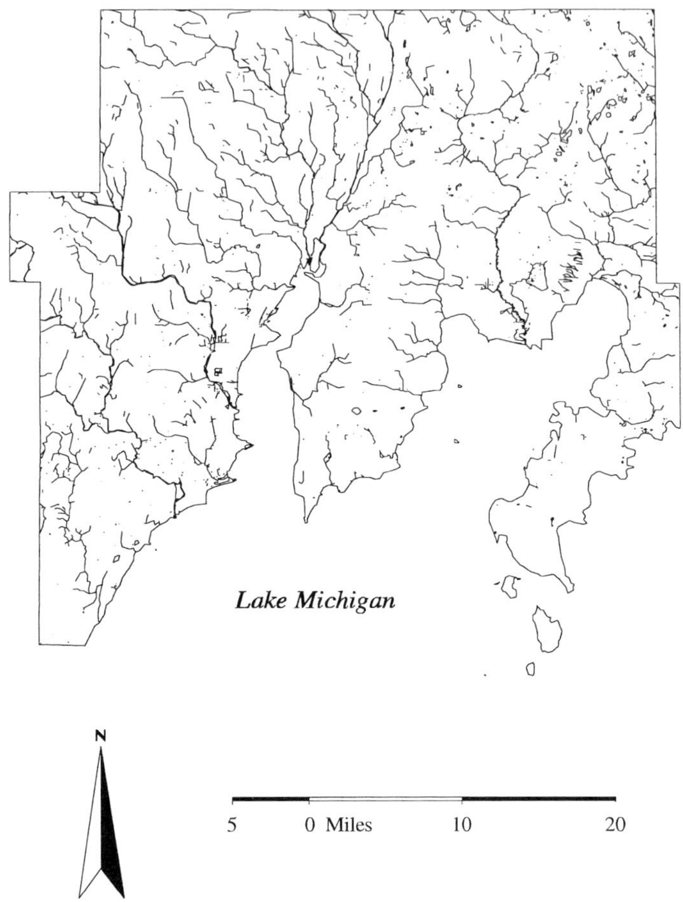

FIGURE 13.4 Hydrological data for Delta County, Michigan (adapted from information available from the Michigan Department of Natural Resources).

Clearly, there is a great deal of information of value to natural resource management and planning efforts that can be obtained from public sources at little or no cost. The exact types of information and available scales vary by location, but an extensive geographic information system for supporting landscape management can be assembled today in a relatively short amount of time at a much lower cost than ever before. There are still many technical challenges to linking information at various scales from various sources in digital form. The process of standardizing the geographical coordinate systems of different data layers is referred to as *georeferencing* and may require adjustment of coordinates on one or more data layers to ensure the compatibility of coordinate systems. Most geographic information systems contain routines to assist in georeferencing and coordinate adjustment, but these procedures must be used with care to avoid introducing excessive errors into the database. These procedures are necessary, however, to the process of geographic information system development and the linkage of different types of spatially explicit information to support resource management.

One example of a geographic information system designed to support landscape-level natural resource management is described in Table 13.1. This information was taken mostly from public sources and was assembled in about a year by a small group of people. The geographic information system containing these data layers was developed as a cooperative effort between university, public agency, and private corporate entities for the Upper Peninsula of Michigan. It provides a means of rapidly linking the various information sources to support landscape-level analyses of policy decisions or the consequences of implementing various management alternatives. Such public information can be linked with private, proprietary information, such as forest inventory records for a particular private ownership, to provide a customized tool for supporting management of private land. Similar efforts are under way in many areas, though access to the information may be limited in some cases.

ERRORS AND ACCURACY

There is a natural tendency to forget that information printed on full-color computer-generated maps is subject to some degree of error. Error in this

TABLE 13.1 Data Layers in a Geographic Information System Designed to Support Sustainable Landscape Management in Upper Michigan.

Information	*Source*
Abiotic Resources	
Base Map	Michigan Department of Natural Resources
Digital Elevation Model	U.S. Geological Survey
Soil Associations	U.S. Natural Resource Conservation Service
Hydrology	Michigan Department of Natural Resources
Biotic Resources	
Ecological Landtypes	U.S. Forest Service
Pre-European-Settlement Forest Cover	Michigan Natural Features Inventory
General Land Office Witness Tree Data	General Land Office Survey Notes
1991 Land Cover	Michigan Department of Natural Resources, Wildlife Division
1992 Forest Inventory and Analysis Data	U.S. Forest Service

context refers to the difference between the presented information and the true situation. It is important to understand the error associated with all information used to support management decision making; this is especially true in regard to maps, which provide one of the major formats for communicating the spatial information that is so critical in landscape management. Error is introduced into the layers of geographic information systems in many ways, several of which are summarized in the following paragraphs; Bolstad and Smith (1992) give a more detailed summary of the types and sources of error that are incorporated into all geographic information systems.

Positional Errors

Positional errors occur when the coordinates of a map feature differ from the true geographic coordinates of the feature. A measure of positional error is the *root mean square error*. Errors in the X-Y coordinate plane are referred to as horizontal errors, and errors in elevation are referred to as vertical errors. The horizontal root mean square error is given by:

$$RMSE_h = \sqrt{\frac{\left(\sum_{i-1}^{n} (x_i - X_i)^2 + \sum_{i=1}^{n} (y_i - Y_i)^2\right)}{\mathrm{n}}} \qquad [13.1]$$

where lowercase letters represent the values of the respective coordinates on the map or stored in the database, uppercase letters represent the true values of the respective coordinates, and n is the number of sample point locations used to evaluate error. The vertical root mean square error is similar, but based on only one coordinate:

$$RMSE_v = \sqrt{\frac{\sum_{i=1}^{n} (z_i - Z_i)^2}{n}} \qquad [13.2]$$

where z_i is the map or database elevation, Z_i is the true elevation, and n is the number of sample points used to evaluate error.

In the United States, national map accuracy standards, which apply to high-quality maps such as U.S. Geological Survey maps, require that 90% of the features on a map have positional errors of less than $1/50^{th}$ inch on maps with scales of less than 1:20,000 (Wolf and Brinker, 1989). On a map with a scale of 1:20,000, $1/50^{th}$, or 0.02 inch, corresponds to 400 inches or 33.3 feet on the ground. On maps with scales greater than 1:20,000, 90% of the points are required to have positional errors of less than $1/30^{th}$ inch. Thompson and Rosenfield (1971) indicate that approximately 83% of U.S. Geological Survey 1:24,000 maps met these standards; the remainder had greater errors in the positional location of well-defined features. Maps of lesser quality can be expected to have greater errors.

Additional positional errors can be introduced into computerized maps if existing paper maps are digitized. Digitizing equipment often has a resolution of about 0.001 inch, which, when coupled with errors introduced by the manual digitizing process, lead to average positional errors of about 0.0025 or 1/400th inch (Bolstad, Gessler, and Lillesand, 1990). On a map with a scale of 1 : 20,000, then, the process of digitizing leads to an average positional error of 50 inches (4.2 feet) on the ground, which is in addition to any errors that may be due to inaccuracies in the map being digitized.

Boundary Errors

The very nature of drawing a line on a map to indicate a boundary between polygons of different properties introduces an error into the location of the boundary. On a typical 1 : 20,000 scale soil map, a line .05 inch in width (the width of a fine drawing instrument) corresponds to a ground distance of 1,000 inches, or 83.3 feet. Even in the case where the boundary between two regions is inherently sharp, there can be considerable uncertainty in the location of the boundary when it is transferred to a digital map. These errors are introduced into digital maps even when the boundary locations are accurately mapped; additional error may be involved in determining the location of the boundary between polygons in the field. Prisley, Gregoire, and Smith (1989) found that the error in estimated polygon areas resulting from uncertainty in boundary location ranged from 5 to 15%. In many situations, boundaries occur across gradients or catenas and are not clearly defined. In these situations, the gradual changes between two different areas in the field result in additional error in the original demarcation.

Classification Errors

Classification errors result when a point or polygon is identified as being one type of entity when it is really another type. A classification error results if a polygon is identified as an upland deciduous forest, but it is really an upland mixed conifer-deciduous forest. Classification errors are usually presented and analyzed using an *error matrix* or, as sometimes referred to, a *confusion matrix* (Table 13.2). In an error matrix, the map classifications are usually presented as the rows and the true classifications are presented as the columns. A number of locations are visited on the ground, and both the map classifications and the true classifications are recorded; this process is known as *ground truthing*. Ground truthing may be a stratified sampling exercise (see Chapter 15), whereby samples are randomly or systematically selected from all categories of map classifications. In this case, the major objective would be to assess

TABLE 13.2 Error Matrix for the Classification of Forest Types from Satellite Imagery; Values Are Numbers of Visited Locations.

	True Forest Type						
	Upland			***Lowland***			
Mapped Forest Type	*Deciduous*	*Mixed*	*Conifer*	*Deciduous*	*Mixed*	*Conifer*	*Total*
Upland							
Deciduous	25	5	0	3	1	0	34
Mixed	4	19	2	0	2	0	27
Conifer	1	6	39	2	1	4	53
Lowland							
Deciduous	4	2	1	25	9	3	44
Mixed	3	6	0	6	18	7	40
Conifer	0	0	2	0	6	25	33
Total	37	38	44	36	37	39	231

errors within individual classes and to examine and possibly improve the classification process. From a map user's perspective, overall classification accuracy may be estimated by randomly or systematically selecting ground truth locations throughout the entire classified area. Bowden and Frayer (1985) present methods for assessing map accuracy and associated precision within a range of sampling designs. The following discussion assumes that simple random sampling (without replacement) was used to select sample locations for evaluating classification accuracy.

In the error matrix, the observations along the main diagonal are correctly classified and the observations off the main diagonal are incorrectly classified. The overall classification accuracy rate is usually expressed as the percentage of the ground truthed samples that are correctly classified on the map. This is calculated by summing the main diagonal elements from the error matrix and dividing by the total number of ground truth observations:

$$\text{Accuracy} = \frac{\sum_{i=1}^{t} \varepsilon_{ii}}{\sum_{i=1}^{t}\sum_{j=1}^{t} \varepsilon_{ij}}(100) \qquad [13.3]$$

where ε_{ij} refers to the number of observations in the ith row and j^{th} column of the error matrix and t is the number of different classes; the numerator is the sum of the diagonal elements and the denominator is the total number of ground truth observations. As noted earlier, error can be assessed on the basis of comparing the classification of selected point locations on the ground with their map classification. Given a representative sample, the proportion of points correctly classified represents the area of the map that is correctly classified. Error could also be evaluated on the basis of polygon comparison, whereby selected polygons are visited and classified; the error matrix is constructed by comparing the map classifications to the classes observed on the ground. The areas of the polygons are then usually used to construct the error matrix, rather than simply the number of polygons. There is an implicit assumption that the polygon boundaries are located without error in using this approach.

In Table 13.2, a simple random sample of point locations were visited and the error matrix was constructed. There were 34 locations classified as upland deciduous forest on the map; when visited on the ground, only 25 of these were actually upland deciduous forest and the remainder were misclassified. There were a total of 231 checked locations, 151 of which were correctly classified. Using Equation [13.3], the overall classification accuracy of this map is estimated to be:

$$\begin{aligned}\text{Accuracy} &= \frac{151}{231}(100)\\ &\approx 65\%\end{aligned} \qquad [13.4]$$

Usually, commercial mapping exercises are required to result in overall classification accuracies of 85% or better. Bolstad and Smith (1992) refer to the accuracy within a row as the *user accuracy* that can be calculated for each polygon class. Errors within the rows are sometimes referred to as *errors of commission.* User accuracy provides an indication of the percentage of area indicated on the map to be in a given category that is actually in the category. It is expressed as the percentage of the sample locations in the class that are classified correctly:

$$\text{User accuracy}_\text{i} = \frac{\varepsilon_{ii}}{\sum_{j=1}^{t} \varepsilon_\text{ij}}(100) \qquad [13.5]$$

For the error matrix in Table 13.2, the user accuracy calculated for the locations on the map classified as upland deciduous forest is given by the number of locations classified as upland deciduous forest that are classified correctly, divided by the total number of locations classified as upland deciduous forest:

$$\begin{aligned}\text{User accuracy}_\text{UDF} &= \frac{25}{34}(100)\\ &\approx 74\%\end{aligned} \qquad [13.6]$$

where the subscript UDF refers to upland deciduous forest.

Bolstad and Smith (1992) refer to the accuracy within the columns as *producer accuracy;* errors within columns are also sometimes referred to as *errors of omission.* Producer accuracy indicates the proportion of area of a certain type that is correctly classified on the map:

$$\text{Producer accuracy}_i = \frac{\varepsilon_{ii}}{\sum_{i=1}^{t} \varepsilon_{ij}}(100) \qquad [13.7]$$

For the example in Table 13.2, the producer accuracy for the upland deciduous forest class is estimated by dividing the number of upland deciduous forest locations that were correctly classified by the total number of upland deciduous forest areas visited on the ground;

$$\begin{aligned}\text{Producer Accuracy}_{\text{UDF}} &= \frac{25}{37}(100) \\ &\approx 68\%\end{aligned} \qquad [13.8]$$

where the subscript UDF, again, refers to upland deciduous forest.

Overall map accuracy can be quite high even if accuracy within some categories is low. Similarly, accuracy within a few categories may be quite high even if overall map accuracy is low. Accuracy is generally higher in maps with fewer, very distinct categories as compared with maps with many, finely distinguished categories.

It is possible to compare error matrices resulting from different classification systems. Congalton, Oderwald, and Mead (1983) present methods to statistically evaluate and compare error matrices. These procedures have great utility in evaluating different classification methods, but they are of less direct utility in a resource assessment exercise in which the emphasis is usually on the interpretation of spatial information rather than the comparison of classification schemes.

Attribute Errors

Attributes of points, lines, or polygons may also contain some error in their values. Forest overstory inventory information, for example, is only rarely based on complete enumeration; it is almost always derived from a sample (see Chapters 8 and 15). The properties of estimates derived from basic sampling designs are well characterized and can be used to derive the variance of the estimate or associated confidence intervals (see Chapter 15). These errors are based on statistical properties of the estimates, and it is a reasonably straightforward process to estimate them for most basic sampling designs. Some attribute errors can, therefore, be quantified during the analysis of field sampling results. Attribute errors can often be quantified for individual polygons or points, whereas the classification errors discussed earlier are difficult to define and estimate for individual polygons without direct observation.

Some types of attribute errors, however, are difficult to quantify for individual points or polygons. Examples include attributes that are derived from published sources or those based on interpretive information derived from a classification system. An example is the interpretive information obtained from soil maps. As noted in Chapter 4, the Natural Resource Conservation Service soil survey reports contain a great deal of interpretive information for each mapping unit concerning properties associated with recreational development, wildlife habitat, building site development, and so on. Confidence limits or variability may not be given for this interpretive information, and, even if variability is given, it is difficult to evaluate for a single polygon. The user accuracy defined in the preceding section can be used to estimate the probability that a polygon with a particular classification is really a member of that class, but there is usually no linkage with the uncertainty of interpretive information associated with that class.

One method of addressing this problem is to examine the classification error matrix to understand the types of classification errors that occur. It may be that a particular class is often confused with classes that have similar interpretive properties. A coniferous wetland, for example, may have a greater likelihood of being misclassified as another wetland type than as a drier upland type. In this case, interpretive information, such as engineering or hydrologic properties, may be similar for the class in question and the classes with which it is likely to be confused, resulting in a lower likelihood of attribute error than classification error. In other cases, however, it will be difficult to place any quantitative value on the likely error rate associated with the attributes of points and polygons. In all cases, if there is a particular area for which it is very important to have precise attribute estimates, some sort of field sampling procedure should be implemented to verify interpretive information from published sources. It is then possible to estimate the variability or place confidence limits on particular attributes associated with the point or polygon in the geographic information system.

Error Propagation

Each layer of information in a geographic information system has an associated error, as discussed earlier. For a layer with classified polygons, such as forest overstory cover types or soil mapping units, the error is usually expressed as the percentage of the area that is correctly classified (Table 13.2). When data layers are combined, the resulting cross-classification also has an associated error rate that may be derived from knowledge of the variability of each layer and the statistical relationship between information in the different layers. The combined error rate may be estimated, as may the error rate of the individual layers, through field sampling and comparison of the field classification with the map classification. Probabilistic estimates of the combined error rates may also be developed. This is accomplished by defining an artificial random variable, I_i, which takes the value of one (1) if polygon i is classified correctly and a value of zero (0) if the polygon is classified incorrectly. Over all classes, the probability that $I_i = 1$ (denoted as $P(I_i = 1)$) is equivalent to the overall classification accuracy in Table 13.2; for a polygon in a given class, the probability that it is correctly classified is equivalent to the user accuracy (Equation [13.5]). The likelihood that the information from combined data layers is all correct may be derived from basic probabilistic relationships. For two independent data layers, the probability of simultaneous correct classification in both data layers is given as:

$$P(I_1 = 1 \text{ and } I_2 = 1) = P(I_2 = 1) \cdot P(I_2 = 1) \qquad [13.9]$$

If the first data layer has an overall classification accuracy of 85% and the second data layer has an overall classification accuracy of 90%, then the estimated proportion of the map combining the data layers where both are correctly classified is simply:

$$\begin{aligned} P(I_1 = 1 \text{ and } I_2 = 1) &= 0.85 \cdot 0.90 \\ &= 0.765 \text{ or } 76.5\% \end{aligned} \qquad [13.10]$$

If two independent data layers each have a 90% classification accuracy, then 81% of the map area will be expected to be correctly classified in both data layers. This can be expanded to any number of independent data layers:

$$P(I_i = 1 \ \forall\ i) = \prod_{i=1}^{n} P(I_i = 1) \qquad [13.11]$$

if n data layers are being combined. The likelihood that at least one data layer contains a classification error increases rapidly as more data layers are combined. Even if all data layers have 90% classification accuracies, the likelihood that the classification in every data layer is correct declines to less than 50% when seven data layers are combined:

$$\begin{aligned} P(I_i = 1 \ \forall\ i, i = 1,2, \ldots ,7) &= \prod_{i=1}^{7} 0.90 \\ &= 0.90^7 \\ &= 0.478 \text{ or } 47.8\% \end{aligned} \qquad [13.12]$$

In practice, information in different data layers may not be independent. Vegetative association, for example, is at least partially dependent on the soil mapping unit. To derive the likelihood that both data layers are correctly classified in this situation requires the specification of the conditional probability that one of the data layers is correct, given that the other is correct:

$$P(I_1 = 1 \text{ and } I_2 = 1) = P(I_2 = 1) \mid I_1 = 1) \cdot P(I_1 = 1) \qquad [13.13]$$

If the overall classification accuracy of the first data layer is 80%, and if the conditional accuracy of the second data layer is 90% (i.e., the second data layer has a 90% classification accuracy if the first data layer is correctly classified), then the estimated combined classification accuracy is:

$$\begin{aligned} P(I_1 = 1 \text{ and } I_2 = 1) &= 0.90 \cdot 0.80 \\ &= 0.72 \text{ or } 72\% \end{aligned} \qquad [13.14]$$

When three data layers are combined, the combined classification accuracy is derived as:

$$\begin{aligned} P(I_1 = 1 \text{ and } I_2 = 1 \text{ and } I_3 = 1) = P(I_3 = 1 \mid I_1 = 1 \text{ and } I_2 = 1) \\ \cdot P(I_2 = 1 \mid I_1 = 1) \cdot P(I_1 = 1) \end{aligned} \qquad [13.15]$$

This expression can be extended to derive the probability of error-free classification for any number of combined data layers. If the layers are not independent, estimation of the conditional probabilities is often difficult. Field sampling to estimate these probabilities involves finding numerous locations for each combination of classes from the different layers in order to evaluate the classification errors. It is important to note that there may be dependencies between data layers that are not obvious from their content; development of a pre-European-settlement forest cover map may utilize soil and landform maps to aid in interpolation between section corners in the General Land Office survey notes. Soil maps, in turn, are developed with the aid of information on landform and current vegetative cover obtained from aerial photographs. Thus, information in several data layers (soil, landform, current vegetative cover, and pre-European-settlement forest cover) may all be linked to a common set of aerial photographs that have been interpreted by different personnel for different reasons.

The probabilistic approach is appropriate when the error rate associated with membership in different classes is being examined. Sometimes it is necessary to combine attributes stored as continuous variables to develop indices or variables to support management decision making. Habitat suitability indices, for example, may be calculated using attributes associated with different data layers (see Chapter 9). When linear combinations of attributes from different layers are calculated, the attributes can be viewed as random variables for the purpose of estimating the variance of the resulting variable. A linear combination is simply a linear function of the different attributes:

$$U = \sum_{i=1}^{t} a_i X_i \qquad [13.16]$$

where U is the combined attribute, X_i is the ith attribute incorporated in the combination, and a_i is the linear coefficient of the ith attribute. If two attributes are being added together, for instance, then both coefficients are equal to one:

(*a*)

FIGURE 13.5 Overlays of geographic information system layers for an area in Delta County, Michigan: (*a*) location of two ecological landtype associations on a peninsula extending into Lake Michigan; (*b*) combination of soil association and the two ecological landtype associations; (*c*) pre-European-settlement forest cover for the southernmost ecological landtype association; and (*d*) 1991 LANDSAT cover classification for the southernmost ecological landtype association.

$$U = X_1 + X_2 \tag{13.17}$$

If one attribute is being subtracted from the other, a_2 is equal to negative 1:

$$U = X_1 - X_2 \tag{13.18}$$

Estimation of the variance of U is dependent on knowledge of the variance associated with each X_i and the covariance (see Chapter 15) between each pair of attributes being combined:

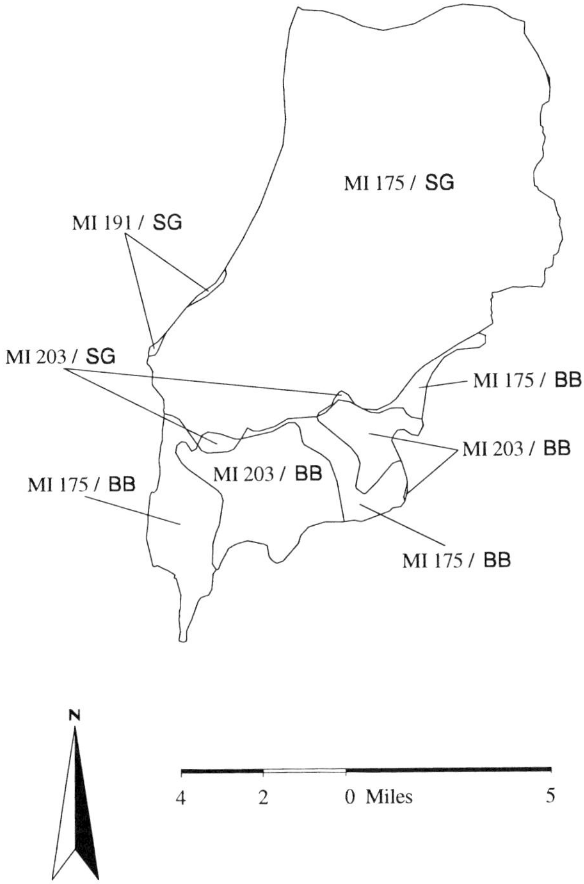

FIGURE 13.5 (*Continued*)

$$\sigma_U^2 = \sum_{i=1}^{t} a_i^2 \, \sigma_i^2 + 2 \sum_{i=2}^{t} \sum_{j=1}^{i-1} a_i a_j \, \sigma_{ij} \qquad [13.19]$$

where σ_U^2 is the variance of the combination of attributes, σ_i^2 is the variance of the ith attribute, and σ_{ij} is the covariance between X_i and X_j. Note that if X_i and X_j are independent, then σ_{ij} is equal to zero and σ_U^2 is equal to the sum of the $a_i^2 \, \sigma_i^2$. For the sum of two attributes, as in Equation [13.17], the variance is equal to:

$$\begin{aligned} \sigma_{X_1+X_2}^2 &= (1)^2\sigma_1^2 + (1)^2\sigma_2^2 + 2(1)(1)\,\sigma_{ij} \\ &= \sigma_1^2 + \sigma_2^2 + 2\,\sigma_{ij} \end{aligned} \qquad [13.20]$$

whereas the variance of the difference between two random variables (Equation [13.18]) is given by:

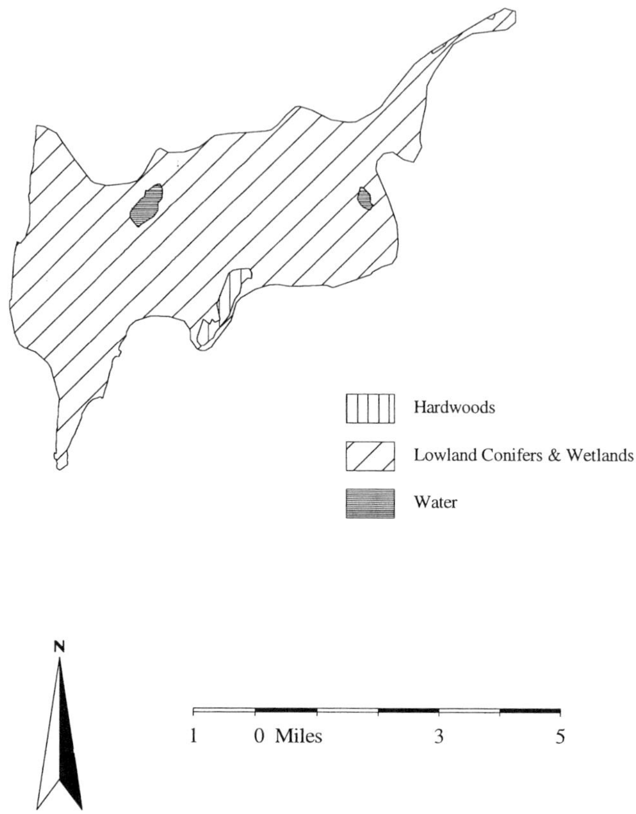

(*c*)

FIGURE 13.5 (*Continued*)

$$\sigma^2_{X_1-X_2} = (1)^2\sigma_1^2 + (-1)^2\sigma_2^2 + 2(1)(-1)\ \sigma_{ij}$$
$$= \sigma_1^2 + \sigma_2^2 - 2\ \sigma_{ij} \qquad [13.21]$$

If X_1 has a variance of 5, X_2 has a variance of 8, and if the covariance between X_1 and X_2 is 2, then the variance of the sum of X_1 and X_2 is given by Equation [13.20]:

$$\sigma^2_{X_1+X_2} = 5 + 8 + 2\ (2)$$
$$= 17 \qquad [13.22]$$

The variance of the difference between X_1 and X_2 (Equation [13.21]) is given by:

$$\sigma^2_{X_1-X_2} = 5 + 8 - 2\ (2)$$
$$= 9 \qquad [13.23]$$

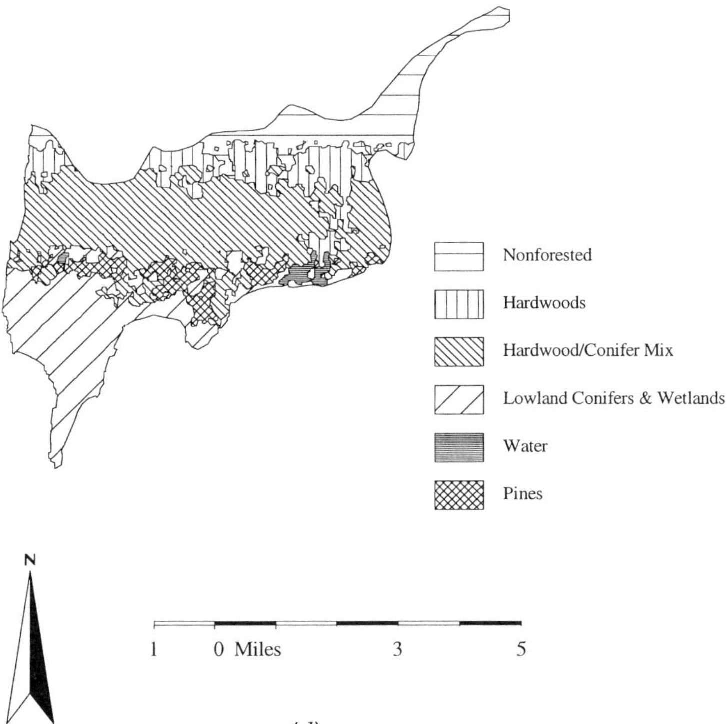

(*d*)

FIGURE 13.5 (*Continued*)

As with any single random variable, the variance of a linear combination of random variables from Equation [13.19] may be used to derive confidence limits for U as described in Chapter 15. Although not being an "error" in the sense of a mistake in classification of a polygon, this is a measure of the uncertainty of the value of a continuous variable derived from different attributes from one or more data layers in the geographic information system.

ANALYSIS

Geographic information systems can greatly speed the production of customized maps of land management compartments and landscapes, which can save a great deal of time and money. But the potential of a geographic information system as an aid to the land management decision-making process goes far beyond the production of maps. Combining information from different data layers to perform analyses of resource characteristics and the consequences of management alternatives can lead to much greater insight than can be derived from simple maps. Many land management agencies invest a great deal of time and money in developing geographic information systems, but experience a natural letdown after the initial push to develop the system. At this point, some within the organization may question the value of the entire effort. Advancement to the stage of utilizing the geographic information system as an analytical tool to support management decision making is often organizationally difficult and not always immediately successful. Once the system is developed, however, the additional value of advancing to the analytical stage of use can be very great.

To *overlay* two data layers is to combine the two sets of information to create a new layer containing polygons that are combinations of those in the original two layers. Figure 13.5 illustrates several examples of the combination of different data layers in a geographic information system. Two ecological landtype associations from a portion of Delta County, Michigan, are illustrated in Figure 13.5a. The soil associations for the two ecological landtype associations are shown in Figure 13.5b; the pre-European-settlement forest cover information and the 1991 Landsat classification of current land cover are shown for the southernmost landtype association in Figures 13.5c and 13.5d, respectively. Using similar procedures, a data layer consisting of point locations could be combined with one describing spatial polygons.

The most basic numeric summary is a *crosstabulation,* which gives the joint distribution of area or an attribute within two or more data layers. Table 13.3 contains a crosstabulation of the area of different forest overstory cover types by ecological landtype association derived from an overlay similar to the one in Figure 13.5. The column or row totals provide the marginal distribution of the respective single data layers. If forest overstory inventory information is available as attributes in the forest cover data layer, a crosstabulation of the volume of different species by ecological landtype association may be easily developed (Table 13.4). Crosstabulations of attributes within a single data layer could be completed in a geographic information system, but it may be simpler to use a database program since this does not require spatial information.

TABLE 13.3 Crosstabulation of Area (Hectares) for Selected Combinations of Landtype Association and Forest Cover Type.

Landtype Association	Forest Cover Type			
	Maple–Birch	*Jack Pine*	*Mixed Pine*	*Spruce–Fir*
A	3,500	175	1,205	525
B	496	78	678	2,586
C	127	5,476	978	276
D	4,387	93	597	175
E	676	1,297	1,543	787

TABLE 13.4 Crosstabulation of Commercial Timber Volume (m^3) for Selected Combinations of Landtype Association and Species.

Landtype Association	Species			
	Sugar Maple	*Quaking Aspen*	*Jack Pine*	*Balsam Fir*
A	140,100	19,200	9,600	9,300
B	9,900	62,000	32,800	66,300
C	1,900	32,500	167,500	33,800
D	175,400	4,700	8,800	4,300
E	13,400	69,200	52,500	17,400

Overlays and crosstabulations are simple descriptions of information combined from two or more layers. Summation is the only mathematical operation used to develop crosstabulations like those in Tables 13.3 and 13.4. It is possible to do calculations or crosstabulations involving more complex combinations of attributes. If the necessary information is stored as attributes, for example, habitat suitability indices for selected wildlife species can be calculated for every location in the landscape. For the example in Chapter 9, the four variables required to estimate the habitat suitability index for fisher are percentage of canopy closure, mean diameter at breast height of overstory trees, tree canopy diversity class, and percentage of the overstory consisting of deciduous species. These variables are either available in most forest inventory information or can be obtained using aerial photo interpretation techniques (Thomasma, 1988). If these variables are stored as attributes of one or more data layers, they can easily be combined into a habitat suitability index in most geographic information system software packages.

More complex analyses of spatial information may be made using specialized routines available in most geographic information system software programs. A data layer containing the location of rivers and streams can be combined with a forest overstory cover layer. Then, summaries of the area of each forest type within some fixed distance, such as 100 meters, of a stream can be developed and economic analyses of the financial consequences of streamside set-aside zones to protect riparian habitat and water quality may be developed. This can be combined with a property ownership data layer to determine the landowners, public or private, who will bear the financial burden of a policy decision to require such set-asides. This information may be used to develop landowner assistance or compensation programs to design policies to protect societal resources, such as water quality, which are equitable for individuals directly impacted by such policies. A geographic information system is a critical component of the analyses required to balance the costs and benefits of such policy decisions.

Geographic information systems are also ideal tools for conducting complex analyses of landscape structure. Calculation of the total length of edge within a landscape, or the length of edge between different types of habitat, such as forest and nonforest areas, can provide critical information regarding habitat availability for some wildlife species. Specialized programs, such as the FRAGSTATS program developed by McGarigal and Marks (1995), can be used to estimate aspects of landscape structure, such as area-perimeter ratios for different classes of polygons or the fractal dimension (see Chapter 7) of polygons in one data layer or in an overlay of two or more data layers.

PROJECTIONS

Analytical procedures such as crosstabulations and overlays provide a great deal of insight into processes operating at the current time within the landscape. This capability can be extended by incorporating a means of projecting resource characteristics into the future. The time increments used for these projections are called *projection intervals.* A forest growth model, for example, may be linked with forest inventory information to project forest composition and structure through time. Harvesting activity may be simulated by altering the structure or composition of the overstory at the beginning of a projection period. The resulting projected changes in leaf area and transpiration may be linked to a hydrologic model to examine changes in stream flow over time as a consequence of harvesting and regrowth of the forest overstory. Climate simulation routines, such as those developed by Desanker and Reed (1992), may be linked to these analyses to estimate the year-to-year variability in water yield under different management alternatives.

Projections supporting the evaluation of management alternatives should be explicitly linked to management goals and objectives. For instance, a management objective may be to manipulate landscape structure to the point where the fractal dimension of the forest-cover-type landscape elements approximates the value found in natural landscapes. To determine the degree to which this objective is achieved, the results of planned vegetation manipulations must be coupled with planned changes in the sizes and shapes of management units under each management alternative being considered. Computational routines can then be used to evaluate the fractal dimension of polygons developed from overlays of land use and vegetation layers at different lengths of time into the future.

Landscape-level impacts of implementing different management alternatives may be estimated through analyses linking models projecting different landscape components. The cumulative impacts of a management alternative specifying 60-year commercial timber rotations on all industrial ownership within a watershed may be compared with an alternative specifying 40-year rotations on highly productive sites and 90-year rotations on sites that are less productive; habitat suitability indices for selected wildlife species or the frequency of streamflow volume exceeding some threshold under each alternative may be estimated. In this way, the degree to which each management goal is satisfied may be evaluated for every management alternative.

SUMMARY

The development of geographic information systems has greatly improved the technology available for manipulation, analysis, and synthesis of spatially explicit information from diverse sources. This capability is important solely from the perspective of the ability to customize maps of natural resources and management areas. Of even greater potential importance is the capability to develop crosstabulations of area or attribute values by combining different sources of information or data layers. The insights developed from such summarizations of current landscape characteristics are of great value in the development and evaluation of management alternatives during the management planning process. Linkage with models for projecting different landscape or resource components through time provides both visual and quantitative analysis of the consequences within the landscape for the various management alternatives contemplated.

APPLICATIONS

13.1 Obtain a soil map for a county in your state and answer the following questions:

a. What is the scale of the map?
b. What is the width of the lines denoting boundaries between soil mapping units? What is the ground distance covered by this line width?
c. Apply a dot grid to the map. What is the average percentage of the total soil map area that is beneath boundary lines denoting the soil mapping units?

13.2 Consider three different data layers (A, B, and C), where A has a classification accuracy of 92%, B has a classification accuracy of 76%, and C has a classification accuracy of 82%.

a. What proportion of an overlay of layers A and C is expected to contain an error in classification?
b. If the classification accuracy of layer B improves to 88% if layer A is classified correctly, what is the classification accuracy of an overlay of layers A and B?

13.3 Consider three continuous variables stored as attributes of polygons in a geographic information system. If the first variable (X_1) has a variance of 12, the second variable (X_2) has a variance of 6, and the covariance between the two variables is 5, what is the variance of the following:

a. $X_1 - X_2$
b. $3 \cdot X_1 + X_2$
c. If there is a third variable (X_3) with a variance of 9 that is independent of both X_1 and X_2, what is the variance of:

$$X_1 + X_2 - X_3$$

13.4 Find examples of models for projecting the following resources through time that are applicable to your region:

a. Forest overstory
b. Hydrologic yield from a watershed
c. Wildlife habitat

13.5 For a forested landscape, describe the analyses and projections necessary to determine the following:

a. Present net worth of forest management activity
b. Wildlife habitat (as indicated by habitat suitability index for a species you select)
c. Streamflow
d. Soil erosion (or stream sedimentation)

Chapter

14

Integrated Environmental Assessment

Formal requirements for the development of environmental impact statements for federal actions were established by the National Environmental Policy Act of 1969. Section 102 (C) of this Act states that all agencies of the Federal Government shall:

> include in every recommendation or report on proposals for legislation and other major federal actions significantly affecting the quality of the human environment, a detailed statement by the responsible official on:
>
> i. the environmental impact of the proposed action,
> ii. any adverse environmental effects which cannot be avoided should the proposal be implemented,
> iii. alternatives to the proposed action,
> iv. the relationship between local short-term uses of man's environment and the maintenance and enhancement of long-term productivity, and,
> v. any irreversible and irretrievable commitments of resources which would be involved in the proposed action should it be implemented.

This legislation not only established the requirement to develop environmental impact statements, but essentially provides a reporting format that is useful in many resource assessment efforts. As noted in previous chapters, some individual states have enacted similar legislation applicable to state agencies. While formal environmental impact statements are usually not required for resource management activity on private land except in special circumstances, many types of management activities on private lands do require permits that depend to some extent on an environmental assessment process. This is particularly true for activities that may impact water quality. In addition, many private organizations are beginning to utilize resource assessments and impact analyses as a means to demonstrate the sustainability of their resource management and utilization activity.

Resource assessment activities may be conducted at many spatial scales. As originally envisioned and as most often encountered, formal environmental impact statements usually deal with site-specific impacts of particular projects, such as the location of highway corridors. The original concept included larger-scale projects, such as reservoir construction or resort development. In resource management applications, there may not be a specific action, such as a construction project, upon which to focus in the evaluation process. Also, the management decision is usually not whether to initiate a specific project, but whether to continue management activity as currently implemented or to change resource allocation in order to better meet management objectives while maintaining ecological integrity and sustainability. In recent years the concept has been extended even further; the State of Minnesota, for example, recently completed a Generic Environmental Impact Statement (GEIS), which addressed the impacts and levels of timber harvesting on a statewide level (Kilgore, Ellefson, and Rose, 1996). In a similar, though usually less extensive, vein, efforts are under way to develop and implement programs to certify that specific land-management organizations are practicing sustainable management. These efforts may include labeling of commercial products to indicate that they were produced in a sustainable manner, with all due consideration to long-term ecosystem health and viability.

Resource assessment documents supporting all of these types of analyses differ in exact outline, depending on the relevant regulations or purposes. Almost all resource assessments, however, address several general topics:

1. The description of the area of concern and associated resource status and condition;
2. A statement of management goals and objectives with a description of possible management alternatives;
3. An analysis of the anticipated success of the different management alternatives in meeting management goals and objectives;
4. An analysis of the possible impacts of each alternative on the resources of the area of concern; and, finally,
5. The selection of a preferred alternative designed to satisfy management goals and objectives to the maximum extent possible while minimizing negative impacts.

DESCRIPTION OF RESOURCE STATUS AND CONDITION

Most of the preceding chapters have concentrated on describing methods of assessing the status and condition of various components of forested landscapes. Along with physical descriptions of current status, a good resource assessment will also describe the limitations placed on the development of management alternatives by the different landscape components. The presence of hydric soils, for example, as noted in Chapters 4 and 6, places significant restrictions on the range of possible management actions. The state of the forest overstory not only places constraints on current management alternatives, but will impact management options for decades into the future. And, of course, the social implications of different management alternatives are extremely important to consider in evaluating their effectiveness in achieving management goals and objectives.

The presentation of material describing the status and condition of different landscape components is a critical element of every resource assessment. Each assessment, however, will not necessarily place the same emphasis on each landscape component. In some cases, water quality or water yield may be of primary importance, as may be the case in the management of a municipal watershed. In other cases, maintaining or enhancing habitat for an endangered species may receive more emphasis. In many situations, the most detailed information that will be presented will concern the forest overstory and the spatial distribution of different vegetative communities in the landscape. These resources are critical for commercial reasons, but also form the essential characteristics of wildlife habitat.

There is no single outline or reporting format that will always be ideal for describing landscape resources and developing management alternatives, but there is a logical progression from the abiotic features of the landscape, to the nonhuman biotic components, to the human impact on, and utilization of, the various resources within the landscape. This general approach will almost always provide an acceptable organization of knowledge about the landscape upon which to base management decisions, though the major emphases will vary from place to place, as some resources are more limiting or more critical to maintaining landscape structure and function. The components listed in Table 14.1 will all be considered and described in most resource assessment efforts. It is important to remember that the most critical landscape elements, and the level of informational detail required for each in order to support management planning, is not constant. Nor is this an exclusive list, as other factors may be critical in the management of some landscapes. As a general starting point, however, Table 14.1 will usually provide an acceptable outline, which may be modified as circumstances demand.

DESCRIPTION OF MANAGEMENT ALTERNATIVES

A *management alternative* is a scenario of resource management activities that satisfies the constraints placed on management by the abiotic, biotic,

TABLE 14.1 Example Organization of a Resource Assessment Report, with Concentration on Evaluation of Landscape Components.

Element	*Section*
I. Introduction	A. Location and Legal Description B. Management Objectives and Goals
II. Background	A. Ownership B. Current Land Use C. Issues, Concerns, and Opportunities
III. Abiotic Resources	A. Geology* i. Sources of Information ii. Data Collection Methods iii. Description iv. Management Constraints B. Soils C. Climate D. Hydrology
IV. Biotic Resources	A. Vegetative Communities B. Forest Overstory C. Wildlife Habitat
V. Social Context	A. Legal and Regulatory Framework B. Historic and Prehistoric Land Use C. Recreation and Aesthetics
VI. Summary	

*Subsections within the sections are not given for every landscape component, but would be similar to those given for geological resources.

and social characteristics of a landscape and that achieves, to some extent, one or more management goals. The development of management alternatives is a complex exercise that involves balancing biological potential, landowner goals and objectives, and social and political realities. The methods and procedures for developing management alternatives are beyond the scope of this text. There are, however, some general requirements that must be met for every resource assessment effort prior to the formulation of management alternatives. Foremost among these is a clear statement of the management goals and objectives for the landscape, or operational unit within the landscape, that is the subject of the management planning effort. A *management goal* is a general statement describing a desired outcome of management activity, such as the maintenance and improvement of water quality. A *management objective* is a specific statement that provides measurable criteria of success in achieving a management goal, such as a 20% decrease in stream sedimentation from erosion due to management activity within a watershed. It is also critical to identify the relative importance of each specified management goal or objective; Table 14.2 contains an example set of statements of management goals along with an indication of the relative importance of each. It is important to note that judgment of the relative importance of different management goals is a subjective procedure, which may become the focus of extensive debate among various constituencies utilizing the landscape.

The description of resource status and condition (Table 14.1) identifies critical processes within the landscape and characterizes the landscape components. The description of resource status and condition also identifies constraints placed on possible management actions by the various landscape components. These constraints may take many forms:

1. Physical constraints, such as those forced by topography or hydrology;
2. Biological constraints, such as those caused by the successional stages of the vegetative communities within the landscape;
3. Financial constraints, such as those resulting from the cost of land ownership; and
4. Legal or regulatory requirements and restrictions, which may be caused by (a) specific landscape components, such as endangered species habitat; (b) permitting procedures, such as those required to ensure the maintenance of water quality; (c) access requirements, which may be mandated by state commercial forest management acts; or (d) any of a number of other regulatory requirements that may be placed on resource management within a particular landscape.

The description of resource status and condition must explicitly identify all of the constraints that must be satisfied by any management alternative to be implemented. In the context of public land management, there may also be a public review period that serves to identify critical social and cultural

TABLE 14.2 Examples of Management Goals and Objectives of a Public Land Management Agency, with the Relative Importance of Each, for an Example Forested Landscape; the Goals and Relative Weights Assigned to Them Will Vary by Organization and Situation.

Goal	*Relative Weight (%)*
1. Protect threatened and endangered species	20
2. Maintain/enhance biodiversity	20
3. Maintain long-term soil productivity and nutrient cycling	15
4. Enhance wildlife habitat	15
5. Achieve acceptable financial returns	15
6. Provide dispersed recreation opportunities	10
7. Produce commercially valuable commodities	5
	100

management issues and that may provide input into the ranking of the importance of different issues identified in the resource assessment effort. Following this, it is necessary to develop a number of viable management alternatives that both satisfy the constraints on management and meet, to some extent, the identified management goals and objectives. The different goals and objectives will not be met to the same extent by all possible alternatives. Each viable management alternative, however, must meet the constraints placed on management by the abiotic, biotic, and social landscape components. Alternatives that satisfy constraints must be evaluated and compared. The alternative which, in the judgment of management personnel, best achieves the management goals and objectives will be identified as the *preferred alternative* during the decision-making process. It is important to emphasize the fact that selection of the preferred alternative is a subjective process. There will be objective input concerning the success of the different alternatives in achieving the specified management goals and objectives, but the final selection of the preferred alternative is based on the judgment of the manager or management team. Resource assessment procedures should be designed to provide complete, high-quality information for managers to consider during the decision-making exercise.

There is usually no formal specification of a fixed number of management alternatives that must be considered during the planning process. There are also no formal procedures for the development of management alternatives to be considered, though in public land management there may be an expectation of formal or informal input by the public into the development of alternatives to be considered. Formal public involvement may take several forms, ranging from participation in public hearings and the provision of written comments to membership on advisory committees, reviewing management alternatives and analyses developed by agency personnel.

One widely used approach to the development of management alternatives for consideration in the planning process is to develop a set of alternatives, each of which optimizes the achievement of one of the specified management goals or objectives. A second set of alternatives is often developed that utilizes some combination of aspects of the previous set to maximize the likelihood that some combination of the management goals will be realized. A final alternative, which is almost always included, is the "business-as-usual" alternative, or a management scenario that simply continues current management practice.

ANALYSIS OF MANAGEMENT ALTERNATIVES

Following the development of the management alternatives to be considered during the decision-making process, it is necessary to analyze each alternative in regard to its:

1. Level of success in achieving each management goal and objective;
2. Level of success in satisfying the constraints placed on management by the abiotic, biotic, and social characteristics of the landscape; and
3. Level of impact on resource status and condition within the landscape, including the impact on landscape processes and functions that will result if the alternative is implemented.

Management is successful only to the degree that the management goals and objectives are achieved. Most successful scenarios will achieve the various management goals and objectives to different degrees. Unfortunately, it is not possible to maximize the achievement of more than one management goal at a time; often, the most successful management scenarios fail to optimize any single management goal.

The constraints placed on management by the abiotic, biotic, and social components of the landscape are hard constraints; a management scenario that contains elements that violate the Clean Water Act, for example, cannot be implemented. A management alternative is not viable if it violates constraints placed on management activity by any of the various landscape components. There will be a wide range of alternatives, however, that do not violate the constraints on management and that achieve differing mixes of the management goals and objectives. These alternatives will also have a wide range of impacts on the landscape components, even though they may all satisfy the hard constraints on management. It is important, therefore, to develop qualitative or quantitative estimates of both the degree of achievement of the different management goals and objectives, as well as the expected impacts on the various landscape components of every considered alternative.

The analysis of management alternatives must include qualitative or quantitative statements of the degree to which each candidate management alternative succeeds in meeting each of the stated management goals and objectives. If a management goal concerns financial returns, for example, financial criteria such as the expected present net worth or rate of return on investment may be calculated for the landscape or operational area in question under each candidate alternative; the management objectives provide specific target levels of performance for the chosen metrics. Table 14.3 contains an example of the type of financial information that may be developed for each candidate management alternative; the particular financial criteria used to support the decision-making process may differ by organization. Realized returns from the selected alternative may be used as a benchmark in future planning efforts.

Analyses may be quite explicit for some resources; if there is a sensitive soil mapping unit, the number of hectares of that mapping unit that will be disturbed under each management alternative should be noted, and any necessary procedures to minimize impacts to long-term soil productivity

TABLE 14.3 Example of Financial Criteria That Might Be Used in the Evaluation of Management Alternatives.

	Management Alternative			
Criteria	*1*	*2*	*3*	*4*
Present Net Worth per Unit Area	$ 100	$ 70	$ 90	$ 130
Internal Rate of Return on Inventory	8%	6%	9%	7%
Value of Inventory per Unit Area at End of Period	$ 300	$ 200	$ 150	$ 400

should be clearly stated. Table 14.4 contains an example of the type of quantitative information that might be developed during the analysis of the impacts of the different management alternatives on soil resources. Similar tables would be developed for measures relative to different landscape components such as water quality, habitat of game and nongame species, and so on. The measures used to quantify performance of the various alternatives should be selected to address the critical management issues and constraints in the landscape being considered. They may vary from place to place and from time to time. In addition, a table indicating the degree of administrative

TABLE 14.4 Example of Quantitative Information That Might Be Utilized in Evaluating the Impacts of Management Alternatives on Soil Resources.

	Management Alternative			
Criteria	*1*	*2*	*3*	*4*
Well-Drained Soils				
Miles of Road Construction	4	1	4	6
Acres of Skid Trails*	13	3	14	16
Creation of Permanent Openings	3	1	3	5
Moderately Well-Drained Soils				
Miles of Road Construction	3	1	4	5
Acres of Skid Trails*	8	3	9	12
Creation of Permanent Openings	2	1	3	3
Poorly Drained Soils				
Miles of Road Construction	1	0	1	2
Acres of Skid Trails*	3	0	3	4
Creation of Permanent Openings	1	0	1	1

*Assumes that 10% of a harvested area will be covered by skid trails.

overhead that would be required to ensure regulatory compliance of each management alternative should be developed; this table, for example, might include the number of water quality permits expected to be required for the implementation of each alternative.

Following the detailed analyses of the impacts of each candidate alternative on the components of the landscape and the degree to which each alternative satisfies the various management goals and objectives, it is necessary to bring all of this together in an easily understood format. If at all possible, a one-page summary table should be developed, detailing the success of each alternative in meeting each management goal or objective and the impacts of each alternative on the various landscape components. This table should be carefully prepared and should rely on solid supporting evidence from the detailed analyses for the individual goals and landscape components. This table should be easily understood and simply presented; it will be a focal point of any reviews of the management planning effort.

There are several ways to present summary information of this type. The best way is usually to simply combine the information for every landscape component into a single summary table. This may result in a complex table comparing alternatives using many different metrics. It is sometimes better to summarize this information; three examples are presented here. These do not constitute an exhaustive list of the possible methods of presentation, but they do provide a range of suggestions that may be modified during a particular resource assessment effort. The first (Table 14.5) relies on the business-as-usual alternative as a baseline. Achievement of management goals and impacts on landscape components are rated as improved (+) or degraded (−) as compared with the business-as-usual scenario, which receives a no change rating (0) for every evaluation criteria. A qualitative indication of the degree of improvement or degradation is indicated by repeating the symbols (e.g.,

TABLE 14.5 Evaluation of Management Alternatives Based on Relative Comparisons with a "Business-as-Usual" Alternative (Alternative 1).

	Management Alternative			
Management Goal	*1*	*2*	*3*	*4*
1. Threatened and Endangered Species	0	2+	1+	1−
2. Biodiversity	0	1−	2+	0
3. Soil Productivity	0	2+	0	1−
4. Wildlife Habitat	0	1+	0	1+
5. Financial Return	0	2−	1−	2+
6. Recreation	0	1−	0	1−
7. Commodity Production	0	3−	1−	2+

+++) or by incorporating a numerical indicator (e.g., 3+ or +3). These numerical values are ordinal values, implying a ranking (3 indicates a greater degree of impact than 2), but not a numerical scale (4 is not necessarily "twice" 2). This type of summary excels at highlighting the differences between current management and possible future management scenarios being considered.

A second type of summary places the qualitative expectation of achievement of each goal, or impact on each landscape component, on an absolute scale rather than a relative scale that is based on comparisons with the business-as-usual alternative. In this case, the business-as-usual alternative is scored as to its impacts or degree of achievement of management goals and all the other alternatives are rated on the same scales. Table 14.6 contains a summary based on the same analyses as the summary in Table 14.5, except that absolute scales are used. It is easy to see that the presentation in Table 14.5 highlights the possible changes from current management, while the presentation in Table 14.6 highlights the degree to which the management goals are obtained by each alternative and the level of expected impacts on the various landscape components.

A third common method of summarizing the analyses of management alternatives is to simply rank each alternative, assigning the alternative that maximizes the achievement of a management goal the value 1, the next best alternative in achieving that management goal the value 2, and so on. Similarly, alternatives that have the least impact on a landscape component would be assigned the value 1, the alternative with the next lowest impact on that component would be assigned the value 2, and so on. Table 14.7 contains a summary of the same analyses described in Tables 14.5 and 14.6, which uses ranking as the method of comparing the alternatives. Some would argue that there could never be ties in ranks between alternatives, but the analyses really

TABLE 14.6 Comparison of Management Alternatives Based on Absolute Evaluation of the Success of Different Alternatives in Achieving Management Goals and Objectives and Impacts on Landscape Components.

	Management Alternative			
Management Goal	*1*	*2*	*3*	*4*
1. Threatened and Endangered Species	2+	4+	3+	1+
2. Biodiversity	2+	1+	4+	2+
3. Soil Productivity	1−	1+	1−	2−
4. Wildlife Habitat	1+	2+	1+	2+
5. Financial Return	3+	1+	2+	5+
6. Recreation	0	1−	0	1−
7. Commodity Production	0	3−	1−	2+

TABLE 14.7 Evaluation of Management Alternatives Utilizing a Ranking Procedure.

	Management Alternative			
Management Goal	*1*	*2*	*3*	*4*
1. Threatened and Endangered Species	3	1	2	4
2. Biodiversity	2.5	4	1	2.5
3. Soil Productivity	2.5	1	2.5	4
4. Wildlife Habitat	3.5	1.5	3.5	1.5
5. Financial Return	2	4	3	1
6. Recreation	1.5	3.5	1.5	3.5
7. Commodity Production	2	4	3	1

do not have the precision to allow absolute demarcation between the candidate alternatives. If two alternatives are considered equal either in their achievement of a management goal or in their impact on a landscape component, a midranking procedure should be used, whereby the ranks are "split" among the tied alternatives (see Table 14.7 for examples). Ranks emphasize the relative performance of the management alternatives, but they do not indicate a numeric scale. In other words, the difference between two alternatives ranked 2 and 3 could be small or it could be large; there is no way to tell from a summary based on ranks.

There are variations of the three methods of summarizing the comparisons among the management alternatives illustrated in Tables 14.5 through 14.7. All are designed to facilitate evaluation and comparison of the alternatives; good summary presentations can be easily read and digested by both managers and interested public participants. Good presentations are short and succinct, preferably no longer than one page, but they are solidly based on extensive, detailed analyses of each management goal and landscape component. The resource assessment provides the basis for these analyses by providing the information needed to evaluate the different aspects of the management alternatives. The final management decision selecting the preferred alternative should be clearly based on the qualitative and quantitative analyses of the group of possible management alternatives.

SELECTION OF A PREFERRED ALTERNATIVE

The preferred alternative is the management alternative that will be implemented. The crux of the management planning process is the selection of the preferred alternative. The resource assessment provides the basic information needed for the development and evaluation of alternatives. This ultimately leads to the selection of the alternative that best achieves the management goals and objectives, satisfies the constraints placed on management by the abiotic, biotic, and social components of the landscape, and minimizes adverse impacts to the landscape components.

The selection of the preferred alternative is based on a combination of the relative importance of the management goals and objectives (Table 14.2) and the summary of the analyses of the management alternatives being considered (e.g., Tables 14.5–14.7). While there is a great deal of qualitative and quantitative information developed during the resource assessment and the analysis of the management alternatives, the final selection of the preferred alternative is a subjective decision that balances the achievement of management goals and objectives with the minimization of impacts on the landscape components and associated ecological processes. There is no way to avoid the fact that this is a judgment call and, thus, different individuals or organizations may not all select the same alternative as the preferred alternative in a given situation.

One cautionary note is in order concerning the management alternative summary information contained in Tables 14.5 through 14.7; there is a great temptation to "overinterpret" this kind of information. In some instances, the selection of the preferred alternative has been justified by some quantitative combination of the summary information. In Table 14.5, for instance, it is possible to sum the values assigned each management alternative, or even to weight the values by the relative importance of the different management goals or landscape components. These are ordinal values, however, and not numerical values. The scale used to assign the values is subjective and is not numerically consistent among management goals or landscape components. In this situation, numerical summation of the values does not provide a valid evaluation of the management alternatives; all it does is provide a false sense of quantitative legitimacy to the process of selecting the preferred alternative.

Summation of the ranks (Table 14.7) is somewhat more justifiable in a quantitative or statistical sense, but the subjective nature of the scoring and ranking process, in most instances, implies that the quantitative information contained in the summation of the ranks is insufficient as a basis for selecting the preferred alternative. There is a great deal of appeal to managers in somehow using the summary information (Tables 14.5–14.7) to develop a single, quantitative score on which to base the selection of the preferred alternative. Unfortunately, this is usually not legitimate. In the final analysis, the selection of the preferred alternative is a subjective process that can be aided by quantitative information, but must be based on professional judgment in attempting to develop a balance among resource issues within the landscape.

SUMMARY

The resource assessment process is just a portion of the management decision-making process, but it is crucial to the success of management planning.

The development of management alternatives is contingent on a thorough assessment of the abiotic, biotic, and social components of forested landscapes. Evaluation of these alternatives, including both their success in achieving management goals and objectives and their impact on the components of the landscape, relies heavily on information developed during the resource assessment process. The final selection of a preferred alternative must be supported by qualitative and quantitative information concerning the status and condition of landscape components. In the final analysis, the management decision-making process is a subjective, human process. It must be defensible and solidly supported by the best available information, but the use of subjective judgment cannot, and should not, be eliminated from the decision-making process.

APPLICATIONS

14.1 Obtain a management plan for a National Forest and answer the following:

a. What critical issues were identified during the planning process?
b. How many alternatives were evaluated? Describe the differences among them.

14.2 Obtain the management plan for a National Park and answer the following:

a. What critical issues were identified during the planning process?
b. Is the preferred alternative the best at meeting all of the management goals? Which goals might have been better met by a different alternative?

14.3 Obtain the management plan from a National Forest, National Park, or other area, such as a state forest. How are the management alternatives compared? Is there a table like one of Tables 14.5–14.7? Are the differences among the alternatives clearly identified? Is it clear why the preferred alternative was selected?

14.4 Obtain the management plans from two or more National Forests. For one landscape component, such as soil, give the criteria used to evaluate the management alternatives in each plan. Why might these differ among the plans? Is the method of summarizing the differences among the management alternatives the same in each plan or are different methods used in the different plans?

PART VI

Supplemental Material

Chapter

15

Basic Statistical Sampling Methods

Conducting any type of field inventory involves several steps. The first is the development of an inventory plan. An *inventory plan* is a document giving a complete description of the objectives of the inventory, the field methods, the analytical methods, and the archival plans for the data (see Chapter 1). Development of an inventory plan includes the specification of the entities and the variables to be observed at each sample location and the methods of selecting samples to be measured. Following the development of the inventory plan, the execution of the field, analytical, reporting, and archival procedures are required to complete the inventory. It may not be necessary to develop a detailed inventory plan for every inventory effort; the formulation of organizational policy on precision requirements, analytical procedures, reporting formats, and archival procedures may be sufficient, though there is always a need to inspect areas to be sampled to see whether any standard methods need to be modified to fit a particular situation.

The following sections provide an overview of basic statistical sampling techniques and the practical aspects of field inventories. A number of recent texts have examined the statistical aspects of forest inventory in greater detail; Avery and Burkhart (1994) and Philip (1994) both provide descriptions of forest inventory methods, DeVries (1986) and Schreuder, Gregoire, and Wood (1993) offer a detailed treatment of the theory supporting forest inventory methods, and Shiver and Borders (1996) present a survey of the practical application of sampling theory in forest inventory. Cochran (1977) provides comprehensive discussion and development of the general statistical theory for most of the methods described here.

SAMPLING VERSUS COMPLETE ENUMERATION

Perfect knowledge about the resource base would greatly improve a manager's ability to make rational management decisions. There is no such thing as perfect knowledge, of course, but it is possible to make measurements of every tree in a watershed, for example, to obtain an estimate of overstory biomass. This would be very expensive and time-consuming, and the effort required to do this is not justified by the value of the information. In some situations, it is not even possible to obtain a complete enumeration, even if it is desired. The Constitution of the United States requires a decennial census of the human population, but there is still considerable uncertainty in the resulting estimates because of problems in locating and obtaining a response from every member of the population. Trees cannot hide or refuse to answer a question, but there can be observational errors and imprecise measurements by even the most conscientious observer. In most natural resource situations, approximate values based on imperfect knowledge will usually suffice, especially if the degree of uncertainty can be quantified.

For these reasons, managers usually utilize statistical *sampling* procedures, in which a representative subset of the resource base is observed and used to estimate the characteristics of the overall population. The *population* is the area or group of individuals for which information is required. The description of the population may be simple, such as the legal description of the operational area to be addressed in the resource assessment. Even here, though, some elements in the assessment of the population will extend to the surrounding landscape. The population could be described as a particular group of organisms, such as the white-tailed deer inhabiting a particular landscape. It could be quite specific, such as the overstory sugar maple trees on some particular tract of land. The definition of the population is dependent on the goals and the objectives of the resource assessment process. There may be different definitions of the relevant population for the various landscape characteristics to be addressed in the assessment process, because there are differences in the appropriate spatial and temporal scales for addressing different ecological processes (see Chapter 1). Each population is made up of individual *sample units,* which are the units in the population that will actually be observed and measured during the sampling process. The sample units could be individual organisms in wildlife sampling efforts, they could be physical entities such as soil pits, or they could be groups of individuals, such as all trees meeting certain selection requirements within a small plot.

The *sample design* describes the method of selecting the individual sample units to be observed from the population. The *sampling frame* is a list or description of the sample units that make up the population to be sampled. In other words, the sample design is really the method of selecting sample units from the sampling frame for observation. In the following discussion, the population and the sampling frame are considered to be synonymous. In reality, this is not necessarily the case, and it is very important to ensure that the sampling frame used to select sample units is actually representative of the population of interest.

There are four basic sampling designs commonly used in resource assessment efforts: simple random sampling, systematic sampling, stratified random sampling, and cluster sampling. DeVries (1986), Schreuder, Gregoire, and Wood (1993), Avery and Burkhart (1994), Philip (1994), and Shiver and Borders (1996) give detailed information on some or all of these basic designs, as well as discussion of more specialized designs. Here, the major sampling designs used in resource assessment efforts are summarized and the necessary computational procedures are illustrated for each.

SIMPLE RANDOM SAMPLING

The most basic sampling design is *simple random sampling,* whereby every possible combination of sample units in the population has an equal probability of being selected for measurement. Note that this is more restrictive than requiring that every sample unit have an equal probability of being selected.

In simple random sampling, individual sample units are independently and randomly selected from the sampling frame for observation. The selection of one sample unit does not alter the likelihood of selecting any other sampling unit. Sampling *without replacement* implies that a sample unit can be observed only once, and once selected for observation, it is removed from the list of possible sample units for the next observation. Sampling *with replacement* means that, following observation, a sample unit is placed back into the list of possible sample units and it may be selected for observation again.

If the sampling frame is a map, simple random sampling may be conducted by randomly selecting Cartesian map coordinates to determine the location of the sample units. In vegetation sampling, these coordinates may be centers of the sample units. In sampling bird populations within some fixed area such as a watershed, county, or other region, the location of observation points or transects may also be randomly selected from a map of the area.

In sampling, measurements of the variables of interest are made for each selected sampling unit. The objective is to estimate the value of a population *parameter*, which is the true value of some characteristic of the population. Examples of parameters in forest overstory sampling include the basal area or volume per hectare. It is usually impossible to know the actual values of population parameters, because it is not possible or practical to measure every element in a population. Following measurement of the sample units, calculations are conducted to obtain *estimates* of the corresponding population parameters. Usually, information concerning the uncertainty of the estimates is also included in the reporting process.

To estimate the mean value of some variable within the population, such as the mean basal area per hectare for the forest overstory, the following formula is used in simple random sampling:

$$\overline{x} = \frac{\sum_{i=1}^{n} x_i}{n} \qquad [15.1]$$

where $\overline{x}$ is the estimate of the population mean, x_i is the value observed for the variable on the ith sample unit, and n is the number of sample units observed from the population.

Table 15.1 contains a set of 11 example observations of basal area per hectare from the overstory of a northern hardwood forest; methods of determining the basal area of the individual sample units are described in Chapter 8. The mean basal area per hectare is calculated by first summing the observed basal areas per hectare to obtain 206.0 m²/ha. This value is then divided by 11, the number of sample units observed from the population, to obtain the mean basal area per hectare:

$$\begin{aligned}\overline{x} &= \frac{206.0}{11} \\ &= 18.7 \text{ m}^2\text{/hectare}\end{aligned} \qquad [15.2]$$

TABLE 15.1 Example Observations of Basal Area per Hectare from the Overstory of a Northern Hardwood Forest in Houghton County, Michigan.

Plot Number	*Basal Area (m²/ha)*
1	31.2
2	13.4
3	11.2
4	15.6
5	26.8
6	22.3
7	13.6
8	22.1
9	15.9
10	22.5
11	11.4
Total	206.0
Mean	18.7 m²/hectare
Variance	44.2 m⁴/hectare²
Standard Deviation	6.6 m²/hectare
Standard Error of the Mean	2.0 m²/hectare
Coefficient of Variation	35%

The estimate of the mean is just that: it is an estimate of the true value of the mean of the population. There is uncertainty in the estimate, because if a different sample of the same number of sample units were to be selected from the population, there is a very high likelihood that the second observed sample mean will have a different value from the first observed sample mean. This uncertainty is due to the number and size of the sample units, but it is also due to the variability of the characteristic of interest within the population. This variability is usually not known, but it may be estimated from the sample observations. The *sample variance* is defined as:

$$s^2 = \frac{\sum_{i=1}^{n} (x_i - \overline{x})^2}{n - 1} \qquad [15.3]$$

The sample variance is in square units of the variable being observed. An equivalent expression, which is easier to calculate by hand or to program in a computer is given by:

$$s^2 = \frac{\sum_{i=1}^{n} x_i^2 - \frac{\left(\sum_{i=1}^{n} x_i\right)^2}{n}}{n - 1} \qquad [15.4]$$

For the data in Table 15.1, the sum of the squared basal areas per hectare is 4,299.7 m^4/ha^2, so the sample variance for the data in the table is given by:

$$\begin{aligned} s^2 &= \frac{4{,}299.7 - \frac{(206.0)^2}{11}}{11 - 1} \\ &= 44.2 \text{ m}^4/\text{ha}^2 \end{aligned} \qquad [15.5]$$

The sample variance is expressed in squared units of the observed variable, which may be difficult to quickly and easily relate to the mean. The *sample standard deviation* is the square root of the variance. For the sample data in Table 15.1, the sample standard deviation is 6.6 m^2/ha, which is the square root of 44.2 m^4/ha^2 from Equation [15.5]. The sample standard deviation is based on the differences between the individual observed values and the estimated mean for the variable of interest. In many resource assessment efforts, it is more important to understand variability in the estimate of the population characteristic of interest, such as the mean basal area per hectare, than it is to understand the variability among the individual sample units. The *standard error of the mean* is equivalent to the standard deviation of the estimated value of the mean, and it is given by:

$$s_{\overline{x}} = \frac{s}{\sqrt{n}} \qquad [15.6]$$

For the data in Table 15.1, the standard error of the mean is given by:

$$s_{\bar{x}} = \frac{6.6}{\sqrt{11}}$$
$$= 2.0 \text{ m}^2\text{/hectare} \qquad [15.7]$$

When sampling without replacement, a relatively large fraction of the population may be sampled. If this occurs, Equation [15.6] will overestimate the true standard error of the mean. This can be seen intuitively, because there is less and less uncertainty in an estimate of the population mean as greater fractions of the total population are sampled. There is still variability among the sample units that is not affected by the proportion of the population sampled. As a practical guide, use of Equation [15.6] is appropriate when less than 5% of the population is sampled. When more than 5% of the population is sampled, a correction is necessary to avoid overestimating the standard error of the mean. This correction is called the *finite population correction factor:*

$$FPC = \frac{N - n}{N}$$
$$= 1 - \frac{n}{N}$$
$$= 1 - f \qquad [15.8]$$

where N is the number of sample units in the entire population and f is the sampling fraction (n/N). An estimate of the standard error of the mean for use in sampling more than 5% of the population without replacement is given by:

$$s_{\bar{x}} = \frac{s}{\sqrt{n}} (\sqrt{1 - f}) \qquad [15.9]$$

The standard deviation and the standard error of the mean are expressed in the same units as the mean. A relative measure of the variability in the population is given by the *coefficient of variation,* which is the ratio of the sample standard deviation to the mean. The coefficient of variation is usually expressed as a percentage:

$$CV = \frac{s}{\bar{x}} (100) \qquad [15.10]$$

where CV is the coefficient of variation. For the example in Table 15.1:

$$CV = \frac{6.6}{18.7} (100)$$
$$\approx 35\% \qquad [15.11]$$

Sometimes it is not the population mean that is of interest in the assessment effort, but the population total. An obvious case in which this is true is when the value of a tract of land is being appraised; it is not the mean value per hectare that is of major interest but, instead, the total value of the entire area. In cases in which estimates of the population total are desired, the population mean is estimated as described earlier and then multiplied by the population size (N) to obtain an estimate of the population total. The standard deviation of the estimate of the population total is obtained by multiplying the population size by the standard error of the mean.

A second special case arises in the estimation of a population proportion, such as the proportion of sample units that contain evidence of some particular disease. In this case, the response variable is an indicator variable that takes the value of one if the sample element possesses the characteristic of interest and the value of zero if it does not. The mean of the indicator values is the estimate of the proportion of the population possessing the characteristic of interest ($\hat{p}$). The standard error of the estimated proportion ($s_{\hat{p}}$) is estimated using the indicator values for each sample unit (zeros or ones) in Equation [15.3] or [15.4] to estimate the sample variance, and then using the resulting standard deviation in the standard error calculation (Equation [15.6] or [15.9] as appropriate).

Error in the estimate of the mean simply refers to the fact that the estimated value is not exactly equal to the true value; it does not imply that mistakes have been made. As noted previously, the variability of the estimated mean is described by the standard error of the mean (Equation [15.6] or [15.9] as appropriate). The distribution of estimated values obtained from samples collected following the simple random sampling protocols for sample unit selection are easily described mathematically. A *statistical distribution* describes the relative frequency of different values of the variable in question. Estimates of the mean from independent samples follow a t-distribution which is dependent on the sample size. For estimates of the mean, the t-distribution can be used to develop a range of values that can be expected to contain the true population value with some known probability. These are referred to as *confidence limits,* which form a *confidence interval* for the estimated value.

The first step in calculating the confidence interval is to specify the probability level desired; a 95% confidence interval is a range of values that have a 95% probability of containing the true mean. Figure 15.1 illustrates the specification of percentile points from a statistical distribution for use in confidence limit calculations. There are one-sided confidence limits, in which the range of values is expressed only as a maximum or minimum value (Figure 15.1a), with the interpretation that the observer is 95% confident that the true value of the population mean is less than (or greater than) the specified value; in two-sided confidence limits, the interpretation is that the observer is 95% confident that the true value lies within some range (Figure 15.1b). The confidence interval is developed from the estimated mean, its standard error, and a percentage point of the t-distribution, which can be found in

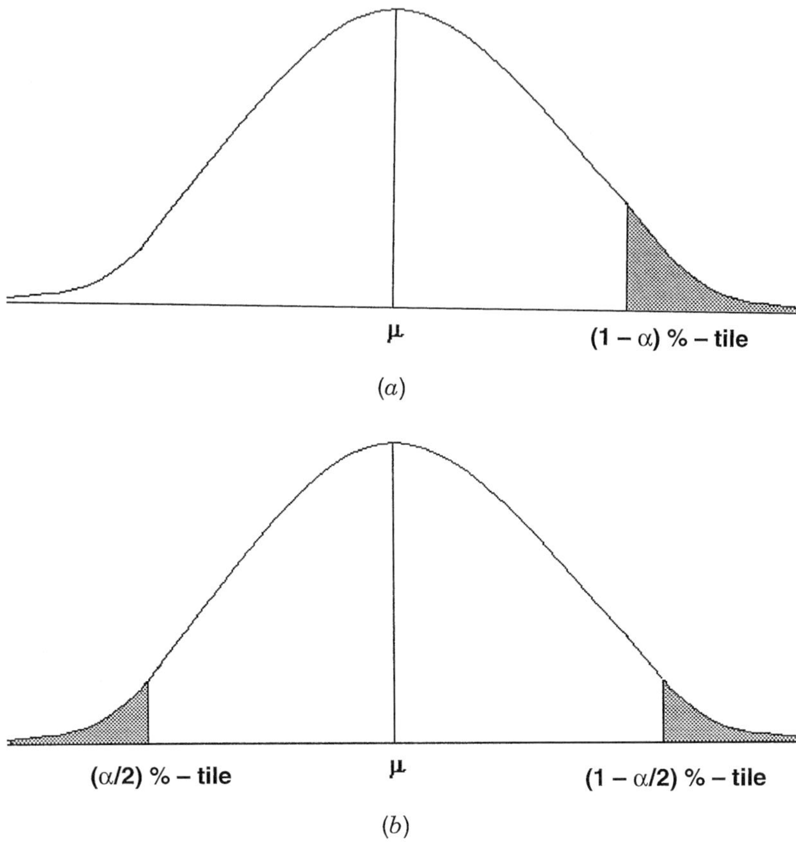

FIGURE 15.1 Specification of percentile points from a statistical distribution for use in estimating confidence limits; (*a*) one-sided $(1 - \alpha)$% confidence limits and (*b*) two-sided $(1 - \alpha)$ % confidence limits.

almost any elementary statistical text or in special compilations of statistical tables (e.g., Rohlf and Sokal, 1969). The confidence interval is calculated as:

$$\text{Confidence interval} = \bar{x} \pm t_{\alpha/2,df}\, s_{\bar{x}} \qquad [15.12]$$

where $t_{\alpha,df}$ is the αth percentage point of a t-distribution having df degrees of freedom. For simple random sampling, the degrees of freedom are obtained by subtracting 1 from the number of observed sample units ($df = n - 1$). The data in Table 15.1 have a mean of 18.7 m^2/ha, a standard error of the mean of 2.0 m^2/ha, and 10 degrees of freedom ($11 - 1 = 10$). To estimate a 95% upper confidence interval, as in Figure 15.1a, the one-sided t-value with 10 degrees of freedom ($t_{.05,10} = 1.812$) is used:

$$\begin{aligned}\text{Confidence interval} &= 18.7 + 1.812\,(2.0)\\ &= 22.3\ \text{m}^2/\text{hectare} \qquad [15.13]\end{aligned}$$

The interpretation of this confidence interval is that the observer is 95% confident that the true value of the population mean is less than 22.3 m^2/ha. A two-sided 95% confidence interval (as in Figure 15.1b) is obtained using the value of the $(\alpha/2)^{th}$ percentage point of the t-distribution, as illustrated in Figure 15.1b. For the example, a two-sided 95% confidence interval for the estimate of the mean is calculated using the two-sided t-value ($t_{.025,10} = 2.228$):

$$\begin{aligned}\text{Confidence interval} &= 18.7 \pm 2.228\ (2.0)\\ &= 14.2,\ 23.2\ \text{m}^2/\text{hectare}\end{aligned} \quad [15.14]$$

The interpretation of this result is that the observer is 95% confident that the true value of the population mean lies between 14.2 and 23.2 m^2/ha. There are many possible confidence intervals that could be calculated, depending on the situation (Table 15.2). As the level of confidence increases, the width of the corresponding confidence interval increases. The reason, intuitively, is that an observer will be more sure that wider intervals contain the true mean. The choice of whether to use one-sided or two-sided confidence intervals depends on the situation and the reason for conducting the inventory. If an observer is conducting an inventory to estimate commercial value for the purpose of making a bid for the purchase of an area, a lower one-sided confidence interval may be more appropriate for establishing an initial bid. If the observer estimates that the mean value per hectare is $1,000 with a 95% lower confidence limit of $900 per hectare, then the observer will be 95% confident that a bid of $900 per hectare will not be in excess of the true value. Of course, the greater the bid, the greater the likelihood of a successful purchase, so there is a trade-off between confidence in not overbidding and the need to be successful in obtaining the purchase. In another case, if the observer is attempting to estimate the mean basal area per hectare in a stand for use in an analytical model, a two-sided confidence interval may be appropriate. In this case, if the observer estimates the mean basal area to be 28.0 m^2/ha and is 95% confident that the true value is between 26.8 and

TABLE 15.2 Example of Different Confidence Limits Calculated for the Data in Table 15.1; the Mean Equals 18.7 m^2/ha, the Standard Error of the Mean Equals 2.0 m^2/ha, and There are Ten Degrees of Freedom.

Confidence Level	$t_{\alpha,df}$	One-Sided Confidence Limit Lower	One-Sided Confidence Limit Upper	$t_{\alpha/2,df}$	Two-Sided Confidence Interval
80%	0.879	16.9	20.5	1.372	16.0, 21.4
90%	1.372	16.0	21.4	1.812	15.1, 22.3
95%	1.812	15.1	22.3	2.228	14.2, 23.2
99%	2.764	13.2	24.2	3.169	12.4, 25.0

29.2 m^2/ha, then the analyses can be performed with the estimated value and the lower and upper confidence limits to obtain a range of likely outcomes.

An important step in planning a sampling effort is the specification of the number of sampling units to be observed. This requires specification of an acceptable error rate. A planner must specify both the acceptable degree of error and the probability level to be associated with the error. In a sampling effort to obtain an estimate of the mean basal area per hectare for the forest overstory, for example, the management information needs might require an estimate within plus or minus 4 m^2/ha of the true mean with a 95% level of confidence. In other words, the sampling effort should be designed to result in an estimated mean basal area per acre that is within 4 m^2/ha of the true mean 95% of the time.

In addition to the error rate and level of confidence, the planner must be able to specify the sample variance expected in the population in order to estimate the minimum number of samples required to achieve the desired error bounds. The expected variance may be estimated from prior sampling efforts in similar populations or, alternatively, a small pilot sampling effort may be conducted within the population of interest to obtain an estimate of the sample variance within the population. Pilot studies have the added advantage of allowing an evaluation of planned field procedures and sampling protocols, as well as providing additional training if inexperienced personnel will be involved.

It is not possible to guarantee that the desired error rate will be achieved during the sampling effort. The methods discussed here provide an indication of the minimum number of sample units that must be observed in order to expect to achieve the desired error rate. Occasionally, the target error rate will be exceeded owing to the vagaries of sampling. The estimate of the minimum required sample size is derived from the same relationship used to develop a confidence interval for the estimate (Equation [15.12]). The target error rate is equivalent to the quantity added and subtracted from the mean to develop the confidence interval:

$$E = t_{\frac{\alpha}{2},df}\, s_{\bar{x}} \qquad [15.15]$$

where E is the target error rate (expressed in the same units as the mean). For extremely large populations or populations sampled with replacement:

$$E = t_{\frac{\alpha}{2},df}\left(\frac{s}{\sqrt{n}}\right) \qquad [15.16]$$

This implies that n can be expressed as a function of the other quantities:

$$n = \frac{t^2_{\frac{\alpha}{2},df}\, s^2}{E^2} \qquad [15.17]$$

Of course, it is not possible to look up $t_{\alpha/2,df}$ without knowing the degrees of freedom. It is important to remember that these methods only provide guidelines for setting the number of sample units to be observed. In practice, the t-value is either set using the degrees of freedom from a pilot study used to obtain an estimate of the variance, or set at an approximate value. For a two-sided 95% level of confidence, for example, the t-value is usually approximated by simply substituting the value 2 into Equation [15.17]. For the data in Table 15.1, if the estimate of basal area per hectare had a target error rate of + 1.0 m^2/ha with a 95% level of confidence, the minimum required number of samples to take in order to expect to achieve this acceptable error rate is:

$$\begin{aligned} n &= \frac{(2)^2\,(44.2)}{1^2} \\ &= 176.8 \approx 177 \text{ sample units} \end{aligned} \qquad [15.18]$$

So a minimum of roughly 177 sample units should be measured to achieve the desired error rate in this example. There is no guarantee that measuring 177 sample units will result in an error rate less than or equal to ± 1 m^2/ha; this is just a guideline, and the greater the number of sample units observed, the lower the expected error in the resulting estimate of the mean.

In some situations, it makes more sense to express the acceptable error rate in relative terms, as a percentage of the mean, rather than in absolute units. When this is the case, dividing both the numerator and the denominator of Equation [15.17] by the mean and expressing the relevant values as percentages results in:

$$n = \frac{t^2_{\frac{\alpha}{2},df}\, CV^2}{A^2} \qquad [15.19]$$

where $t_{\alpha/2,df}$ and n are as defined above, A is the acceptable error rate in percentage, and CV is the coefficient of variation (Equation [15.10]). For the example in Table 15.1, if the acceptable error rate in the estimate of the mean basal area per hectare is 5%, then since the coefficient of variation is 35% (Equation [15.11]), the minimum number of sample units to observe in order to expect to achieve the 5 % error bound is:

$$\begin{aligned} n &= \frac{(2^2)\,(35^2)}{5^2} \\ &= 196 \text{ sample units} \end{aligned} \qquad [15.20]$$

In sampling a finite population without replacement, Equations [15.17] and [15.19] will sometimes dramatically overestimate the required number of sample units to observe if the finite population correction (Equation [15.8]) is not used to correct the standard error of the mean. In this case:

$$E = t_{\frac{\alpha}{2},df}\left(\frac{s}{\sqrt{n}}\right)\left(\sqrt{1-f}\right) \qquad [15.21]$$

where E, $t_{\alpha/2,df}$, s, and n are as defined earlier for Equation [15.17] and $(1 - f)$ is the finite population correction factor as defined in Equation [15.8]. This relationship implies that the minimum required number of sample units to observe to achieve a given error rate in a finite population sampled without replacement is given by:

$$n = \frac{t^2_{\frac{\alpha}{2},df}\; N\; s^2}{E^2\; N + t^2_{\frac{\alpha}{2},df}\; s^2} \qquad [15.22]$$

For the data in Table 15.1, if we assume that the total population size is 100 sample units and approximate the two-sided t-value as 2, then the estimated minimum number of sample units required to obtain a target error rate of $\pm$ 1.0 m^2/ha with a 95% level of confidence is given by:

$$\begin{aligned} n &= \frac{2^2\,(100)\,(44.2)}{1^2\,(100) + 2^2\,(44.2)} \\ &= 63.9 \approx 64 \text{ sample units} \end{aligned} \qquad [15.23]$$

Thus, in a large population, Equation [15.18] indicated a required sample size of at least 177 sample units to allow the expectation of achieving the acceptable error rate, whereas, in a small population of 100 sample units, a minimum of about 64 sample units are required in order to expect to achieve the same target error rate.

Equation [15.22] can be modified to allow the acceptable error rate to be expressed as a percentage of the mean. As in Equation [15.19], dividing both the numerator and the denominator by the mean and converting to percentage yields:

$$n = \frac{t^2_{\frac{\alpha}{2},df}\; N\; CV^2}{A^2\; N + t^2_{\frac{\alpha}{2},df}\; CV^2} \qquad [15.24]$$

where A is the acceptable error rate expressed as a percentage of the mean, CV is the coefficient of variation of the characteristic of interest, N is the total number of sample units in the population, and $t_{\alpha/2,df}$ is the (α/2)-percentile point of the two-sided t-distribution with df degrees of freedom (Figure 15.1b). For the example in Table 15.1, again assuming a total population size (N) of 100 sample units, the minimum number of samples required to achieve a 5% error rate in the estimation of the mean basal area per hectare is given by:

$$\begin{aligned} n &= \frac{(2^2)\,(100)\,(35^2)}{(5^2)\,(100) + (2^2)\,(35^2)} \\ &= 66.2 \approx 66 \text{ sample units} \end{aligned} \qquad [15.25]$$

which is much less than the 196 samples required to achieve the same expected error rate in an infinite population (Equation [15.20]).

The best practical procedure to determine the required sample size is usually to proceed in an iterative fashion through the following steps:

1. Use Equation [15.17] or Equation [15.19] (as appropriate, depending on how the acceptable error rate is expressed) to obtain an estimate of the minimum number of sample units required to allow the expectation of achieving the desired error rate in an infinite population.
2. If the estimated minimum number of sample units obtained in Step 1 is less than 5% of the population size, then proceed with the sampling.
3. If the estimated minimum number of sample units obtained in Step 1 is more than 5% of the population size, then recalculate the minimum required number of sample units using Equation [15.22] or Equation [15.24], depending on which is appropriate.

Again, the 5% cutoff for using the finite population correction factor is an arbitrary dividing point. A value of 5% is usually set because when less than 5% of the population is sampled, the finite population correction factor is very small and has little practical impact. This is especially true in obtaining estimates of the minimum required number of sample units since the indicated number is an approximation anyway.

SYSTEMATIC SAMPLING

Simple random sampling is sometimes a difficult sampling design to implement in field resource assessment situations. Random selection of sample units to be observed provides statistical rigor, but it may introduce a great deal of difficulty into the field data collection process. In many field assessment situations, the exact location of samples is not known with a great deal of precision, although this is changing with the introduction of Global Positioning System receivers. In many forest overstory inventories, sample location is determined through the use of a hand compass to identify direction and pacing to identify distance. Random selection of sample units implies that the sample units may be scattered throughout the area to be inventoried. It is sometimes difficult to plan an efficient route to travel from sample unit to sample unit; it is also sometimes difficult to quickly and easily determine a compass bearing and the distance from one sample unit to another. This additional planning and travel in the field add to the cost of implementing simple random sampling in field assessment operations.

Systematic random sampling is an alternative sampling design in which the initial sample unit is randomly selected from a subset of the population and all additional sample units are established in a systematic fashion. From an initial random start in one corner of the area of interest, for example, additional sample units may be established on a 100 m × 100 m grid throughout the rest of the area. Sample units are usually established on a square or rectangular grid, because it is relatively easy to move in straight lines in the field using a hand compass. Every sample unit in the population still has an equal probability of being selected for observation, but only fixed combinations of sample units may be selected for sampling. This approach is opposed to simple random sampling whereby every combination of sample units is equally likely. In many applications, the population is divided into square or rectangular areas using a systematic grid and the sample units are established at the centers of the grid units.

There are several advantages to systematic random sampling over simple random sampling in field settings:

1. Systematic sampling is much easier and more efficient to implement in the field because the route of travel between sample units is fixed.
2. The sample units are dispersed throughout the whole area of interest, whereas in simple random sampling it is possible to get random coordinates that result in large areas of the population being unsampled.
3. Dispersion of the sample units throughout the population has the added advantage of forcing the field crews to travel throughout the area. This can facilitate other objectives of the field procedures, such as verification of soil maps, the refinement of the boundaries in overstory community maps derived from aerial photos, and so on.

The primary disadvantage of systematic random sampling is a statistical consideration; since not all combinations of sample units are equally likely, it is not possible to obtain a truly valid estimate of the sample variance. This is because there is, at most, one randomly selected observation; all of the other observations follow from the selection of the initial sample unit. Mendenhall, Ott, and Schaeffer (1971) note that for most situations encountered in field sampling, the variance may be estimated using the same procedures as in simple random sampling without serious consequences. The exception is a situation in which there is a spatial periodicity to the variables being observed. Spatial periodicities may occur when forests have become established on old shoreline ridges, along lakes, or on other physical features where there are succeeding ridges of higher elevation and drier conditions. The population may also be periodic in alluvial or fluvial soils where there may be systematic spatial variation in soil texture. There are also culturally induced periodicities, as noted by Shiver and Borders (1996), that may be introduced during site preparation procedures when residual debris from the previous overstory is concentrated into linear windrows. In periodic populations, the sample variance obtained using the simple random sampling procedures (Equation [15.3]) is an underestimate of the true variance of the estimated mean. Avery and Burkhart (1994) recommend that systematic sampling be used in forest inventory applications if the goal is only to estimate the population mean. If determination of the precision of the resulting estimate is important, they recommend that simple random sampling be used.

In systematic sampling the sample units are located at fixed intervals along grid lines; the location of the grid lines can impact the level of potential problems with periodicity in the population. If grid lines are established so that the observers travel across the topography (i.e., up and down hills, perpendicular to the topographic contour lines), then the possibility of having a majority of sample units fall into a single elevational setting or soil unit is minimized. Traveling across the topography also increases the likelihood that the field crews will encounter boundaries between overstory cover types, soil units, or vegetative associations that may be associated with elevation or hydrology. In most situations, supplemental observations, such as verification of existing soil or cover maps, will be facilitated if travel is up- and downslope as much as possible.

STRATIFIED RANDOM SAMPLING

If there are extreme differences within a population resulting from historic, physiographic, hydrologic, or other factors that impact the ecological processes or characteristics of interest, *stratification,* or the division of the population into relatively homogeneous subpopulations, or strata, may be advisable. *Stratified random sampling* may be used to obtain estimates of the variables of interest both within strata and for the overall population. Strata definition should be based on factors relevant to the processes and characteristics of interest. Delineation may be based on such things as overstory vegetation or soil mapping unit if a relatively small area is being examined. Delineation could also be based on larger-scale characteristics, such as watershed boundaries, if larger areas are being examined. *A priori* definition of strata is not possible without knowledge of the context and purpose of the sampling exercise. Strata delineations may be developed from existing information such as soil maps, topographic maps, aerial photographs, or other types of information contained in a geographic information system (Chapter 13).

The following discussion assumes that the strata are delineated prior to field sampling (*prestratification*) and that the number of sample units in each strata is known. An alternative procedure, whereby the strata are defined beforehand, but delineated during field sampling, is referred to as *poststratification.*

In a prestratification situation, the number of sample units in the population (N) is known, as is the number of sample units in each stratum (N_h). The sum of the population sizes in the individual strata equals the number of sample units in the entire population:

$$N = \sum_{h=1}^{k} N_h \qquad [15.26]$$

Either simple random sampling or systematic random sampling is performed within each stratum. The sample mean is estimated for each stratum as if independent sampling procedures were applied in each stratum:

$$\bar{y}_h = \frac{\sum_{i=1}^{n_h} y_{hi}}{n_h} \qquad [15.27]$$

where $\bar{y}_h$ is the sample mean for strata h and y_{hi} is the ith sample unit in strata h (i = 1, 2, . . . , n_h). Table 15.3 contains an example from a small population with only two strata. There are six samples of total aboveground biomass per hectare in the first stratum ($n_1 = 6$) and eight samples in the second stratum ($n_2 = 8$). The mean of the samples in the first stratum ($\bar{y}_1$) is 273.7 Mg/ha and the mean of the samples from the second stratum ($\bar{y}_2$) is 221.0 Mg/ha.

The sample variance is also estimated for each stratum:

$$s_h^2 = \frac{\sum_{i=1}^{n_h} y_{hi}^2 - \frac{\left(\sum_{i=1}^{n_h} y_{hi}\right)^2}{n_h}}{n_h - 1} \qquad [15.28]$$

TABLE 15.3 Observed Values of Aboveground Biomass (Mg/Hectare) of the Forest Overstory in a Northern Hardwood Forest Divided into Two Strata.

Plot Number	*Stratum 1*	*Stratum 2*
1	282	202
2	257	227
3	293	215
4	272	235
5	275	219
6	263	222
7	–	240
8	–	208
n_h	6	8
N_h	50	70
$\bar{x}_h$	273.7	221.0
s_h	13.0	12.9
$s_{\bar{x}_h}$	5.0	4.3

where s_h^2 is the sample variance for strata h and other notation is as previously defined. The sample standard deviation (s_h) may be estimated for stratum h by taking the square root of the sample variance for the stratum. For the data in Table 15.3, the sample standard deviation for the first stratum is 13.0 Mg/ha and the sample standard deviation for the second stratum is 12.9 Mg/ha.

The standard error of the sample mean for strata h may be estimated as follows:

$$s_{\bar{y}_h} = \left(\frac{s_h}{\sqrt{n_h}}\right)\left(\sqrt{\frac{N_h - n_h}{N_h}}\right) \quad [15.29]$$

Note that the finite population correction factor is included in Equation [15.29]. For the data in Table 15.3, if the total number of sample units in the first stratum (N_1) is 50 and the total number of sample units in the second stratum (N_2) is 70, then the standard error of the mean of the first stratum is:

$$\begin{aligned} s_{\bar{y}_1} &= \left(\frac{13.0}{\sqrt{6}}\right)\left(\sqrt{\frac{50 - 6}{50}}\right) \\ &= 5.0 \text{ Mg/hectare} \end{aligned} \quad [15.30]$$

and the standard error of the mean for the second stratum is:

$$\begin{aligned} s_{\bar{y}_2} &= \left(\frac{12.9}{\sqrt{8}}\right)\left(\sqrt{\frac{70 - 8}{70}}\right) \\ &= 4.3 \text{ Mg/hectare} \end{aligned} \quad [15.31]$$

If less than 5% of the sample units in stratum h are observed, or if stratum h is sampled with replacement, then the finite population correction factor may be ignored and the standard error of the mean for stratum h is simply:

$$s_{\bar{y}_h} = \frac{s_h}{\sqrt{n_h}} \quad [15.32]$$

As noted earlier, stratified random sampling is used when estimates for the entire population across all strata are desired in addition to estimates for individual strata within the population. The overall population mean is estimated by weighting the sample means for the individual strata by the respective population sizes within the strata:

$$\bar{y}_s = \frac{1}{N}\sum_{h=1}^{k} N_h\,\bar{y}_h \quad [15.33]$$

For the data in Table 15.3, the estimate of the mean of the entire population is given by:

$$\bar{y}_s = \frac{(50)\,(273.7) + (70)\,(221.0)}{(50 + 70)}$$
$$= 243.0 \text{ Mg/hectare} \qquad [15.34]$$

The standard error of the mean for the entire population is estimated in a similar fashion:

$$s^2_{\bar{y}_s} = \sum_{h=1}^{k} \left[\frac{N_h}{N}\right]^2 s^2_{\bar{y}_h} \qquad [15.35]$$

Note that, if appropriate, the finite population correction factors for each stratum are incorporated into the estimates of the standard errors of the mean in the individual strata ($s_{\bar{y}_h}$). For the data in Table 15.3, the standard error of the estimated mean for the entire population is given by:

$$s^2_{\bar{y}_s} = \left[\frac{50}{120}\right]^2 \cdot 5.0^2 + \left[\frac{70}{120}\right]^2 \cdot 4.3^2$$
$$= 4.34 + 6.29$$
$$= 10.6 \text{ Mg/hectare} \qquad [15.36]$$

and

$$s_{\bar{y}_s} = 3.3 \text{ Mg/hectare} \qquad [15.36a]$$

The confidence interval for the overall population mean is calculated as in Equation [15.12], with the appropriate substitutions:

$$\text{Confidence interval} = \bar{y}_s \pm t_{\alpha/2,df}\, s_{\bar{y}_s} \qquad [15.37]$$

The degrees of freedom used to calculate the confidence limits for the overall estimate of the population mean in stratified random sampling is the sum of the degrees of freedom of the individual strata:

$$\text{df}_s = \sum_{i=1}^{k} (n_h - 1) \qquad [15.38]$$

For the data in Table 15.3, the degrees of freedom for calculation of confidence intervals for the estimate of the population mean is 12 (12 = (6 − 1) + (8 − 1)). This leads to a value for $t_{\alpha/2,df}$ of 2.179 for two-sided 95% confidence intervals. Using the estimates from Equations [15.34] and

[15.36], two-sided 95% confidence intervals for the estimate of the population mean in this example are given by:

$$\begin{aligned}\text{Confidence interval} &= 243.0 \pm 2.179\ (3.3)\\ &= 235.8,\ 250.2\ \text{Mg/hectare}\end{aligned} \qquad [15.39]$$

Again, the interpretation here is that we are 95% confident that the true mean of the population lies between 235.8 and 250.2 Mg/ha. The advantage of stratification can be seen in this example by calculating the estimated mean and confidence limits for the mean as if the entire sample were a single, simple random sample of the population. Doing this results in an estimated mean of 243.6 Mg/ha, which is not very different from the stratified random sampling estimate of the mean, but a two-sided 95% confidence interval for the simple random sampling mean is 226.4 to 260.8 Mg/ha, which is considerably wider, implying less confidence in the estimate, than that from stratified random sampling given in Equation [15.39].

One obvious question in the implementation of stratified random sampling is the method of determining the number of sample units to be observed in each stratum. In some situations, it may be necessary to specify error bounds to be achieved within each stratum. In this case, the methods for estimating the required sample size in simple random sampling (Equations [15.17], [15.19], [15.22], or [15.24]) may be applied to each of the strata. In other cases, it may not be appropriate to place bounds on the sampling error in the individual strata. This would be the case, for example, when the primary goal is to estimate either the population mean or the population total across all strata, with the estimation of the mean or total within the individual strata as a secondary objective. In this case, the overall sample size across all strata may be estimated as described in subsequent paragraphs (Equations [15.46], [15.48], [15.50] and [15.52]), but the number of sample units to be observed in each strata must still be determined. There are two basic methods to do this. The first method is called *proportional allocation,* and here the samples are allocated among the strata in proportion to the relative size of the stratum in the population; more samples are allocated to larger strata than to smaller strata.

$$n_h = \frac{N_h}{N}\,(n) \qquad [15.40]$$

For the data in Table 15.3, if 48 observations are to be made in the two strata, the proportional allocation of samples between the strata would result in the following number of samples in each stratum:

$$\begin{aligned} n_1 &= \frac{50}{120}\,(48)\\ &= 20\ \text{sample units}\end{aligned} \qquad [15.41]$$

and

$$n_2 = \frac{70}{120}(48)$$
$$= 28 \text{ sample units} \qquad [15.41a]$$

A second method of allocating the sample units to the individual strata is based on both the size and the expected sample variance of the respective strata. More variable strata are sampled more intensively than less variable strata of the same size, and larger strata are allocated more samples than smaller strata with the same variability. This method of allocating samples among the strata is termed *optimal allocation* by many authors, but it is sometimes referred to as *Neyman allocation*.

$$n_h = \frac{N_h\, s_h}{\sum_{h=1}^{k} N_h\, s_h}(n) \qquad [15.42]$$

Equation [15.42] requires estimates of the sample variance within the individual strata, but use of this method will result in a more precise estimate of the population mean or total in comparison with proportional allocation (Cochran, 1977). The improvement in precision over proportional allocation depends on the difference in variability for the characteristic of interest among the strata; if the sample variance is equal in all strata, then optimal allocation is equivalent to proportional allocation. This can be seen in the example data in Table 15.3. The standard deviation in the first stratum is 13.0 Mg/ha, and the standard deviation in the second stratum is 12.9 Mg/ha. If 48 sample units are to be allocated among the strata, then the number of sample units to place in each stratum is given by:

$$n_1 = \frac{(50)\,(13.0)}{(50)\,(13.0) + (70)\,(12.9)}(48)$$
$$= 20.1 \approx 20 \text{ sample units} \qquad [15.43]$$

and

$$n_2 = \frac{(70)\,(12.9)}{(50)\,(13.0) + (70)\,(12.9)}(48)$$
$$= 27.9 \approx 28 \text{ sample units} \qquad [15.43a]$$

Based on optimal allocation, therefore, 20 sample units would be allocated to stratum 1 and 28 sample units would be allocated to stratum 2. For this example, these results are the same as those derived by using proportional allocation because of the similarity in strata standard deviations. If the standard deviation of stratum 1 had been twice as large as it is, optimal allocation would have allocated 28 samples to stratum 1 and 20 to stratum 2 because of the greater variability of stratum 1.

In some situations, the relative cost of establishing and measuring the sample units may differ among the strata. For example, in denser overstories, usually more individual trees are measured in each sample unit, topography may be such that travel among the sample units is more difficult in some strata, and there are obviously other situations in which the relative cost of sampling differs among the sample units. There is a third method that incorporates these differences in the cost of sampling among the strata into the allocation of sampling units among the strata. This method is also sometimes referred to as *optimal allocation.*

$$n_h = \frac{N_h\ s_h/\sqrt{C_h}}{\sum_{h=1}^{k} N_h\ s_h/\sqrt{C_h}}(n) \qquad [15.44]$$

where C_h is either the actual or relative cost of establishing and measuring a sample unit in stratum h; the other terms are as previously defined. If the actual or relative cost for establishing and measuring the sample units is equal in all strata, Equation [15.44] is equivalent to Equation [15.42]. For the example data in Table 15.3, assuming that there are 48 sample units to be allocated between the two strata, the costs of measuring a sample unit in the two strata may differ because of vegetation density, ease of access, or other factors. For purposes of illustration, if the cost of measuring a sample unit in stratum 1 is two-thirds the cost of measuring a sample unit in stratum 2, then the number of sample units to allocate to stratum 1 is given by:

$$\begin{aligned} n_1 &= \frac{(50)\ (13.0)/(\sqrt{2/3})}{((50)\ (13.0)/(\sqrt{2/3})) + (70)\ (12.9)}(48) \\ &= 22.5 \approx 22 \text{ sample units} \end{aligned} \qquad [15.45]$$

and

$$\begin{aligned} n_2 &= \frac{(70)\ (12.9)}{((50)\ (13.0)/(\sqrt{2/3})) + (70)\ (12.9)}(48) \\ &= 25.5 \approx 26 \text{ sample units} \end{aligned} \qquad [15.45a]$$

In this case, more sample units are allocated to stratum 1 because they are cheaper to measure.

The sample size needed among all strata to estimate the overall population mean or total depends on whether proportional or optimal allocation is being used to allocate the sample units among the strata. Because optimal allocation results in a more precise estimate of the mean or total, usually fewer total samples will be required to achieve the same overall error rate in the population estimates. This is qualified by the fact that reasonable estimates of the sample variances of the individual strata are required to achieve this increased precision.

If proportional allocation is used, and if the sampling is with replacement or of low intensity, an estimate of the overall sample size (n) required to achieve some specified error bound (E) is given by:

$$n = \frac{t^2_{\alpha/2,df} \sum_{h=1}^{k} N_h \, s_h^2}{E^2 \, N} \qquad [15.46]$$

In application, because the true number of degrees of freedom cannot be known prior to determining n, an approximate value is used in Equation [15.46], as was done with simple random sampling. For the data in Table 15.3, if the goal is to estimate the average biomass per hectare within 2.0 Mg/ha with 95% confidence (i.e., the value of 2 is used to approximate $t_{\alpha/2,df}$), then the total number of required samples in the population is:

$$\begin{aligned} n &= \frac{(2^2)\,[(50)\,(5.0^2) + (70)\,(4.3^2)]}{(2.0^2)\,(120)} \\ &= 21.2 \approx 21 \text{ sample units} \end{aligned} \qquad [15.47]$$

If the sampling is without replacement, the overall sample size needed to obtain a specified error bound is given by:

$$n = \frac{N \sum_{h=1}^{k} N_h \, s_h^2}{\left(\dfrac{N \cdot E}{t_{\alpha/2,df}}\right)^2 + \sum_{h=1}^{k} N_h \, s_h^2} \qquad [15.48]$$

Again, to estimate the population mean to within 2.0 Mg/ha with 95% confidence, and using the data in Table 15.3, the number of sample units to observe in the population is:

$$\begin{aligned} n &= \frac{(120)\,[(50)\,(5.0^2) + (70)\,(4.3^2)]}{\left[\dfrac{(120)\,(2.0)}{2}\right]^2 + [(50)\,(5.0^2) + (70)\,(4.3^2)]} \\ &= 18.0 \text{ sample units} \end{aligned} \qquad [15.49]$$

If using optimal allocation and sampling is with replacement or of low intensity, an estimate of the overall sample size required to achieve some specified error bounds on $\bar{y}_s$ is given by:

$$n = \frac{t^2_{\alpha/2,df} \left[\sum_{h=1}^{k} N_h \, s_h\right]^2}{N^2 \, E^2} \qquad [15.50]$$

In using optimal allocation in the situation represented in Table 15.3, to achieve an acceptable error rate of 2.0 Mg/ha requires the following minimum number of sample units:

$$n = \frac{(2^2)\,[(50)\,(5.0) + (70)\,(4.3)]^2}{(120^2)\,(2.0^2)}$$
$$= 21.1 \approx 21 \text{ sample units} \qquad [15.51]$$

which is essentially the same number of sample units required to achieve the same expected error rate in this example population when proportional allocation is used (Equation [15.47]). The number of sample units required for equivalent error rates is always less with optimal allocation of the sample units among the strata, but the advantage of optimal allocation depends on the relative variation in the individual strata. In the example (Table 15.3), the variability is similar in the two strata, which reduces the advantage of optimal allocation over proportional allocation.

If optimal allocation is used and sampling is without replacement, the estimated required sample size is:

$$n = \frac{\left[\sum_{h=1}^{k} N_h\, s_h\right]^2}{\left[\dfrac{N \cdot E}{t_{\alpha/2,df}}\right]^2 + \sum_{h=1}^{k} N_h\, s_h^2} \qquad [15.52]$$

For the example data in Table 15.3, the number of samples required to estimate the mean biomass per hectare within 2.0 Mg/ha with 95% confidence is:

$$n = \frac{[(50)\,(5.0) + (70)\,(4.3)]^2}{\left[\dfrac{(120^2)\,(2.0^2)}{2^2}\right] + [(50)\,(5.0^2) + (70)\,(4.3^2)]}$$
$$= \frac{303{,}601}{14{,}400 + 2544.3}$$
$$= 17.9 \approx 18 \text{ sample units} \qquad [15.53]$$

which is essentially the same number of samples required if proportional sampling without replacement was being used. Again, the similarity of these results is due to the similarity in the sample variances in the two strata in this example. Optimal allocation is never less efficient than proportional allocation (i.e., for an equal number of samples, the resulting variance with optimal allocation will be less than or equal to the variance resulting from proportional allocation), but in some situations there is little gain in efficiency achieved by using optimal allocation.

In many actual applications of stratified random sampling, the overall population size is known, but the sizes of the subpopulations of each stratum are estimated. This is the case, for example, when vegetation maps are developed from aerial photographs to prestratify the area prior to field sampling. In this situation, the population sizes within the strata (N_h) are estimates rather than true values. This is an example of what is called *double sampling* or *two-phase sampling*. The first phase estimates the strata sizes, and the second phase estimates the population mean or total. If the estimates of the strata sizes are of high precision, this has little impact on the resulting precision of the stratified random sampling estimates of the overall population mean and total (Avery and Burkhart, 1994; Shiver and Borders, 1996). The estimates of the mean and variance within each strata are obtained in the same manner as those from Equations [15.27] and [15.28]. In some cases, the stratum to which a sample unit belongs cannot be identified prior to observation. For example, a soil mapping unit may consist predominantly of one soil series with a second series commonly occurring as inclusions (Chapter 4). Observation is required at each point for which series identification is needed. Another example is a situation in which a limited number of understory vegetative associations (Chapter 7) are known to occur in a given overstory type on a given physiographic feature. It is not possible to make an *a priori* determination of the vegetative association occurring at a given location. Note that in both of these examples, the strata are defined prior to sampling, but the stratum membership of the sample units is not determined until sampling occurs. This is sometimes referred to as *poststratification*. The number of observations falling in each stratum (n_h) and the total number of sample units observed in the population (n) are substituted for N_h and N, respectively, in Equation [15.33] to estimate the population mean:

$$\bar{y}_{ps} = \frac{1}{n} \sum_{h=1}^{k} n_h \, \bar{y}_h \qquad [15.54]$$

The variance of the mean in a poststratified sample is given by:

$$s^2_{\bar{y}_{ps}} = \left(\frac{N - n}{N}\right) \sum_{h=1}^{k} \left(\frac{n_h}{n}\right)^2 \frac{s_h^2}{n_h} + \frac{1}{n^2} \sum_{h=1}^{k} \left(\frac{n - n_h}{n}\right) s_h^2 \qquad [15.55]$$

Note that the first term of this expression is roughly equivalent to the variance of the mean in a prestratified sample (Equation [15.35]), with n_h and n substituted for N_h and N, respectively, and the population-level finite population correction factor is included. The second term may be thought of as the penalty for imprecise knowledge of the strata sizes resulting from poststratification. As the sample size (n) increases, the second term is reduced, because better information concerning the true strata sizes is obtained.

It is possible to extend two-phase sampling to more phases; satellite imagery may be used to develop broad classifications, aerial photograph interpretation may provide more detailed information, followed by ground sampling, for example, to provide a three-phase sampling design. Frayer and associates (1979) presented methodology extending the results from the two-phase sampling discussed earlier to designs with greater numbers of phases.

RATIO AND REGRESSION ESTIMATION

A second type of double sampling occurs when observations of an easy-to-measure, or *auxiliary,* variable are made on a large number of sample units and observations of a more-difficult-to-measure variable that is of primary interest (the *primary* variable) are made on a subset of these same sample units. The relationship between the two variables is used to obtain more precise estimates of the primary variable for the overall population. An example of this is readily available with the use of aerial photography; aerial photography of an area is obtained, and photo interpretation techniques are used to estimate overstory cover and density on a large number of photo interpretation plots. A subset of these plots is visited on the ground, and more detailed information on forest structure, including such variables as basal area per hectare, is estimated on each ground plot. The relationship between basal area per hectare (the primary variable) and the photo interpretation estimates (the auxiliary variable) on the subset of plots is determined and used to obtain a more precise estimate of basal area per hectare based on the photo interpretation estimates of cover in the larger area. This type of sampling can lead to more precise estimates of the primary variable for a given cost of sampling if (1) there is a strong linear correlation between the primary and auxiliary variables and (2) the auxiliary variable is much easier or cheaper to measure than the primary variable.

Ratio estimation is one method of obtaining double sampling estimates that are more precise and cost-effective than simple or systematic random sampling estimates. Ratio estimation is advantageous if the auxiliary and primary variables are strongly, linearly correlated and if their relationship passes through the origin (i.e., if the value of the primary variable is zero when the value of the auxiliary variable is zero). In ratio estimation, as discussed here, both the primary and auxiliary variables are observed on a small number of samples (n_s), and the auxiliary variable is measured on an additional set of sample units ($n_l = n_s + n'$). The *ratio-of-means* estimator for the mean of the primary variable is given by:

$$\overline{y}_{rm} = \left[\frac{\overline{y}_s}{\overline{x}_s}\right] \overline{x}_l = R\,\overline{x}_l \qquad [15.56]$$

where $\overline{y}_{rm}$ is the ratio-of-means estimate of the population mean of the primary variable, $\overline{y}_s$ is the mean of the primary variable in the small sample (n_s), $\overline{x}_s$ is the mean of the auxiliary variable in the small sample, and $\overline{x}_l$ is the mean of the auxiliary variable in the larger sample. The ratio-of-means estimate is a biased estimate of the population mean of the primary variable, with the bias decreasing as the sample size increases. In the special case in which the empirical relationship between the primary and auxiliary variables passes exactly through the origin, $\overline{y}_{rm}$ is unbiased. Thus, when sample sizes are large and the data support the assumption that the primary variable equals zero when the auxiliary variable equals zero, the bias in the ratio-of-means estimate is minimized.

Table 15.4 contains an example data set illustrating the application of ratio-of-means estimation in double sampling. This example was generated during an exercise to estimate the aboveground biomass per hectare of a stand of northern hardwoods. The basal area per hectare, which is rapidly estimated in the field, is the auxiliary variable that was observed on 20 sample units. On a subset of these (n_s = 10), the diameters and heights of each tree were measured and recorded, which allowed estimation of the aboveground biomass per hectare (the primary variable). The sample size in this example is small, and there may be problems with bias in the resulting estimates.

The mean of the auxiliary variable on the small sample is 20.7 m²/ha, the mean of the auxiliary variable on the large sample is 20.0 m²/ha, and the mean of the primary variable on the small sample is 175.4 Mg/ha. This results in a ratio-of-means estimate of the aboveground biomass per hectare of:

$$\begin{aligned}\overline{y}_{rm} &= \left[\frac{175.4}{20.7}\right](20.0)\\ &= (8.47)\,(20.0)\\ &= 169.5 \text{ Mg/hectare}\end{aligned} \qquad [15.57]$$

It is difficult to estimate the variance of the ratio-of-means estimate. Schreuder, Gregoire, and Wood (1993) provide a summary of several alternative methods of estimating the variance of a ratio-of-means estimate, and Sukhatme et al. (1984) and Cochran (1977) provide a more detailed theoretical treatment of these issues. DeVries (1986) derived the following expression for the variance of $\overline{y}_{rm}$:

$$s^2_{\overline{y}_{rm}} = \frac{s^2_y + R^2\, s^2_{x_l} - 2\, R\, s_{xy}}{n_s} + \frac{2\, R\, s_{xy} - R^2\, s^2_{x_l}}{n_l} - \frac{s^2_y}{N} \qquad [15.58]$$

where N is the total number of sample units in the population, n_s is the number of sample units where both the primary and the auxiliary variable were observed, s^2_y is the sample variance of the primary variable in the small sample, $s^2_{x_l}$ is the sample variance of the auxiliary variable in the large sample, s_{xy} is the covariance between the primary and auxiliary variables in the small sample, and R is the ratio of the means of the primary and auxiliary variables in the small sample. This approximation of the variance of $\overline{y}_{rm}$ tends to

TABLE 15.4 Observed Values of Basal Area (m^2/Hectare) and Aboveground Biomass (Mg/Hectare) from a Northern Hardwood Stand in Northern Michigan; Heights and Diameters of Each Tree were Measured on Every Other Plot in Order to Estimate Aboveground Biomass.

Plot Number	*Basal Area (m^2/ha)*	*Aboveground Biomass (Mg/ha)*
1	13	117
2	20	–
3	25	205
4	13	–
5	31	267
6	16	–
7	20	186
8	13	–
9	25	210
10	20	–
11	18	142
12	25	–
13	20	166
14	17	–
15	20	176
16	29	–
17	13	113
18	22	–
19	22	172
20	18	–
Small Sample Mean	20.7	175.4
Small Sample Variance	30.2	2,121.8
Large Sample Mean	20.0	–
Large Sample Variance	27.1	–

underestimate the true variance and is a poor approximation in cases in which the bias of $\overline{y}_{rm}$ is not small. Schreuder, Gregoire, and Wood (1993) discuss alternative estimators of the variance of $\overline{y}_{rm}$ for application in situations in which it is critical to obtain the best available estimate of $s^2_{\overline{y}_{rm}}$.

The *covariance* between the auxiliary and primary variables is a measure of the joint variability of the two variables. In the small sample it is obtained by:

$$s_{xy} = \frac{\sum_{i=1}^{n_s} (x_i - \overline{x})(y_i - \overline{y})}{n_s - 1} \qquad [15.59]$$

An equivalent alternative expression for calculating s_{xy} is given by:

$$s_{xy} = \frac{\sum_{i=1}^{n_s} x_i y_i - \frac{\left(\sum_{i=1}^{n_s} x_i\right)\left(\sum_{i=1}^{n_s} y_i\right)}{n_s}}{n_s - 1} \qquad [15.60]$$

For the example in Table 15.4, there are 10 observations in the small sample, the sum of the observed values of the auxiliary variable (basal area) was 207.0, and the sum of the observations of the primary variable was 1,754.0. The sum of the cross-products (the observed basal areas multiplied by the corresponding biomass) is 38,542. This results in an estimated covariance between the primary and auxiliary variables of:

$$\begin{aligned} s_{xy} &= \frac{38{,}542 - \frac{(207.0)\,(1{,}754.0)}{10}}{9} \\ &= 248.2 \end{aligned} \qquad [15.61]$$

If the sample units in Table 15.4 were randomly located in a 16 ha area, with the restriction that there was no more than one sample per 0.25 ha and sampling was without replacement, we can estimate the variance of $\overline{y}_{rm}$ using Equation [15.58]. This information implies that the overall number of sample units in the population is 64 (N). Recall that the number of sample units in which both the primary and auxiliary variable were observed is 10 (n_s), the variance of the auxiliary variable (basal area) in the large sample is 27.1 m^4/ha^2, the variance of the primary variable (aboveground biomass) in the small sample is 2,121.8 Mg^2/ha^2, the covariance between the primary and auxiliary variables in the small sample is 248.2 Mg^2/ha^2, and the ratio of the means (from Equation [15.57]) is 8.47. This results in an estimated variance of $\overline{y}_{rm}$ of:

$$\begin{aligned} s^2_{\overline{y}_{rm}} &= \frac{2{,}121.8 + (8.47)^2\,27.1 - 2\,(8.47)\,248.2}{10} \\ &\quad + \frac{2\,(8.47)\,248.2 - (8.47)^2\,27.1}{20} - \frac{2{,}121.8}{64} \\ &= -\,13.9 + 113.0 - 33.2 \\ &= 65.9\ \text{Mg}^2/\text{hectare}^2 \end{aligned} \qquad [15.62]$$

and

$$s_{\overline{y}_{rm}} = 8.1\ \text{Mg/hectare} \qquad [15.62a]$$

If the small sample were analyzed as a simple random sample, the resulting standard error of the mean would be 14.6, which is greater than the standard error of $\overline{y}_{rm}$ obtained in Equation [15.62]. Schreuder, Gregoire, and Wood

(1993) note that a sufficient condition for ratio-of-means estimation to result in a smaller standard error of the mean than simple random sampling (i.e., be more efficient) is for the correlation between the auxiliary and primary variable (r) to satisfy the following condition:

$$r > \frac{CV_X}{2\ CV_Y} \qquad [15.63]$$

where CV_X and CV_Y are the coefficients of variation of the auxiliary and primary variables, respectively. The covariance and the standard deviations of the auxiliary and primary variables are used to calculate the *correlation* between the two variables. Correlation is a measure of the linear relationship between two variables and takes values in the range of $-1 \leq r \leq 1$. Correlation values of -1 or 1 imply a perfect linear relationship with either a negative or positive slope, respectively. A correlation of zero implies statistical independence of the two variables. Correlation is calculated as:

$$r = \frac{s_{xy}}{s_{x_s}\ s_{y_s}} \qquad [15.64]$$

where s_{xy} is the covariance between the primary and auxiliary variables (Equation [15.59]) and s_{x_s} and s_{y_s} are the sample standard deviations for each of the two variables in the small sample. For the example in Table 15.4, the correlation between basal area per hectare and aboveground biomass per hectare in the small sample is:

$$\begin{aligned} r &= \frac{248.2}{(5.5)\ (46.1)} \\ &= 0.98 \end{aligned} \qquad [15.65]$$

The coefficient of variation of basal area per hectare in the small sample is 26.5%, and the coefficient of variation of aboveground biomass in the small sample is 26.3%. The condition in Equation [15.63], therefore, is satisfied in the example data:

$$0.98 > \frac{26.5}{(2)\ (26.3)} = 0.50 \qquad [15.66]$$

and the standard error of the mean from ratio-of-means estimation should be less than that from simple random sampling, as indeed happened.

A second type of ratio estimation uses the *mean-of-ratios* procedure to estimate the mean of the primary variable in the population. In mean-of-ratios estimation, the ratio between the primary and the auxiliary variable is calculated for every sample unit in the small sample; these are averaged and then multiplied by the mean of the auxiliary variable in the large population:

$$\bar{y}_{mr} = \bar{x}_l \sum_{i=1}^{n_s} \left[\frac{y_i}{x_i} \right] \qquad [15.67]$$

Ratios are very variable, as is nicely illustrated by Johnson et al., (1995) using a data set consisting of baseball batting averages. Murthy (1967) noted that the variance of $\bar{y}_{mr}$ and the variance of $\bar{y}_{rm}$ are approximately equal, but $\bar{y}_{mr}$ is not consistent. In this case, consistency means that the bias in the estimator is reduced as the sample size increases; this is true for ratio-of-means estimates but is not true for mean-of-ratio estimates. Mean-of-ratios estimates are not often used because of this instability. An exception is a situation in which the likelihood of sampling is proportional to the value of the auxiliary variable, as may be true in variable radius plot sampling of the forest overstory (see Chapter 8). In this special case, $\bar{y}_{mr}$ is unbiased and may be highly efficient.

For the example in Table 15.4, it is reasonable to expect that the above-ground biomass of the forest overstory will be zero when the basal area per hectare is zero. In this case ratio estimation is appropriate, but this expectation is not true for every situation. *Regression estimation* is a method of obtaining double sampling estimates when the relationship between the primary and secondary variables cannot be assumed *a priori* to pass through the origin. Regression estimation assumes a linear relationship between the primary and auxiliary variables that may or may not pass through the origin. Regression estimation, therefore, may be thought of as a more general procedure than ratio estimation.

The data structure for regression estimation is identical to that for ratio estimation; there is an auxiliary variable that is relatively easy or inexpensive to measure. The auxiliary variable must be linearly related to the primary characteristic of interest that is more difficult or expensive to observe. Commonly, the auxiliary variable is observed on all of the measured sample units, but the primary variable is observed only on a subset of the sample units. The regression estimator of the primary variable for the population is given by:

$$\bar{y}_{\text{reg}} = \bar{y}_s + b\,(\bar{x}_l - \bar{x}_s) \qquad [15.68]$$

where the subscript s denotes means from the small sample, the subscript l denotes means from the larger sample, y is the primary variable, x is the auxiliary variable, and b is equivalent to the slope of a simple linear regression line relating y to x. The slope b is estimated as in simple linear regression:

$$b = \frac{s_{xy}}{s_{x_s}^2} \qquad [15.69]$$

where $s_{x_s}^2$ is the sample variance of the auxiliary variable in the small sample and s_{xy} is the covariance between the auxiliary and the primary variables in the small sample (Equation [15.59]). Note that the data in Table 15.4 can also be analyzed using regression estimation if the investigator is unwilling to make the assumption that the relationship between basal area per hectare and aboveground biomass passes through the origin. Using the values of the covariance between basal area per hectare and aboveground biomass (248.2) and the variance of basal area per hectare in the small sample (30.2), b is calculated as:

$$\begin{aligned} b &= \frac{248.2}{30.2} \\ &= 8.22 \end{aligned} \qquad [15.70]$$

The regression estimator of the mean aboveground biomass is calculated using b, the mean basal area per hectare in the small sample (20.7, Table 15.4), the mean basal area per hectare in the large sample (20.0, Table 15.4), and the mean aboveground biomass in the small sample (175.4, Table 15.4):

$$\begin{aligned} \bar{y}_{\text{reg}} &= 175.4 + 8.22\ (20.0 - 20.7) \\ &= 169.6 \text{ Mg/hectare} \end{aligned} \qquad [15.71]$$

which is very similar to the ratio-of-means estimate (169.5 Mg/hectare, Equation [15.57]). The similarity between these two estimates is indicative of the validity of the assumption that the linear relationship between basal area per hectare and aboveground biomass passes through the origin.

DeVries (1986) gives the following expression for estimating the variance of $\bar{y}_{\text{reg}}$

$$s^2_{\bar{y}_{\text{reg}}} = \frac{s^2_y}{n_s} - \frac{s^2_y\ (n_l - n_s)\ r^2}{(n_l)\ (n_s)} \qquad [15.72]$$

where s^2_y is the sample variance of y in the small sample, n_s and n_l are the number of observations in the small and large samples, respectively, and r is the correlation between the auxiliary and primary variables in the small sample. For the example in Table 15.4, the sample variance of aboveground biomass in the small sample is 2,121.8 and the correlation between basal area per hectare and aboveground biomass is 0.98 (Equation [15.65]). This results in an estimated variance for $\bar{y}_{\text{reg}}$ of:

$$\begin{aligned} s^2_{\bar{y}_{\text{reg}}} &= \frac{2{,}121.8}{10} - \frac{2{,}121.8\ (20 - 10)\ 0.98^2}{(20)\ (10)} \\ &= 212.18 - 101.89 \\ &= 110.29 \text{ Mg}^2\text{/hectare}^2 \end{aligned} \qquad [15.73]$$

and

$$s_{\bar{y}_{\text{reg}}} = 10.50 \text{ Mg/hectare} \qquad [15.73a]$$

which is greater than the standard error of $\bar{y}_{rm}$ (8.1 Mg/hectare) given in Equation [15.62]. One should keep in mind that the estimated variances of both the ratio and the regression estimates are approximations.

One final topic to consider in regression estimation is the estimation of the required number of sample units in the large and small samples to achieve some specified acceptable error rate. Loetsch and Haller (1964) present methods to estimate the required sample size that are based on the variance of the primary variable, the correlation between the auxiliary and the primary variables, and the ratio of costs between plots where both the primary and auxiliary variables are observed and plots where only the auxiliary variable is observed. For the large sample:

$$n_l = \frac{t^2_{\frac{\alpha}{2},df}\left[s_y^2\left(\sqrt{\frac{C_s}{C_l}}\right) r\,(\sqrt{1 - r^2}) + s_y^2\, r^2\right]}{E^2} \qquad [15.74]$$

where C_s is the cost of observing a sample unit in the small sample (measuring both the primary and auxiliary variables), C_l is a cost of an observation in the large sample (measuring only the auxiliary variable), and all other variables are as defined before. Note that the ratio between C_s and C_l appears in Equation [15.74], which means that an investigator does not have to explicitly define the two costs, but can instead use the ratio of the costs to estimate the required sample size.

For the example in Table 15.4, if we assume that it costs three times as much to measure both aboveground biomass and basal area per hectare as it does to measure basal area per hectare alone, and noting from Equation [15.65] that the correlation between basal area per hectare and aboveground biomass is 0.98 and the sample variance of aboveground biomass from Table 15.4 is 2,121.8, then the number of large sample observations required to estimate the aboveground biomass to within ±20.0 Mg/ha with 95% confidence is given by:

$$\begin{aligned} n_l &= \frac{2^2\left[2{,}121.8\left(\sqrt{3}\right)(0.98)\left(\sqrt{1 - 0.98^2}\right) + 2{,}121.8\,(0.98)^2\right]}{20^2} \\ &= \frac{4 \cdot [716.70 + 2{,}037.78]}{400} \\ &= 27.5 \approx 28 \text{ sample units for the large sample} \end{aligned} \qquad [15.75]$$

Loetsch and Haller (1964) give the following expression to estimate the required number of observations in the small sample:

$$n_s = \frac{t^2_{\frac{\alpha}{2},df}\left[s_y^2\left(\sqrt{\frac{C_l}{C_s}}\right) r\left(\sqrt{1 - r^2}\right) + s_y^2\,(1 - r^2)\right]}{E^2} \qquad [15.76]$$

So, to complete the preceding example, the number of required observations in the small sample, where both the auxiliary and the primary variables are observed, is given by:

$$
\begin{aligned}
n_s &= \frac{2^2 \cdot \left[2{,}121.8\left(\sqrt{\frac{1}{3}}\right)(0.98)\sqrt{1-0.98^2} + 2{,}121.8\,(1-0.98^2)\right]}{20^2} \\
&= \frac{4 \cdot [238.90 + 84.02]}{400} \\
&= 3.2 \approx 3 \text{ sample units in the small sample}
\end{aligned}
\qquad [15.77]
$$

This implies that slightly greater precision could have been obtained by measuring aboveground biomass on fewer samples (3 instead of 10) while increasing the number of observations of basal area alone (from 10 to 25). Note that, assuming three units of cost are required to measure both basal area per hectare and aboveground biomass versus one unit of cost for measuring basal area alone, the cost of sampling to obtain the data in Table 15.4 was 40 units of cost $(10 \cdot 1 + 10 \cdot 3)$, whereas the cost to obtain approximately the same precision using the sample sizes from Equations [15.75] and [15.77] would only be 34 units of cost $(25 \cdot 1 + 3 \cdot 3)$.

CLUSTER SAMPLING

The final major sampling design to be examined is *cluster sampling.* The objective in cluster sampling is to obtain estimates of a characteristic of the individual sample units, such as the individual animals or plants. Cluster sampling is most appropriate when the individual sample units in the population fall into groups, or clusters, and it is logistically easier to select the clusters for sampling rather than the individual sample units. The clusters are randomly selected from the population, and observations are made on all sample units in each sampled cluster. All sample units in the population have an equal chance of being selected, but all combinations of sample units are not equally likely. Likewise, the clusters are not stratified in the sense of homogeneity, as are the strata in stratified random sampling, although in some applications the within-cluster variation is less than the between-cluster variation. Examples of clustered populations arise in social settings: all workers with offices on a particular floor or within a certain wing of an office building, or all families living on a block or in a small neighborhood of a city or town. Clusters naturally arise in ecological settings as well. Cluster sampling may be advantageous in wildlife populations, in which clusters may be defined as family units or social units within a population (young in a litter or individuals in a colony, for example). Plant communities also lend themselves to cluster sampling; if the objective is to estimate some characteristic of the individual plants, then the clusters may be defined as the collection of individual plants in a plot or quadrat.

The cluster sampling estimator of the population mean takes the form of a ratio estimate:

$$\overline{y}_c = \frac{\sum_{i=1}^{n} y_i}{\sum_{i=1}^{n} m_i} \qquad [15.78]$$

where $\overline{y}_c$ is the cluster sampling estimate of the population mean, y_i is the total value of the characteristic of interest over all the sample elements in the ith sampled cluster, and m_i is the number of sample elements in the ith sampled cluster.

Table 15.5 contains data from an application of cluster sampling. The objective of the exercise is to estimate the average height of sugar maple seedlings growing beneath a forest canopy; the taller the seedlings, the more effectively they shade the forest floor and therefore affect the regeneration of other species. The clusters are 1 m^2 (1 m $\times$ 1 m) quadrats. The mean height of the individual sugar maple seedlings in the sample is:

$$\begin{aligned}\overline{y}_c &= \frac{5{,}441}{391}\\ &= 13.9 \text{ cm}\end{aligned} \qquad [15.79]$$

TABLE 15.5 Observed Values of the Number of Sugar Maple Seedlings on 1 m^2 Quadrats in a 1 ha Area and the Total Heights (cm) of the Seedlings Located within Each Quadrat.

Quadrat	*Number of Seedlings*	*Total Height (cm)*
1	20	204
2	38	574
3	27	327
4	29	365
5	35	689
6	58	957
7	32	294
8	47	569
9	16	218
10	22	352
11	39	472
12	28	420

Note that Equation [15.78] utilizes the total value of the characteristic of interest for each cluster. In some instances, it may be more efficient to observe the average value of the characteristic of interest for the elements in a cluster. In this case, the average value for the elements in a cluster can be multiplied by the number of elements in the cluster to obtain the total for the cluster, and then Equation [15.78] may be utilized as presented.

The calculation of the variance of $\overline{y}_c$ is a two-step process. In the first step, the variance of the cluster totals is given by:

$$s_C^2 = \frac{\sum_{i=1}^{n} (y_i - m_i\, \overline{y}_C)^2}{n - 1} \qquad [15.80]$$

which can be calculated as:

$$s_C^2 = \frac{\sum_{i=1}^{n} y_i^2 - 2\, \overline{y}_c \sum_{i=1}^{n} y_i\, m_i + \overline{y}_C^2 \sum_{i=1}^{n} m_i^2}{n - 1} \qquad [15.81]$$

This value is then adjusted by the finite population correction factor and the respective population sizes to obtain the variance of $\overline{y}_c$:

$$s_{\overline{y}_c}^2 = \frac{N - n}{n\, N\, \overline{M}^2}\, s_C^2 \qquad [15.82]$$

where N is the total number of clusters in the population, n is the number of sampled clusters, and $\overline{M}$ is the average number of elements per cluster. The average number of elements per cluster is rarely known beforehand, and this quantity is usually approximated in the calculations by the average number of elements in the observed clusters. For the data in Table 15.5, the variance of the total height in the clusters is calculated using Equation [15.81], which requires the value of $\overline{y}_c$ from Equation [15.79], the sum of the squared total heights (2,982,625), the sum of the squared number of elements in the clusters (14,281), and the sum of the cross products between the total heights and the numbers of elements (202,478):

$$\begin{aligned} s_C^2 &= \frac{2{,}982{,}625 - 2\,(13.9)\,202{,}478 + 13.9^2\,(14{,}281)}{12 - 1} \\ &= 10{,}269.9 \text{ cm}^2 \end{aligned} \qquad [15.83]$$

Assuming further that N = 10,000 (1 ha) and knowing that the average number of elements per cluster in Table 15.5 is 32.6, the variance of $\overline{y}_c$ is given by:

$$\begin{aligned} s_{\overline{y}_c}^2 &= \frac{10{,}000 - 12}{(12)\,(10{,}000)\,(32.6^2)}\,(10{,}269.9) \\ &= 0.8 \text{ cm}^2 \end{aligned} \qquad [15.84]$$

and

$$s_{\overline{y}_c} = 0.9 \text{ cm} \qquad [15.84a]$$

The approximate number of clusters required in order to estimate $\overline{y}_c$ within some acceptable error limits can be calculated as follows:

$$n = \frac{N\, s_C^2}{\dfrac{N\, E^2\, \overline{M}^2}{t^2_{\frac{\alpha}{2},df}} + s_c^2} \qquad [15.85]$$

where E is the acceptable bounds on the error of $\overline{y}_c$ and all other variables are as previously defined. For the example in Table 15.5, if the acceptable error on the mean estimated height of sugar maple seedlings with 95 % confidence is 1 cm, then the number of clusters that must be sampled is given by:

$$\begin{aligned} n &= \frac{(10{,}000)\ (10{,}269.9)}{\dfrac{(10{,}000)\ (1^2)\ (32.6^2)}{2^2} + 10{,}269.9} \\ &= \frac{10{,}000 \cdot 10{,}269.9}{2{,}656{,}900 + 10{,}269.9} \\ &= 38.5 \approx 39 \text{ sampled clusters} \end{aligned} \qquad [15.86]$$

Cluster sampling may be extended to include sampling at several stages. Two-stage cluster sampling, for example, is implemented when clusters are sampled randomly from the population and then, instead of every element in each cluster being observed, a second random sample of elements is observed in each cluster. Two-stage sampling could have been implemented in the example in Table 15.5 by randomly selecting 1 m^2 quadrats and then, instead of measuring the heights of all sugar maple seedlings in each quadrat, measuring the heights of a randomly selected subset (e.g., 10%) of the seedlings in each quadrat. Another example of two-stage cluster sampling would occur if the different vegetation units in a forested landscape were identified as clusters and observations on randomly selected quadrats were performed within selected vegetation units. In this case, the objective would be to estimate some characteristic of individual quadrats across the entire landscape. Cochran (1977) and others provide details of two-stage cluster sampling. Frayer and associates (1979) extended two-stage sampling techniques to situations in which there were more than two stages of sampling.

SUMMARY

It is usually impossible, both logistically and financially, to obtain a complete census of a population of interest during resource assessment efforts. Some form of sampling must be implemented to obtain estimates of the relevant characteristics of the population. There are four basic sampling designs used in resource assessment efforts in forested landscapes: simple random sampling, systematic sampling, stratified random sampling, and cluster sampling. In addition, ratio and regression estimation, which use auxiliary variables to provide greater precision in the estimates of the primary variable of interest, are often suited to resource assessment objectives. More than one of these sampling designs may be implemented during an integrated resource assessment effort; different designs may be used for different landscape components, depending on the temporal and spatial scales of the processes of interest.

APPLICATIONS

15.1 Twelve observations of diameter and height of eastern white pine trees were recorded. Estimate the mean, variance, standard deviation, and standard error of the mean for each of the two variables. Estimate both the covariance and the correlation between diameter and height for this sample of white pine.

Tree	*Diameter (cm)*	*Height (m)*
1	40	22
2	62	29
3	37	21
4	25	20
5	48	25
6	36	23
7	57	32
8	52	30
9	45	27
10	39	22
11	44	25
12	33	24

15.2 For the simple random sampling example data in Table 15.1, estimate the total basal area (m^2) for the population if the sample population was from a 16 hectare area. Give the standard error and a 95% confidence interval for this estimate.

15.3 The following data were collected as a simple random sample of a population of individual members of a herbaceous species. If the individual plants were flowering or had already flowered at the time of observation, a value of 1 was recorded; if the plants had not yet flowered, a value of 0 was recorded. Estimate the proportion of the population that had flowered at the time of observation. Give the standard error and a 90% confidence interval for this estimate.

Plant	*Flowered?*	*Plant*	*Flowered?*	*Plant*	*Flowered?*
1	0	11	1	21	1
2	1	12	0	22	1
3	0	13	0	23	1
4	1	14	1	24	1
5	1	15	0	25	1
6	1	16	1	26	0
7	0	17	1	27	0
8	1	18	1	28	1
9	0	19	1	29	1
10	1	20	0	30	0

15.4 A 1,100 ha watershed was divided into three main vegetative communities. A survey was undertaken to estimate the average number of bird species per hectare in the watershed. Sample locations were established in each vegetative type, and the numbers of different species observed in 30-minute time intervals were recorded. It is assumed that each sampling location represents about one hectare. Estimate the average number of bird species per hectare from the following data, give the standard error of this estimate, determine how many samples would be needed to estimate the number of bird species per hectare to within ± 5 with 95% confidence, and, using optimal allocation, determine how many observations to allocate to each vegetative type.

	Community 1	*Community 2*	*Community 3*
		Area (ha)	
	664	247	189
Observation			
1	12	4	16
2	16	8	23
3	9	12	17
4	14	7	21
5	13	6	24
6	19	11	–
7	12	8	–
8	20	–	–
9	15	–	–
10	18	–	–

15.5 Aerial photography was used to estimate canopy closure for 100 0.25 ha sample locations in a 1,000 ha watershed. The average canopy closure for the watershed was 87% with a standard deviation of 5%. Ten of these locations were then visited on the ground, and aboveground biomass was recorded. Use regression estimation to obtain an estimate of the average aboveground biomass per hectare in the watershed and give the standard error of this estimate.

Ground Plot	*Canopy Closure (%)*	*Aboveground Biomass (Mg/ha)*
1	83	177
2	68	125
3	97	188
4	78	133
5	88	195
6	72	142
7	92	180
8	87	189
9	79	171
10	84	168

15.6 A survey of white-water rafters at a given recreation area was undertaken to determine the average distance traveled by individuals to arrive at the site. Professional guides were assigned to accompany groups of rafters, and among these groups cluster sampling was undertaken to estimate the average distance traveled. Groups of rafters were randomly selected, and all individuals in a group were asked how far they had traveled from home. From these data, estimate the average distance traveled by people to participate in this activity. Give the standard error of this estimate.

Group	*Number of Individuals*	*Average Distance Traveled (km)*
1	6	90
2	8	157
3	5	64
4	9	236
5	4	122
6	7	189
7	8	45
8	6	347

15.7 A series of 10 m^2 quadrats were placed in a 16 hectare vegetation unit. The goal was to estimate the fraction of saplings of overstory tree species experiencing browsing of the terminal bud by herbivores. This is an example of cluster sampling with the intent to estimate a proportion. The data are given in the following table:

Plot	*Total Number of Saplings*	*Number of Browsed Saplings*
1	16	8
2	26	10
3	12	2
4	38	26
5	18	8
6	52	36
7	19	11
8	9	0

Use cluster sampling to estimate the proportion of the saplings experiencing browsing of the terminal bud. Give the standard deviation of this estimated proportion.

Appendices

APPENDIX A Measurement Units and Approximate Conversion Factors*

To Convert From:	*Divide By:*	*To Convert To/From:*	*Divide By:*	*To Convert To:*
Length				
millimeter	10	centimeter	0.1	millimeter
millimeter	1,000	meter	0.001	millimeter
centimeter	100	meter	0.01	centimeter
meter	1,000	kilometer	0.001	meter
inch	12	foot	0.08333	inch
inch	36	yard	0.02778	inch
foot	66	chain	0.01515	foot
foot	5,280	mile	0.0001894	foot
chain	80	mile	0.0125	chain
link	100	chain	0.01	link
inch	7.92	link	0.1263	inch
millimeter	25.4	inch	0.03937	millimeter
centimeter	2.54	inch	0.3937	centimeter
foot	3.28	meter	0.3048	foot
kilometer	1.609	mile	0.6214	kilometer
Area				
centimeter^2	10,000	meter^2	0.0001	centimeter^2
meter^2	10,000	hectare	0.0001	meter^2

APPENDIX A (Continued)

To Convert From:	*Divide By:*	*To Convert To/From:*	*Divide By:*	*To Convert To:*
$inch^2$	144	$foot^2$	0.006944	$inch^2$
$foot^2$	43,560	acre	0.000023	$foot^2$
$chain^2$	10	acre	0.10	$chain^2$
$centimeter^2$	6.452	$inch^2$	0.1550	$centimeter^2$
$foot^2$	10.758	$meter^2$	0.09295	$foot^2$
acre	2.471	hectare	0.405	acre
$kilometer^2$	2.5889	$mile^2$	0.3863	$kilometer^2$
$foot^2$/acre	4.3555	$meter^2$/hectare	0.2296	$foot^2$/acre
Volume				
$millimeter^3$	1,000	$centimeter^3$	0.001	$millimeter^3$
$centimeter^3$	10^6	$meter^3$	0.000001	$centimeter^3$
$inch^3$	1,728	$foot^3$	0.0005787	$inch^3$
$inch^3$	144	board foot	0.006944	$inch^3$
$foot^3$	43,560	acre foot	0.000023	$foot^3$
$foot^3$	128	cord	0.0078125	$foot^3$
board foot	6–7**	$foot^3$ roundwood	0.14–0.17**	board foot
$foot^3$ wood	72**	cord	0.0139**	$foot^3$ wood
board foot	500**	cord	0.002**	board foot
$foot^3$	35.315	$meter^3$	0.02832	$foot^3$
$meter^3$ wood	2.0388**	cord	0.4905**	$meter^3$ wood
$meter^3$	1,233.48	acre foot	0.0008107	$meter^3$
$feet^3$/acre	14.2978	$meter^3$/hectare	0.06994	$feet^3$/acre
Mass				
gram	1,000	kilogram	0.001	gram
kilogram	1,000	ton	0.001	kilogram
ounce	16	pound	0.0625	ounce
pound	2,000	ton	0.0005	pound
gram	28.3495	ounce	0.03527	gram
pound	2.2046	kilogram	0.4536	pound
ton (E)	1.1023	ton (SI)	0.9072	ton (E)
pound/acre	1.1208	kg/hectare	0.8922	pound/acre
ton/acre	2.2417	ton/hectare	0.4461	ton/acre

*Conversion factors are given to at least 1 : 1,000 accuracy.

**Indicates an approximate average conversion factor.

APPENDIX B Common and Scientific Species Names*

Common Name	*Genus*	*Species*
Plants		
Balsam Fir	*Abies*	*balsamea*
Red Maple	*Acer*	*rubrum*
Sugar Maple	*Acer*	*saccharum*
Wild Chives	*Allium*	*schoenoprasum*
Red Alder	*Alnus*	*rubra*
Green Spleenwort	*Asplenium*	*trichomanes-ramosum*
Yellow Birch	*Betula*	*alleghaniensis*
Paper Birch	*Betula*	*papyrifera*
Northern Reedgrass	*Calamagrostis*	*lacustris*
Walking Fern	*Camptosorus*	*rhizophyllus*
Eastern Redcedar	*Juniperus*	*virginiana*
Farwell's Water-milfoil	*Myriophyllum*	*farwellii*
White Spruce	*Picea*	*glauca*
Black Spruce	*Picea*	*mariana*
Jack Pine	*Pinus*	*banksiana*
Red Pine	*Pinus*	*resinosa*
Eastern White Pine	*Pinus*	*strobus*
Trembling Aspen	*Populus*	*tremuloides*
Bigtooth Aspen	*Populus*	*grandidentata*
Douglas Fir	*Pseudotsuga*	*menziesii*
Northern Red Oak	*Quercus*	*rubra*
Rayless Mountain Ragwort	*Senecio*	*indecorus*
Common Dandelion	*Taraxacum*	*officionale*
Northern White-cedar	*Thuja*	*occidentalis*
Eastern Hemlock	*Tsuga*	*canadensis*
Mammals		
Moose	*Alces*	*alces*
Gray Wolf	*Canis*	*lupus*
Gunnison Prairie Dog	*Cynomys*	*gunnisoni*
Porcupine	*Erethizon*	*dorsatum*
Northwest Chipmunk	*Eutamias*	*amoensis*
Least Chipmunk	*Eutamias*	*minimus*
Snowshoe (Varying) Hare	*Lepus*	*americanus*
Marten	*Martes*	*americana*

APPENDIX B (Continued)

Common Name	*Genus*	*Species*
Mammals (*Continued*)		
Fisher	*Martes*	*pennanti*
Pine Mouse	*Microtus*	*pinetorum*
Mule Deer	*Odocoileus*	*hemionus*
White-tailed Deer	*Odocoileus*	*virginianus*
Richardson Ground Squirrel	*Spermophilous*	*richardsonii*
Cottontail Rabbit	*Sylvilagus*	*floridanus*
Pocket Gopher	*Thomomys*	*spp.*
Birds		
Boreal Owl	*Aegolius*	*funereus*
Sharp-tailed Sparrow	*Ammodramus*	*nelsoni*
American Widgeon	*Anas*	*americana*
Ruffed Grouse	*Bonasa*	*umbellus*
Pine Siskin	*Carduelis*	*pinus*
American Goldfinch	*Carduelis*	*tristis*
Sage Grouse	*Centrocercus*	*urophosianus*
Bobwhite Quail	*Colinus*	*virginianus*
American Crow	*Corvus*	*brachyrhynchos*
Blue Jay	*Cyanocitta*	*cristata*
Blue Grouse	*Dendragapus*	*obscurus*
Brewer Blackbird	*Euphagus*	*cyanocephalus*
Common Loon	*Gavia*	*immer*
Bald Eagle	*Haliaeetus*	*leucocephalus*
Valley Quail	*Lophortyx*	*californica*
Wild Turkey	*Meleagris*	*gallopavo*
Osprey	*Pandion*	*haliaetus*
Black-capped Chickadee	*Parus*	*atricapillus*
English Sparrow	*Passer*	*domesticus*
Hungarian (Gray) Partridge	*Perdix*	*perdix*
Ring-necked Pheasant	*Phasianus*	*colchicus*
Black-backed Woodpecker	*Picoides*	*arcticus*
Green-tailed Towhee	*Pipilo*	*chlorurus*
Blue-gray Gnatcatcher	*Polioptila*	*caerulea*
Vesper Sparrow	*Pooecetes*	*gramineus*
American Redstart	*Setophaga*	*ruticilla*

APPENDIX B (Continued)

Common Name	*Genus*	*Species*
Birds (*Continued*)		
Brewer Sparrow	*Spizella*	*breweri*
Chipping Sparrow	*Spizella*	*passerina*
Barred Owl	*Strix*	*varia*
American Robin	*Turdus*	*migratorius*
Sharp-tailed Grouse	*Tympanuchus*	*phasianellus*
Yellow-headed Blackbird	*Xanthocephalus*	*xanthocephalus*
Reptiles and Amphibians		
Eastern Box Turtle	*Terrapene*	*carolina carolina*
Fish		
Lake Sturgeon	*Acipenser*	*fulvescens*
Arthropods		
American Burying Beetle	*Nicrophorus*	*americanus*

*Nomenclature follows Corbett and Hill (1980), Clements (1981), and USDA Soil Conservation Service (1982).

References

Alatolo, R. V. 1991. Problems in the measurement of evenness in ecology. *Oikos* 37:199–204.

Albert, D. A., S. R. Denton, and B. V. Barnes. 1986. *Regional landscape ecosystems of Michigan.* Ann Arbor: School of Natural Resources, University of Michigan.

Allen. A. W. 1983. *Habitat suitability index models: Fisher.* U.S. Department of the Interior, Fish and Wildlife Service. Publication FWS/OBS-82/10.45.

Allen, T. F. H., and T. W. Hoekstra. 1992. *Toward a unified ecology.* New York: Columbia University Press.

Anderberg, M. R. 1973. *Cluster analysis for applications.* New York: Academic Press.

Antenucci, J. C. 1991. *Geographic information systems: A guide to the technology.* New York: Van Nostrand Reinhold.

Aronoff, S. 1989. *Geographic information systems.* Ottawa: WDL Publications.

Avery, T. E., and H. E. Burkhart. 1994. *Forest measurements.* 4th ed. New York: McGraw-Hill.

Bailey, R. L. 1994. *Ecoregions of the United States.* (Map; 1 : 7,500,000). Washington, DC: USDA Forest Service.

Baldwin, J. L. 1983. *Climatic atlas of the United States.* Asheville, NC: U.S. Department of Commerce, Environmental Science Services Administration.

Basharin, G. P. 1959. On a statistical estimate for the entropy of a sequence of independent random variables. *Theory Prob. Appl.* 4:333–336.

Beers, T. W., and C. I. Miller. 1964. Point sampling research results, theory, and applications. *Res. Bull. 786.* Purdue University, Agricultural Experiment Station, Lafayette, IN.

Bell, G., A. Kerr, D. McNickle, and R. Woollons. 1996. Accuracy of the line intersect method of post-logging sampling under orientation bias. *Forest Ecology and Manage.* 84:23–28.

Belyea, H. C. 1931. *Forest Measurement.* New York: John Wiley & Sons.

Belyea, H. C. 1953. Log rules with special reference to the Scribner and Doyle. *South. Lumberman* 187:276–285.

Benyus, J. M., R. R. Buech, and M. D. Nelson. 1992. *Wildlife in the Upper Great Lakes Region: A community profile.* Research Paper NC-301. Washington, DC: USDA Forest Service.

Bergen, S. D., J. L. Fridley, M. A. Ganter, and P. Schiess. 1995. Predicting the visual effect of forest operations. *J. Forestry* 93(2):33–37.

Bitterlich, W. 1947. Die Winkelzahlmessing. *Allg. Forest. Holzwirtsch, Ztg.* 58:94–96.

Black, P. E. 1991. *Watershed hydrology.* Englewood Cliffs, NJ: Prentice-Hall.

Bolstad, B. V. and J. L. Smith. 1992. Errors in GIS. *J. Forestry* 90(11):21–29.

Bolstad, B. V., P. Gessler, and T. M. Lillesand. 1990. Positional uncertainty in manually digitized map data. *J. Geog. Info. Syst.* 4:399–412.

Botkin, D. B., J. F. Janak, and J. R. Wallis. 1972. Some ecological consequences of a computer model of forest growth. *J. Ecology* 60:849–873.

Bourdeau, P. F. 1953. A test of random versus systematic ecological sampling. *Ecology* 34:499–512.

Bourdo, E. A., Jr. 1956. A review of the General Land Office Survey and of its use in quantitative studies of former forests. *Ecology* 37:745–768.

Bowden, D. C., and W. E. Frayer. 1985. *Estimation of the accuracy of mapping systems.* Ford Forestry Center Research Note No. 37. Houghton School of Forestry and Wood Products, Michigan Technological University.

Boyce, S. G. 1995. *Landscape forestry.* New York: John Wiley & Sons.

Brady, N. C. 1990. *The nature and properties of soils.* 10th ed. New York: MacMillan.

Braun-Blanquet, J. 1928. *Pflanzensoziologie. Grundzuge der Vegetationskunde.* Berlin: Springer.

Bray, J. R., and J. T. Curtis. 1957. An ordination of the upland forest communities of southern Wisconsin. *Ecol. Monogr.* 27:325–349.

Briggs, R. D., and R. C. Lemin. 1992. Delineation of climatic regions in Maine. *Can. J. For. Res.* 22:801–811.

Brillouin, L. 1962. *Science and information theory.* 2d ed. New York: Academic Press.

Brinson, M. M. 1993. *A hydrogeomorphic classification for wetlands.* U.S. Army Corps of Engineers, Waterways Experiment Station, Wetland Research Program Technical Report WRP-DE-4.

Brockman, C. F., and L. C. Merriam, Jr. 1979. *Recreational use of wild lands.* 3d ed. New York: McGraw-Hill.

Brooks, K. N., P. F. Ffolliot, H. M. Gregersen, and J. L. Thames. 1991. *Hydrology and the management of watersheds.* Ames: Iowa State University Press.

Brooks, R. T., D. R. Dickson, W. B. Burkman, I. Millers, M. Miller-Weeks, E. Cooter, and L. Smith. 1992. *Forest health monitoring in New England: 1990 Annual Report.* Resource Bulletin NE-125, Radnor, PA: USDA Forest Service, Northeastern Forest Experiment Station.

Brown, G. W. 1985. *Forestry and water quality.* 2d ed. Corvallis: Oregon State University.

Brown, J. K. 1974. *Handbook for inventorying downed woody material.* General Technical Report INT-16. Ogden, UT: USDA Forest Service, Intermountain Forest and Range Experiment Station.

Brown, J. K., and P. J. Roussopoulos. 1974. Estimating biases in the planar intersect method for estimating volumes of small fuels. *Forest Sci.* 20:350–356.

Buongiorno, J., S. Dahir, H. C. Lu, and C. R. Lin. 1994. Tree size diversity and economic returns in uneven-aged forest stands. *Forest Sci.* 40:83–103.

Burkhart, H. E., J. P. Barrett, and H. G. Lund. 1984. Timber inventory. In *Forestry handbook.* 2d ed., edited by K. F. Wenger. New York: John Wiley & Sons.

Burnham, K. P., D. K. Anderson, and J. L. Lake. 1980. *Estimation of density from line transect sampling of biological populations.* Wildlife Monographs 72.

Burton, A. J., K. S. Pregitzer, and D. D. Reed. 1991a. Leaf area and foliar biomass relationships in northern hardwood forests located along an 800 km acid deposition gradient. *Forest Sci.* 37:1041–1059.

Burton, A. J., C. W. Ramm, K. S. Pregitzer, and D. D. Reed. 1991b. Use of multivariate methods in forest research site selection. *Can. J. For. Res.* 21:1573–1580.

Calus, J. K. 1996. *Soil survey of Oceana County, Michigan.* USDA Natural Resources Conservation Service.

Campbell, J. B. 1996. *Introduction to remote sensing.* 2d ed. New York: Guilford Press.

Campbell, S. M., and D. B. Kittredge. 1996. Ecosystem-based management on multiple NIPF ownerships. *J. Forestry* 94:24–29.

Carleton, T. J., R. K. Jones, and G. Pierpont. 1985. The prediction of understory vegetation by environmental factors for the purpose of site classification in forestry: An example from northern Ontario using residual ordination analysis. *Can. J. For. Res.* 15:1099–1108.

Carlile, D. W., J. R. Skalski, J. E. Batker, J. M. Thomas, and V. I. Cullinan. 1989. Determination of ecological scale. *Landscape Ecology* 2:203–213.

Cathey, H. M. 1990. USDA Plant hardiness zone map. Miscellaneous Publication Number 1475. Washington, DC: U.S.D.A., Agricultural Research Service.

Chow, V. T., ed. 1964. *Handbook of applied hydrology.* New York: McGraw-Hill.

Clapham, A. R. 1932. The form of the observational unit in quantitative ecology. *J. Ecology* 20:192–197.

Clark, J. F. 1906. The measurement of saw logs. *Forestry Quarterly* 4(2):79–93.

Clark, R. N., and G. H. Stankey. 1979. *The recreation opportunity spectrum: A framework for planning, management, and research.* General Technical Report PNW-98. Portland, OR: U.S.D.A. Forest Service, Pacific Northwest Forest and Range Experiment Station.

Clarke, K. C. 1996. *Getting started with Geographic Information Systems.* Upper Saddle River, NJ: Prentice-Hall.

Clayton, L. 1984. *Pleistocene geology of the Superior region, Wisconsin.* Circular 46. Madison, WI: Wisconsin Geological and Natural Historical Survey.

Clements, F. E. 1904. Development and structure of vegetation. *Rep. Bot. Survey of Nebraska,* 7.

Clements, F. E. 1905. *Research methods in ecology.* Lincoln, NE: University Publishing Co.

Clements, F. E. 1916. Plant succession. *Carnegie Inst. Wash. Publ.* 242:1–512.

Clements, J. F. 1981. *Birds of the world: A checklist.* New York: Facts on file.

Clesceri, L. S., A. E. Greenberg, and R. Rhodes Trussell eds. 1989 Standard methods for the examination of water & wastewater. 17th edition. American Public Health Assoc., American Water Works Assoc. & Water Pollution Control Federation. Washington, D.C.

Cochran, W. G. 1977. *Sampling techniques.* 3d ed. New York: John Wiley & Sons.

Coffman, M., E. Alyanak, J. Kotar, and J. Ferris. 1984. *Field guide: Habitat classification for the Upper Peninsula of Michigan and Northeastern Wisconsin.* Houghton: Cooperative for Research on Forest Soils, Department of Forestry, Michigan Technological University.

Collins, M. E., B. J. Carter, B. G. Gladfelter and R. J. Southard eds. 1995. Pedological perspectives in archaeological research. Soil Science Society of America, Madison Wis. Special Publication Number 44. 157p.

Congalton, R. G., and K. Green. 1992. The ABCs of GIS. *J. Forestry* 90(11):13–20.

Congalton, R. G., R. G. Oderwald, and R. A. Mead. 1983. Assessing Landsat classification accuracy using discrete multivariate analysis statistical techniques. *Photogramm. Eng. and Rem. Sens.* 49:1671–1678.

Cook, M. 1717. The manner of raising, ordering, and improving forest-trees. 2d. ed.

Corbet, G. B., and J. E. Hill. 1980. *A world list of mammalian species.* London: British Museum of Natural History.

Cowardin, L., V. Carter, F. Golet, and E. LaRoe. 1979. *Classification of wetlands and deepwater habitats of the United States.* Rep. No. FWS/OBS-79/31. Washington DC: U.S. Department of Interior, U.S. Fish and Wildlife Service.

Cremer, K. W., C. J. Borough, F. H. McKinnell, and P. R. Carter. 1982. Effects of stocking and thinning on wind damage in plantations. *New Zealand J. Forest Sci.* 12:245–268.

Cressie, N. A. C. 1993. *Statistics for spatial data.* Rev. ed. New York: John Wiley & Sons.

Critchfield, H. J. 1974. *General climatology.* 3d. ed. Englewood Cliffs, NJ: Prentice-Hall.

Cubbage, F. W. 1991. Public regulation of private forestry. *J. Forestry* 89:31–35.

Cubbage, F. W. 1995. Regulation of private forest practices. *J. Forestry* 93:14–20.

Cubbage, F. W., J. O'Laughlin, and C. H. Bullock. 1993. *Forest resources policy.* New York: John Wiley & Sons.

Cullinan, V. I., and J. M. Thomas. 1992. A comparison of quantitative methods for examining landscape pattern and scale. *Landscape Ecology* 7:211–227.

Czekanowski, J. 1913. *Zarys Metod Statystycznyck.* Warsaw.

Dana, S. T., and S. K. Fairfax. 1980. *Forest and range policy: Its development in the United States.* 2d ed. New York: McGraw-Hill.

Daubenmire, R. 1952. Forest vegetation of northern Idaho and adjacent Washington, and its bearing on concepts of vegetation classification. *Ecol. Monogr.* 22:301–330.

Davis, K. P. 1966. *Forest management: Regulation and valuation.* 2d ed. New York: McGraw-Hill.

Davis, M. G. 1989. Insights from paleoecology on global change. *Bull. Ecol. Soc. Amer.* 70:222–228.

DeGraaf, R. M., M. Yamasaki, W. B. Leak, and J. W. Lanier. 1992. *New England wildlife: Management of forested habitats.* General Technical Report NE-144. Radnor, PA: USDA Forest Service, Northeastern Forest Experiment Station.

Denevan, W. M. 1992. The pristine myth: The landscape of the Americas in 1492. *In* The Americas before and after 1492: Current geographical research (K. W. Butzer, guest editor). *Ann. Amer. Assoc. Geographers* 82:369–385.

Desanker, P. V. and D. D. Reed. 1992. A stochastic model for stimulating daily growing season weather variables for input into ecological models. In *Proceedings of the 1991 symposium on systems analysis in forest resources, Charleston, SC,* pp. 1–10. Asheville, NC: USDA Forest Service.

DeVries, P. G. 1986. *Sampling theory for forest inventory.* New York: Springer-Verlag.

Digby, P. G. N., and R. A. Kempton. 1987. *Multivariate analysis of ecological communities.* London: Chapman and Hall.

Dilworth, J. R., and J. F. Bell. 1984. *Variable probability sampling.* Corvallis: Oregon State University Bookstores.

Dissmeyer, G. E. 1994. *Evaluating the effectiveness of forestry best management practices in meeting water quality goals or standards.* Miscellaneous Publication No. 1520. Washington, DC: USDA Forest Service.

Dissmeyer, G. E., and G. R. Foster. 1984. *A guide for predicting sheet and rill erosion on forest land.* Technical Publication R8-TP-6 Atlanta, GA: USDA Forest Service, Southern Region.

Duff, D. 1993. *Holmes' principles of physical geology.* 4th ed. New York: Chapman and Hall.

Eichenlaub, V. L., J. R. Harmon, F. V. Nurnberger, and H. J. Stolle. 1990. *The climatic atlas of Michigan.* Notre Dame, IN: University of Notre Dame Press.

Ellefson, P. V. 1992. *Forest resources policy: Process, participants, and programs.* New York: McGraw-Hill.

Ellefson, P. V., and A. S. Cheng. 1994. State forest practice programs. *J. Forestry* 92:34–37.

Ellenberg, H. 1956. *Aufgaten und Methoden der Vegetationskunde.* Stuttgart: Eugen Ulmer.

Erickson, M. D., D. D. Reed, and G. D. Mroz. 1990. An economic analysis of alternative cutting methods in northern hardwoods: 32-year results. *North. J. Appl. For.* 7:153–158.

Evelyn, J. A. 1664. *Sylva, or a discourse of forest-trees, and the propogation of timber on His Majesties dominions.*

Evers, D. C. 1992. *A guide to Michigan's endangered wildlife.* Ann Arbor: University of Michigan Press.

FAO (Food and Agriculture Organization of the United Nations). 1992. *Yearbook of forest products.* Forestry Series 25. Rome: FAO.

FAO (Food and Agriculture Organization of the United Nations). 1995. *State of the world's forests.* Rome: FAO.

Farrand, W. R. and D. L. Bell, cartogs. 1982. *Quaternary geology of Northern Michigan.* Lansing, MI: Michigan Department of Natural Resources.

Federal Interagency Committee for Wetland Delineation. 1989. *Federal manual for identifying and delineating jurisdictional wetlands.* Washington, DC: U.S. Army Corps of Engineers, U.S. Environmental Protection Agency, U.S. Fish and Wildlife Service, and U.S. Soil Conservation Service.

Fincher, J., and M.-L. Smith. 1993. *A discriminant-function approach to ecological site classification in Northern New England.* Research Paper NE-686. Radnor, PA: USDA Forest Service, Northeastern Forest Experiment Station.

Finley, R. W. 1976. *Original vegetation cover of Wisconsin from the U.S. General Land Office Notes* (map). St. Paul, MN: USDA Forest Service, North Central Forest Experiment Station.

Finney, D. J. 1948. Random and systematic sampling in timber surveys. *Forestry* 22:64–99.

Finney, D. J. 1950. An example of periodic variation in forest sampling. *Forestry* 23:96–111.

Frayer, W. E., D. C. Bowden, G. E. Dixon, F. A. Graybill, S. Jeyaratnam, D. C. Johnston, B. M. Kent, V. J. LaBau, and E. Roberts. 1979. *Multi-level sampling designs for resource inventories.* Fort Collins: Department of Forest and Wood Sciences, Colorado State University.

Freese, F. 1962. *Elementary forest sampling.* Agriculture Handbook No. 232. U.S. Department of Agriculture, Forest Service.

Frelich, L. E., and C. G. Lorimer. 1991. Natural disturbance regimes in hemlock-hardwood forests of the upper Great Lakes region. *Ecol. Monogr.* 61:145–164.

Gates, C. E., W. H. Marshall, and D. P. Olson. 1968. Line transect method of estimating grouse population densities. *Biometrics* 24:135–145.

Gates, D. M. 1972. *Man and his environment: Climate.* New York: Harper & Row.

Gates, D. M. 1990. Climate change and forests. *Tree Phys.* 7:1–5.

Gates, D. M. 1993. *Climate change and its biological consequences.* Sunderland, MA: Sinauer Associates.

Gevorkiantz, S. R., and L. P. Olsen. 1955. *Composite volume tables for timber and their application in the Lake States.* Technical Bulletin No. 1104. Washington, DC: U.S. Department of Agriculture, Forest Service.

Gleason, H. A. 1917. The structure and development of the association. *Bull. Torrey Bot. Club* 43:463–481.

Gleason, H. A. 1920. Some applications of the quadrat method. *Bull. Torrey Botan. Club* 47:21–33.

Gleason, H. A. 1926. The individualistic concept of the plant association. Contrib. New York Bot. Garden, No. 279.

Gleason, H. A. 1939. The individualistic concept of the plant association (with discussion). *Am. Midland Nat.* 21:92–110.

Graf, W. L., ed. 1987. *Geomorphic systems of North America*. Boulder, CO: Geological Society of America.

Graves, H. S. 1906. *Forest mensuration*. New York: John Wiley & Sons.

Greenacre, M. J. 1984. *Theory and application of correspondence analysis*. London: Academic Press.

Greene, J. L., and W. C. Siegel. 1994. *The status and impact of state and local regulations on private timber supply*. General Technical Report RM-255. Fort Collins, CO: U.S. Department of Agriculture.

Grieg-Smith, P. 1983. *Quantitative plant ecology*. 3d ed. Studies in Ecology (D. J. Anderson, P. Grieg-Smith, and F. A. Pitelka, eds.), vol. 9. Berkeley: University of California Press.

Grosenbaugh, L. R. 1952. *Shortcuts for cruisers and scalers*. Occasional Paper No. 126. USDA forest Service, Southern Forest Experiment Station.

Haines-Young, R. H., D. R. Green, and S. Cousins, eds. 1993. *Landscape ecology and geographic information systems*. London: Taylor and Francis.

Hall, R. B. W. 1991. A re-examination of the use of interpoint distances and least squares in mapping trees. *Ecology* 72:2286–2289.

Hansen, J., I. Fung, A. Lacis, D. Rind, G. Russell, S. Lebedoff, R. Reudy, and P. Stone. 1988. Global climate changes as forecast by the CISS-3-D model. *J. Geophys. Res.* 93:9341–9364.

Hansen, M. H., T. Frieswyk, J. F. Glover, and J. F. Kelly. 1992. *The Eastwide forest inventory data base: Users manual*. General Technical Report NC-151. St. Paul, MN: USDA Forest Service, North Central Forest Experiment Station.

Hansen, P. L., R. D. Pfister, K. Boggs, B. J. Cook, J. Joy, and D. K. Hinkley. 1995. *Classification and management of Montana's riparian and wetland sites*. Miscellaneous Publication No. 54. Missoula: Montana Forest and Conservation Experiment Station, School of Forestry, University of Montana.

Hasel, A. A. 1938. Sampling error in timber surveys. *J. Agric. Res.* 57:715–736.

Hendee, J., G. Stankey, and R. Lucas. 1990. *Wilderness management*. 2d ed. Golden, CO: North American Press.

Hengeveld, H. G. 1990. Global climate change: Implications for air temperature and water supply in Canada. *Trans. of the Amer. Fish. Soc.* 119:176–182.

Hewlett, J. D. 1982. *Principles of forest hydrology*. Atlanta: University of Georgia Press.

Hewlett, J. D., and W. L. Nutter. 1969. *An outline of forest hydrology*. Athens: University of Georgia Press.

Hickman, C. A., W. C. Siegel, and C. A. Martus. 1991. Forest practice regulatory developments in the East. In *Proceedings of the 1991 Society of American Foresters National Convention*. pp. 327–332. Bethesda, MD: Society of American Foresters.

Hill, M. O. 1973. The intensity of spatial pattern in plant communities. *J. Ecol.* 61:225–235.

Hix, D. M. 1988. Multifactor classification and analysis of upland hardwood forest ecosystems of the Kickapoo River watershed, southwestern Wisconsin. *Can. J. For. Res.* 18:1405–1415.

Hodgson, R. W., and R. L. Thayer. 1980. Implied human influence reduces landscape beauty. *Landscape Plan.* 71:171–179.

Holdridge, L. R. 1947. Determinations of world formulations from simple climatic data. *Science* 105:367–368.

Host, G., S. Westin, W. Cole, and K. S. Pregitzer. 1989. *BIOMASS: An interactive program to calculate aboveground biomass of common tree species of the Lake States forests.* General Technical Report NC-127. St. Paul, MN: USDA, Forest Service, North Central Forest Experiment Station.

Hunt, C. B. 1967. *Physiography of the United States.* San Francisco: W. H. Freeman and Co.

Hunter, M. L., Jr. 1990. *Wildlife, forests, and forestry: Principles of managing forests for biological diversity.* Englewood Cliffs, NJ: Prentice-Hall.

Husch, B., C. I. Miller, and T. W. Beers. 1983. *Forest mensuration.* 3d ed. New York: John Wiley & Sons.

Iles, K., and M. Fall. 1988. Can an angle gauge really evaluate "borderline trees" accurately in variable plot sampling? *Can. J. For. Res.* 18:774–781.

Jaccard, P. 1901. Étude comparative de la distribution florale dans une portion des Alpes et du Jura. *Bull. Soc. Vaud. Sc. Nat.* 37:547–579.

Jaccard, P. 1912. The distribution of flora of the alpine zone. *New Phytol.* 11:37–50.

Jaccard, P. 1928. Die statistisch-floristische Methode als Grundlage der Pflanzensoziologie. *Abderhalden, Handb. bio. Arbeitsmeth.* 11:165–202.

Jarvis, P. G., and J. W. Leverenz. 1983. Productivity of temperate, deciduous, and evergreen forests. In *Ecosystem processes: Mineral cycling, productivity, and man's influence.* Vol. 13D of *Physiological plant ecology: New series,* (O. L. Lange, and M. Aragno eds.). New York: Springer-Verlag.

Johnston, R. G., S. D. Schroder, and A. R. Mallawaaratchy. 1995. Statistical artifacts in the ratio of discrete quantities. *The American Statistician.* 49:285–290.

Jolly, G. M. 1965. Explicit estimates from capture-recapture data with both death and immigration stochastic model. *Biometrika* 52:225–247.

Jones, E. A., D. D. Reed, P. J. Cattelino, and G. D. Mroz. 1991. Seasonal shoot growth of planted red pine predicted from air temperature degree days and soil water potential. *Forest Ecology and Manage.* 46:201–214.

Jones, E. A., D. D. Reed, and P. V. Desanker. 1994. Ecological implications of projected climate change scenarios in forest ecosystems of central North America. *Agric. and Forest Meteor.* 72:31–46.

Jones, G. T. 1995. The careful timber harvest: A guide to logging esthetics. *J. Forestry* 93(2):12–15.

Jubenville, A., and B. W. Twight. 1993. *Outdoor recreation management: theory and practice.* State College, PA: Venture Publishing.

Karl, T. R., and R. W. Knight. 1985. *Atlas of monthly Palmer drought severity indices (1931–1983) for the contiguous United States.* U.S. Department of Commerce, Weather Bureau Historical Climatological Series 3-11.

Kenkel, N. C. 1988. Patterns of self-thinning in jack pine: Testing the random mortality hypothesis. *Ecology* 69:1017–1024.

Kent, M., and P. Coker. 1992. *Vegetation description and analysis: A practical approach.* London: Belhaven Press.

Kessler, B. L. 1992. Glossary of GIS terms. *J. Forestry* 90(11):37–45.

Keys, J., Jr., C. Carpenter, S. Hooks, F. Koenig, W. H. McNab, W. Russell, and M. L. Smith. 1995. *Ecological units of the eastern United States: First approximation (map and booklet of map unit tables).* Presentation Scale 1 : 3,500,000, colored. Atlanta, GA: U.S. Department of Agriculture, Forest Service.

Kilgore, M. A., P. V. Ellefson, and G. A. Rose. 1996. Innovative forestry initiatives: Minnesota prepares for the future. *J. Forestry* 94:21–25.

King, A. W., W. R. Emanuel, and R. V. O'Neill. 1990. Models of forest dynamics: problems of temporal scale. In *Process modeling of forest growth responses to environmental stress* (R. K. Dixon, R. S. Meldahl, G. A. Ruark, and W. G. Warren, eds.), pp. 241–248. Portland, OR: Timber Press.

King, D. A. 1986. Tree form, height growth and susceptibility to wind damage in Acer saccharum. *Ecology* 67:980–990.

Kozlowski, T. T., P. J. Kramer, and S. G. Pallardy. 1991. *The physiological ecology of woody plants.* San Diego: Academic Press.

Krummel, J. R., R. H. Gardner, G. Sugihara, and R. V. O'Neill. 1987. Landscape patterns in a disturbed environment. *Oikos* 48:321–324.

Kulow, D. L. 1966. Comparison of forest sampling designs. *J. Forestry* 64:469–474.

Landsberg, J. J. 1986. *Physiological ecology of forest production.* London: Academic Press.

Leach, G., and R. Mearns. 1988. *Beyond the woodfuel crisis: People, land, and trees in Africa.* London: Earthscan.

Lebreton, J.-D., K. P. Burnham, J. Clobert, and D. R. Anderson. 1992. Modeling survival and testing biological hypotheses using marked animals: A unified approach with case studies. *Eco. Monogr.* 62:67–118.

Lee, K. N. 1993. *Compass and gyroscope: Integrating science and politics for the environment.* Washington, DC: Island Press.

Lee, R. 1980. *Forest hydrology.* New York: Columbia University Press.

Leemans, R., and W. Cramer. 1990. The *IIASA climate database for land area on a grid of 0.5° resolution.* Report WP-41. Laxenburg, Austria: International Institute for Applied Systems Analysis.

Lessard, V., D. D. Reed, and N. Monkevich. 1994. Comparing N-tree distance sampling with point and plot sampling in Northern Michigan forest types. *North. J. Appl. For.* 11:12–16.

Liechty, H. O., D. D. Reed, G. D. Mroz, and M. F. Jurgensen. 1990. Comparison of cation leaching in five northern hardwood stands along an air pollution gradient using a throughfall modeling approach. In *Proceedings of the 19th International Union of Forest Research Organizations World Congress, Division I, Montreal, Canada* 1:295–306. Vienna: International Union of Forest Research Organizations.

Lillesand, T. M., and R. W. Kiefer. 1994. *Remote sensing and image interpretation.* 3d ed. New York: John Wiley & Sons.

Lincoln, F. C. 1930. *Calculating waterfowl abundance on the basis of banding returns.* Circular U.S. Department of Agriculture Number 118, pp. 1–4.

Lloyd, D. W., and M. A. MacLeod. 1993. Regulation and perceived compliance. *J. Forestry* 91:41–47.

Loetsch, F., and K. E. Haller. 1964. *Forest inventory*. Vol. I of *Statistics of forest inventory and information from aerial photographs*. Munich: BLV Verlagsgesellschaft.

Lorimer, C. G. 1977. The presettlement forest and natural disturbance cycle of northeastern Maine. *Ecology* 58:139–148.

Lorimer, C. G. 1980. Age structure and disturbance history of a southern Appalachian virgin forest. *Ecology* 61:1169–1184.

Lorimer, C. G. 1985. Methodological considerations in the analysis of forest disturbance history. *Can. J. For. Res.* 15:200–213.

Lucas, O. W. R. 1991. *The design of forest landscapes*. New York: Oxford University Press.

Lucas, R. C., W. R. Bacon, W. Beardsley, P. J. Brown, H. H. Christensen, R. N. Clark, D. N. Cole, B. L. Driver, W. C. Fischer, H. Holloway, D. A. King, W. F. LaPage, J. E. Meier, G. W. Orning, J. Schultz, G. H. Stankey, and P. D. Weingart. 1984. Outdoor recreation management. In *Forestry handbook* (K. F. Wenger, ed.), pp. 801–886. New York: John Wiley & Sons.

Ludwig, J. A., and J. F. Reynolds. 1988. *Statistical ecology: A primer on methods and computing*. New York: John Wiley & Sons.

Luxmoore, R. J. 1983. Water budget of an eastern deciduous forest stand. *Soil Sci. Soc. Am. J.* 47:785–791.

Lyon, G. W. 1986. Economics of all-aged northern hardwood stands. In *The northern hardwood resource: Management and potential* (G. D. Mroz and D. D. Reeds, eds.), pp. 308–320. Houghton: School of Forestry and Wood Products, Michigan Technological University.

Lyon, J. G. 1993. *Practical handbook for wetland identification and delineation*. Ann Arbor, MI: Lewis Publishers. 157 p.

Magill, A. W. 1992. *Managed and natural landscapes: What do people like?* Research Paper PSW-RP-203. Washington, DC: USDA Forest Service.

Magill, A. W. 1994. What people see in managed and natural landscapes. *J. Forestry* 92(9)12–16.

Magurran, A. 1988. *Ecological diversity and its measurement*. London: Croom Helm.

Manabe, S., and R. Wetherald. 1986. Reduction in summer wetness by an increase in atmospheric carbon dioxide. *Science* 232:626–628.

Manwood, J. 1598. *A treatise and discourse of the lawes of the forrest*.

Margalef, D. R. 1958. Information theory in ecology. *General Systems* 3:36–71.

Martin, A. C., H. S. Zim, and A. L. Nelson. 1951. *American wildlife and plants: A guide to wildlife food habits*. New York: McGraw-Hill Book Company.

Martin, P. S. 1984. Prehistoric overkill: the global model. In *Quarternary extinctions* (P. S. Martin and R. G. Klein, eds.), pp. 354–403. Tucson: The University of Arizona Press.

Martus, C. E., H. L. Haney, Jr., and W. C. Siegel. 1995. Local forest regulatory ordinances: Trends in the Eastern United States. *J. Forestry* 93:27–31.

McGarigal, K., and B. J. Marks. 1995. *FRAGSTATS: Spatial pattern analysis program for quantifying landscape structure*. General Technical Report PNW-

GTR-351. Portland, OR: USDA Forest Service, Pacific Northwest Forest Experiment Station.

McIlveen, R. 1992. *Fundamentals of weather and climate.* London: Chapman and Hall.

McLaren, B. E., and R. O. Peterson. 1994. Wolves, moose, and tree rings on Isle Royale. *Science* 266:1555–1558.

McNab, W. H., and P. E. Avers, comps. 1994. *Ecological subregions of the United States: Section descriptions.* Administrative Publication WO-WSA-5. Washington, DC: USDA Forest Service.

Mendenhall, W., L. Ott, and R. L. Schaeffer. 1971. *Elementary survey sampling.* Belmont, CA: Duxbury Press.

Mesavage, C., and L. R. Grosenbaugh. 1956. Efficiency of several cruising designs on small tracts in North Arkansas. *J. Forestry* 54:569–576.

Michigan Department of Natural Resources. 1994a. *Michigan's special animals: Endangered, threatened, special concern, and probably extirpated.* Lansing: Michigan Department of Natural Resources.

Michigan Department of Natural Resources. 1994b. *Water quality management practices on forest land.* Lansing: Michigan Department of Natural Resources.

Millers, I., D. LaChance, W. G. Burkman, and D. C. Allen. 1991. *North American sugar maple decline project: Organization and field methods.* General Technical Report NE-154. Radnor, PA: USDA Forest Service, Northeastern Forest Experiment Station.

Milne, B. T. 1988. Measuring the fractal geometry of landscapes. *Appl. Math. and Computation* 27:67–79.

Milne, B. T. 1990. Lessons from applying fractal models to landscape pattern. In *Quantitative methods in landscape ecology: The analysis and interpretation of landscape heterogeneity* (Turner, M. G. and R. H. Gardner, eds), pp. 129–238. New York: Springer-Verlag.

Minnesota Department of Natural Resources. 1994. *Visual quality best management practices for forest management in Minnesota.* St. Paul, MN: Minnesota Department of Natural Resources.

Mitsch, W. J., and J. G. Gosselink. 1993. *Wetlands.* 2d ed. New York: Van Nostrand Reinhold.

Mroz, G. D., M. Anderson, J. N. Bruhn, P. J. Cattelino, R. Janke, M. F. Jurgensen, E. J. Reed, and C. C. Trettin. 1983. ELF communications system environmental monitoring program: Herbaceous plant cover and trees. In *Compilation of the 1983 Annual Reports of the Navy ELF Communications System Ecological Monitoring Program.* Technical Report E06549-8, pp. 1–145. Chicago: IIT Research Institute.

Mudrey, M. G., B. A. Brown, D. M. Mickelson, and M. D. Johnson. 1982. *Bedrock geological map of Wisconsin* (map; 1 : 500,000). Madison: Wisconsin Geological and Natural Historical Survey.

Mueller-Dombois, D., and H. Ellenberg. 1974. *Aims and methods of vegetation ecology.* New York: John Wiley & Sons.

Murthy, M. N. 1967. Sampling theory and methods. Calcutta, India: Statistical Publishing.

National Park Service. 1993. *Federal historic preservation laws.* Washington, DC: U.S. Department of the Interior, National Park Service, Cultural Resources Programs.

Northern Hardwood and Pine Manufacturer's Association. 1976. *Official grading rules for northern hardwood and softwood logs and tie cuts.* Green Bay, WI: Northern Hardwood and Pine Manufacturer's Association.

Oderwald, R. G. 1981a. Comparison of point and plot sampling basal area estimators. *Forest Sci.* 27:42–48.

Oderwald, R. G. 1981b. Point and plot sampling—the relationship. *J. Forestry* 79:377–378.

Odum, E. P. 1971. *Fundamentals of ecology.* 3d ed. Philadelphia: W. B. Saunders.

Ollinger, S. V. 1995. *Modeling physical and chemical climate of the northeastern United States for a geographic information system.* General Technical Report NE-191. Radnor, PA: USDA Forest Service, Northeastern Forest Experiment Station.

O'Regan, W. G., and M. N. Palley. 1965. A computer technique for the study of forest sampling methods. *Forest Sci.* 11:99–113.

Otis, D. L., K. P. Burnham, G. C. White, and D. R. Anderson. 1978. Statistical inference from capture data on closed populations. *Wildlife Monographs* 62.

Overpeck, J. T., P. J. Bartlein, and T. Webb III. 1991. Potential magnitude of future vegetation change in eastern North America: Comparisons with the past. *Science* 254:692–695.

Palmer, W. C. 1965. *Meteorological drought.* Research Paper No. 45, U.S. Department of Commerce, Weather Bureau.

Patton, D. R. 1992. *Wildlife habitat relationships in forested ecosystems.* Portland, OR: Timber Press.

Pennington, M. 1983. Efficient estimators of abundance for fish and plankton surveys. *Biometrics* 39:281–286.

Perala, D. A. 1985. Predicting red pine shoot growth using growing degree days. *Forest Sci.* 31:913–925.

Petersen, C. G. J. 1896. The yearly immigration of young plaice into the Limfjord from the German Sea. *Rep. Danish Biol. Sta.* 6:1–48.

Philip, M. S. 1994. *Measuring trees and forests.* 2d ed. Wallingford, UK: CAB International.

Pickford, S. G., and J. W. Hazard. 1978. Simulation studies on line intersect sampling of forest residue. *Forest Sci.* 24:469–483.

Pielou, E. C. 1966. Shannon's formula as a measure of specific diversity: Its use and misuse. *American Naturalist* 100:463–465.

Pielou, E. C. 1975. *Ecological diversity.* New York: John Wiley & Sons.

Pielou, E. C. 1977. *Mathematical ecology.* 2d ed. New York: John Wiley & Sons.

Pielou, E. C. 1991. *After the Ice Age: The return of life to glaciated North America.* Chicago: University of Chicago Press.

Pierce, L. L., and S. W. Running. 1988. Rapid estimation of coniferous forest leaf area index using a portable integrating radiometer. *Ecology* 69:1762–1767.

Pollock, K. H., J. D. Nichols, C. Brownie, and J. E. Hines. 1990. *Statistical inference for capture-recapture experiments.* Wildlife Monographs. Washington, DC: The Wildlife Society.

Population Reference Bureau. 1991. *1991 world population data sheet.* Washington, DC: Population Reference Bureau.

Postel, S. L., G. C. Daily, and P. R. Ehrlich. 1996. Human appropriation of renewable fresh water. *Science* 271:785–788.

Pregitzer, K. S. and B. V. Barnes. 1984. Classification and comparison of upland hardwood and conifer systems of the Cyrus H. McCormick Experimental Forest, upper Michigan. *Can. J. For. Res.* 14:362–375.

Prisley, S., T. Gregoire, and J. Smith. 1989. The mean and variance of area estimates computed in an arc-node geographic information system. *Photogramm. Eng. and Remote Sens.* 55:1601–1612.

Pritchett, W. L. and R. F. Fisher. 1987. *Properties and management of forest soils.* 2d ed. New York: John Wiley & Sons.

Range Term Glossary Committee. 1964. *A glossary of terms used in range management.* Portland, OR: American Society of Range Management.

Rast, E. D., D. L. Sonderman, and G. L. Gammon. 1973. *A guide to hardwood log grading.* General Technical Report NE-1. Broommall, PA: USDA Forest Service, Northeastern Forest Experiment Station.

Rauscher, H. M. 1984. *Homogeneous macroclimatic zones of the Lake States.* Research Paper NC-240. St. Paul, MN: USDA Forest Service, North Central Forest Experiment Station.

Recknagel, A. B., J. Bentley, Jr., and C. H. Guise. 1926. *Forest management.* New York: John Wiley & Sons.

Reed, D. D., E. A. Jones, and P. V. Desanker. 1994. Potential impacts of projected climate change scenarios on individual tree species in central North America. In *Proceedings of the 13th International Congress of Biometeorology,* pp. 284–294. Calgary, AB: Ste-Anne-de-Bellevue, Quebec: The International Society of Biometeorology.

Reed, D. D., H. O. Liechty, and A. J. Burton. 1989. A simple procedure for mapping tree locations in forest stands. *Forest Sci.* 35:657–662.

Riitters, K. H., R. V. O'Neill, C. T. Hunsaker, J. D. Wickham, D. H. Yankee, S. P. Timmins, K. B. Jones, and B. L. Jackson. 1995. A factor analysis of landscape pattern and structure metrics. *Landscape Ecol.* 10:23–39.

Ripley, B. D. 1978. Spectral analysis and the analysis of patterns in plant communities. *J. Ecol.* 66:965–981.

Ripley, B. D. 1983. *Spatial statistics.* New York: John Wiley & Sons.

Ritter, D. F. 1986. *Process geomorphology.* Dubuque, IA: William C. Brown.

Rohlf, F. J., and J. W. Archie. 1978. Least-squares mapping using interpoint distances. *Ecology* 59:126–132.

Rohlf, F. J., and R. R. Sokal. 1969. *Statistical tables.* San Francisco: W. H. Freeman and Company.

Root, T. 1988. Environmental factors associated with avian distributional boundaries. *J. Biogeogr.* 15:489–505.

Sample, V. A., ed. 1994. *Remote sensing and geographic information systems in ecosystem management.* Washington, DC: Island Press.

Scalabrelli, G., and G. A. Couvillon. 1986. The effect of temperature and bud type on rest completion and the GDH °C requirement for budbreak in "Redhaven" peach. *J. Am. Soc. Hortic. Sci.* 111:537–540.

Schlesinger, M., and Z. Zhao. 1988. *Seasonal climatic changes induced by doubled* CO_2 as simulated by the atmospheric GCM/mixed layer ocean model. Corvallis: Climate Research Institute, Oregon State University.

Schneider, S. H. 1989. The greenhouse effect: Science and policy. *Science* 243:771–781.

Schreuder, H. T., T. G. Gregoire, and G. B. Wood. 1993. *Sampling methods for multiresource forest inventory.* New York: John Wiley & Sons.

Schuh, D. 1995. Managing esthetic values. *J. Forestry* 93(2):20–25.

Schwenner, C. 1991. *Soil survey of Houghton County area, Michigan.* USDA Soil Conservation Service.

Seagall, T., and S. E. Wilson, cartogs. 1987. *Bedrock geology of Michigan* (map; 1 : 500,000). Lansing: Michigan Geological Survey Division.

Seber, G. A. F. 1965. A note on the multiple-recapture census. *Biometrika* 52:249–259.

Seber, G. A. F. 1982. *The estimation of animal abundance and related parameters.* 2d ed. London: Charles Griffin and Company, Limited.

Seber, G. A. F. 1986. A review of estimating animal abundance. *Biometrics* 42:267–292.

Seber, G. A. F. 1992. A review of estimating animal abundance. II. *Intl. Stat. Rev.* 60:129–166.

Sedjo, R. A., A. C. Wiseman, D. J. Brooks, and K. S. Lyon. 1994. *Global forest products trade: The consequences of domestic forest land-use policy.* Discussion paper 94-13. Washington, DC: Resources for the Future.

Shannon, C. E., and W. Weaver. 1949. *The mathematical theory of communication.* Urbana: University of Illinois Press.

Sheppard, S. R. J. 1989. *Visual simulation: A user's guide for architects, engineers, and planners.* New York: Van Nostrand Reinhold.

Shiver, B. D., and B. E. Borders. 1996. *Sampling techniques for forest resource inventory.* New York: John Wiley & Sons.

Shugart, H. H. 1984. *A theory of forest dynamics.* New York: Springer-Verlag.

Simpson, E. H. 1949. Measurement of diversity. *Nature* 163:688.

Slover, B. L. 1996. A music of opinions: Collaborative planning for the Charles C. Deam Wilderness. *J. Forestry* 94(5):18–23.

Smith, D. M. 1986. *The practice of silviculture.* 8th ed. New York: John Wiley & Sons.

Smith, H. F. 1938. An empirical law describing heterogeneity in the yields of agricultural crops. *J. Agric. Sci.* 28:1–23.

Smith, N. J., and K. Iles. 1988. A graphical depiction of multivariate similarity among sample plots. *Can. J. For. Res.* 18:467–472.

Society of American Foresters. 1992. Red-cockaded woodpecker protection and habitat management on private lands. *J. Forestry* 90:38–39.

Sorensen, T. 1948. A method of establishing groups of equal amplitude in plant sociology based on similarity of species content. *Det Kong. Danske Vidensk. Selsk. Biol. Skr.* (Copenhagen) 5:1–34.

Spies, T. A., and B. V. Barnes. 1985. A multifactor ecological classification of the northern hardwood and conifer ecosystems of Sylvania Recreation Area, Upper Peninsula, Michigan. *Can. J. For. Res.* 15:949–960.

Spurr, S. H., and B. V. Barnes. 1980. *Forest ecology.* 3d ed. New York: Ronald Press.

Stott, C. B. 1961. Log and tree grading for wood quality. Parts 1–V. *Forest control by forest inventory,* Nos. 84, 85, 86, 87, and 91. Milwaukee, WI: USDA Forest Service, Region 9.

Sukhatme, P. V., B. V. Sukhatme, S. Sukhatme, and C. Asok. 1984. *Sampling theory of surveys with applications.* Ames: Iowa State University Press.

Sukwong, S., W. E. Frayer, and E. W. Mogren. 1971. Generalized comparisons of the precision of fixed-radius and variable-radius plots for basal-area estimates. *Forest Sci.* 17:263–271.

Thelin, G. P., and R. J. Pike. 1991. *Landforms of the conterminous United States:* A digital shaded-relief portrayal. Geological Survey, Washington, DC: U.S. Department of the Interior.

Thomasma, L. E. 1988. A test of the Habitat Suitability Index Model for the fisher (Martes pennanti) in the Upper Peninsula of Michigan. Master's thesis, School of Forestry and Wood Products, Michigan Technological University, Houghton.

Thompson, M. M., and G. H. Rosenfeld. 1971. How accurate is that map? *Surv. and Map.* 31:57–64.

Thornthwaite, C. W., and J. R. Mather. 1955. The water balance. *Publications in Climatology* 8:1–104. Drexel Institute Laboratory of Climatology, Centerton, NJ.

Timber Producers Association of Michigan and Wisconsin. 1988. *Official grading rules for northern hardwood and softwood logs and tie cuts.* Rhinelander, WI: Timber Producers Association of Michigan and Wisconsin, Inc.

Tiner, R. W. 1993. The primary indicators method: A practical approach to wetland recognition and delineation in the United States. *Wetlands* 13:50–64.

Trygg, J. W. 1969. *Composite map of United States land surveyors' original plats and field notes.* Ely, MN: J. W. Trygg.

Turner, M. G. 1987. Spatial simulation of landscape changes in Georgia. *Landscape Ecology* 1:29–36.

Turner, M. G., V. H. Dale, and R. H. Gardner. 1989. Predicting across scales: Theory development and testing. *Landscape Ecology* 3:245–252.

USDA Forest Service. 1983. *The principal laws relating to Forest Service activities.* Agricultural Handbook No. 453. Washington, DC: U.S. Department of Agriculture, Forest Service.

USDA Soil Conservation Service. 1975. *Soil taxonomy: A basic system of soil classification for making and interpreting soil surveys.* Agricultural Handbook No. 436. Washington, DC: USDA Soil Conservation Service.

USDA Soil Conservation Service. 1982. *National list of scientific plant names.* SCS-TP-159. Washington, DC: USDA Soil Conservation Service.

USDA Soil Conservation Service. 1992. *Keys to soil taxonomy.* 5th ed. SMSS Technical Monograph 19. Blacksburg, VA: Pocahontas Press.

U.S. Department of the Interior. 1993. *Federal historic preservation laws.* Washington, DC: U.S. Department of the Interior, National Park Service, Cultural Resources Program.

U.S. Environmental Protection Agency. 1980. 40 CFR Part 230. Guidelines for the specification of disposal sites for dredged or fill material. *Federal Register,* Vol. 45, No. 249.

U.S. Environmental Protection Agency. 1988. 40 CFR Part 232. 404 program definitions; exempt activities not requiring 404 permits. *Federal Register,* Vol. 53, No. 108.

U.S. Environmental Protection Agency. 1994. *National water quality inventory: 1992 report to Congress.* Report EPA841-R-94-001. U.S. Environmental Protection Agency, Office of Water.

U.S. Fish and Wildlife Service. 1980. Habitat as a basis for environmental assessment. *Ecological Services Manual 101.* Washington, DC: U.S. Department of Interior, Fish and Wildlife Service, Division of Ecological Services. Government Printing Office.

U.S. Fish and Wildlife Service. 1981. Standards for the development of suitability index models. *Ecological Services Manual 103.* Washington, DC: U.S. Department of Interior, Fish and Wildlife Service, Division of Ecological Services. Government Printing Office.

U.S. Geological Survey. 1966. *National atlas of the United States [Geology]* (1 : 7,500,000 map). Reston, VA: U.S. Department of the Interior, Geological Survey.

Van Wagner, C. E. 1968. The line intersect method in forest fuel sampling. *Forest Sci.* 14:20–26.

Vernon, M. D. 1968. *The psychology of perception.* Baltimore: Penguin Books.

Warren, W. G., and P. F. Olsen. 1964. A line transect technique for assessing logging waste. *Forest Sci.* 10:267–276.

Webster, J. R. 1979. Hierarchical organization of ecosystems. In *Theoretical systems ecology* (E. Halfon, ed.), pp. 119–129. New York: Academic Press.

Whitney, G. D. 1992. *Soil survey of Chippewa County, Michigan.* USDA., Soil Conservation Service.

Whitney, G. G. 1994. *From coastal wilderness to fruited plain: A history of environmental change in temperate North America, 1500 to the present.* Cambridge, UK: Cambridge University Press.

Whittaker, R. H. 1972. Evolution and measurement of species diversity. *Taxon* 21:213–251.

Wiant, H. V., Jr., D. O. Yandle, and R. Andreas. 1984. Is BAF 10 a good choice for point sampling? *North J. Appl. For.* 2:23–24.

Wiens, J. A. 1992. What is landscape ecology, really? *Landscape Ecology* 7:149–150.

Wischmeier, W. H., and D. D. Smith. 1978. *Predicting rainfall erosion loss: A guide to conservation planning.* Agricultural Handbook No. 537. Washington, DC: U.S. Department of Agriculture.

Wisconsin Department of Natural Resources. 1990. *Silviculture and forest aesthetics handbook.* Madison: Wisconsin Department of Natural Resources.

Wolf, P. R., and R. C. Brinker. 1989. *Elementary surveying.* 8th ed. New York: Harper & Row.

Woudenberg, S. W., and T. O. Farrenkopf. 1995. *The Westwide forest inventory data base: User's manual.* General Technical Report INT-GTR-317. Ogden, UT: USDA Forest Service, Intermountain Research Station.

Wright, H. E., Jr. 1987. *Roadside geology of Vermont and New Hampshire.* Missoula MT: Mountain Press Publishing Co.

Wright, H. E., Jr. 1990. *Roadside geology of Pennsylvania.* Missoula MT: Mountain Press Publishing Co.

Zahner, R., and C. E. Grier. 1990. Concept for a model to assess the impact of climate on the growth of the southern pines. In *Process modeling of forest growth responses to environmental stress* (R. K. Dixon, R. S. Meldahl, G. A. Ruark, and W. G. Warren, eds.), pp. 383–392. Portland, OR: Timber Press.

Zahner, R., and R. K. Myers. 1986. Assessing the impact of drought on forest health. In *Proceedings of the Society of American Foresters Annual Meeting,* pp. 227–234. Birmingham, AL: Bethesda, MD: Society of American Foresters.

Zube, E. H., J. L. Sell, and J. G. Taylor. 1982. Landscape perception: Research, application, and theory. *Landscape Plan.* 9:1–33.

Index

Abiotic environment, 15
Advanced regeneration, 125
Aesthetic management zones, 255
Aesthetics, 244
Albic, 49
Alluvial, 27
Angle guage, 185
Animal unit conversion factors, 205
Animal unit month, 204
Argillic, 49
Aspect, 34
Assessments, 299
Attribute errors, 286
Attributes, 275

Basal area factor, 182
Basal area table, 193
Best Management Practices, 108
Biomass, 165
BMP's, 108
Board foot, 145
Boundary errors, 283
Bray-Curtis index, 133
Buffer strips, 107
Bulk density, 43

Capture-recapture sampling, 216
Catchment, 85
Cation exchange capacity, 48
Chapman estimator, 217
Classification errors, 283
Clean Water Act, 235
Climate, 16, 62
  change, 81
  classification, 78
Closed population, 206
Coarse woody debris, 54
Coefficient of variation, 320
Colluvial, 27
Commercial Forest Act, 239
Community, 5
Community similarity, 132
  Bray-Curtis index, 133
  Czekanowski's index, 133
  Ellenberg's index, 133
  Euclidean distance, 134
  Gleason's index, 133

Community similarity (*cont.*)
  Jaccard's index, 132
  Mueller-Dombois, 132
  Relative Euclidean distance, 134
  Sorenson's index, 132
Community structure, 220
Confidence intervals, 321
Confidence limits, 321
Confusion matrix, 283
Continuous variable, 275
Cooling degree day, 67
Cord, 144
Correlation, 343
Covariance, 341
Crosstabulation, 291
Crown classification, 169
  codominant, 169
  dominant, 169
  intermediate, 169
  suppressed, 169
Crown condition, 170
  dieback, 171
  discoloration, 171
  needle retention, 171
  transparency, 171
Czekanowski's index, 133

Degree day, 65
Detection function, 212
Diameter distribution, 193
Digital elevation model, 277
Digitizing, 277
Discrete variables, 275
Drainage basin, 85
Drainage density, 101

Ecological process, 5
Ecosystem, 5
Ecosystem similarity, 136
Elevation, 32
Ellenberg's index, 133
Eluviated, 42
Endangered species, 219, 234
Environmental impact statement, 233
Eolian, 30
Ephemeral stream, 98
Erosion, 107
Erosion factor, 53
Error, 283
  comission, 285
  matrix, 283
  omission, 286
  propagation, 287
Estimation characteristic, 175
Euclidean distance, 134
Evapotranspiration, 71
  moisture deficit, 72
  Palmer drought severity index, 72
  potential, 71
  Thornthwaite's procedure, 72
Expected freeze dates, 65

Federal Insecticide, Fungicide, and Rodenticide Act, 235
Federal Land Policy and Management Act, 229
Federal management agencies, 228
Federal Water Pollution Control Act, 235, 237
Field capacity, 44
Filter strips, 107
Finite population correction factor, 320
Fish and Wildlife Coordination Act, 235
Fixed area plots, 177
Fluvial, 27
Food Security Act, 236
Forest floor, 42
Forest overstory, 143
Forest and Rangeland Renewable Resources Planning Act, 229
Forest Stewardship Act, 236
Fragipan, 42
FRAGSTATS, 294
Frost free period, 65

General Land Office, 115, 263

Geographic Information System, 20, 257, 273
  attributes, 275
  continuous variable, 275
  discrete variables, 275
  errors and accuracy, 281
    attribute errors, 286
    boundary errors, 283
    classification errors, 283
    confusion matrix, 283
    error matrix, 283
    error propagation, 287
    errors of comission, 285
    errors of omission, 286
    ground truthing, 283
    positional errors, 282
    producer accuracy, 285
    root mean square error, 282
    user accuracy, 285
  lines, 275
  nominal, 275
  ordinal, 275
  points, 275
  polygons, 275
  projections, 295
  raster-based, 275
  vector-based, 275
Geology, 15, 25
Geomorphology, 25
Georeferencing, 280
GIS Projections, 295
Glaciers, 28
  drift, 28
  drumlins, 29
  esker, 29
  glaciofluvial sediment, 29
  lacustrine, 29
  moraines, 28
  nonstratified drift, 28
  outwash, 29
  outwash plains, 29
  stratified drift, 29
  till, 28
  till plains, 28
Gleason's index, 133
Government Land Office Public Land Survey, 263
Gravitational water, 44
Ground truthing, 283
Growing degree day, 66
Growing season length, 65
Guilds, 197

Habitat, 5
  assessment, 197
  series, 118
  suitability indices, 201
  type, 118
Heating degree day, 67
Histic, 49
Historic land use, 261, 266
Historic preservation, 263
Holdridge life zones, 80
Hydrologic response, 99
Hydrology, 16, 84

Illuviated horizons, 42
Indian Self-Determination and Educational Assistance Act, 232
Indicator species, 197
Individual species surveys, 124
Infiltration, 85
  capacity, 47
  rate, 47
Integrated environmental assessment, 21, 299
Interception, 85
Intermittent stream, 98
Inventory plan, 315
Inventory planning, 172

Jaccard's index, 132
Jurisdictional wetland, 103

K factor, 53

Landforms, 25
Landscape, 5
  components, 11

Landscape (*cont.*)
structure, 137
average patch size, 137
fractal dimension, 140
perimeter, area ratio, 139
perimeter length, 139
Large scale processes, 7
Layers, 276
Legal and regulatory environment, 19
Legislation, 227
Clean Water Act, 235
Endangered Species Act, 234
environmental impact statement, 233
Federal Insecticide, Fungicide, and Rodenticide Act, 235
Federal Land Policy and Management Act, 229
federal management agencies, 228
Federal Water Pollution Control Act, 235, 237
Fish and Wildlife Coordination Act, 235
Food Security Act, 236
Forest and Rangeland Renewable Resources Planning Act, 229
Forest Stewardship Act, 236
Indian Self-Determination and Educational Assistance Act, 232
Multiple Use Sustainable Yield Act, 229
National Environmental Policy Act, 233
National Forest Management Act, 229
regulation, 227
Renewable Resources Extension Act, 236
Sikes Act, 235
takings, 234
Tax Reform Act, 236
threatened species, 234
Water Quality Act, 235, 237
Wilderness Act, 235
Wild and Scenic Rivers Act, 235
Light, 73
insolation, 73
light extinction curve, 74
photosynthetically active radiation, 73
Lincoln's index, 217
Line intercept sampling, 55
Lines, 275
Line transect sampling, 211
Live crown ratio, 169
Local laws, 240
Loess, 30
Log, 150
cull, 150
faces, 156
grading, 156
grading defects, 159
interior defects, 150
rule, 146
Doyle log rule, 146
International 1/4″ log rule, 146
Scribner Decimal C log rule, 146
scaling, 150
gross scale, 150
gross volume, 150
net scale, 150
net volume, 150
scaling defects, 150
crook deduction, 155
crotch deduction, 155
diameter deduction, 154
pie cut, 153
squared defect method, 152
sweep, 150
side defects, 150

Management:
alternatives, 301, 304
goals, 302
objectives, 302
preferred alternative, 309

Manning equation, 95
Marine, 28
Matric potential, 44
Maximum retentive capacity, 44
Mean of ratios estimator, 343
Mineral rights, 35
Moisture, 67
Moisture deficit, 72
Mollic, 49
Mueller-Dombois, 132
Multiple use sustainable yield act, 229

National Environmental Policy Act, 21, 233
National Forest Management Act, 229
National Historic Landmarks, 263
National Historic Preservation Act, 263
National Register of Historic Places, 263
Natural Features Inventory, 220
Nominal, 275

Ochric, 49
Open population, 206
Optical scanning, 277
Ordinal, 275
Overland flow, 47
Overlay, 291
Overrun, 156

Palmer drought severity index, 72
Pellet counts, 211
Percent scale, 165
Perennial stream, 98
Permanent wilting point, 44
Petersen's index, 217
Photosynthetically active radiation, 73
Physiography, 25
Points, 275
Point sampling, 181
Polygons, 275
Population, 5, 174, 316
  density, 207
Positional errors, 282
Potential evapotranspiration, 89
Precipitation, 67
  intensity, 70
  measurement, 88
  return period, 71
Prehistoric land use, 20, 261, 264
Primary indicator method, 105
Producer accuracy, 285

Quadrat and transect sampling, 207
Quadrats, 123

Rangelands, 204
Raster-based, 275
Ratio estimator, 339
Ratio of means estimator, 339
Recreation, 19, 244
  concentrated, 249
  dispersed, 253
  opportunity spectrum, 247
  visitor day, 246
Regeneration surveys, 124
Regression estimator, 344
Relaskop, 185
Relative Euclidean distance, 134
Renewable Resources Extension Act, 236
Resource, 5
  assessment, 5
  inventory, 172
  management, 5
  monitoring, 173
  planning, 5
Riparian area, 101
Root mean square error, 282

Sample, 174
  design, 174, 316
  standard deviation, 319
  units, 174, 316
  variance, 319

Sampling, 316
  characteristic expansion factor, 175
  characteristics, 175
  cluster sampling, 347
  double sampling, 338
  frame, 174, 316
  simple random sampling, 317
  stratified random sampling, 329
    Neyman allocation, 334
    optimal allocation, 335
    poststratification, 329
    proportional allocation, 333, 338
  systematic random sampling, 327
  two-phase sampling, 338
  units, 175
Sapling surveys, 125
Scale, 5
Sediment, 107
Sedimentation, 107
Seedling surveys, 125
Sikes Act, 235
Site, 5
Site index, 52
Slope, 34
Small scale processes, 7
Soil, 15, 38
  acidity, 48
  aerobic, 46
  alkalinity, 48
  anaerobic, 46
  bulk density, 43
  coarse fragment, 39
  compaction, 46
  defined, 38
  epipedon, 49
    albic, 49
    argillic, 49
    histic, 49
    mollic, 49
    ochric, 49
    spodic, 49
  erosion, 53
  Gley, 46
  horizon, 41
    eluviated horizons, 42
    forest floor, 42
    fragipan, 42
    illuviated horizons, 42
  mapping units, 50
  mottles, 45
  permeability, 46
  pH, 48
  pit, 41
  profile, 41
  redoximorphic features, 46
  structure, 40
    granular, 40
    massive, 40
    platy, 40
    single grained, 40
  survey, 49
  taxonomy, 49
  texture, 39
    clay, 39
    coarse fragment, 39
    sand, 39
    silt, 39
  water, 43
    drainage class, 45
    field capacity, 44
    gravitational, 44
    infiltration capacity, 47
    infiltration rate, 47
    macropores, 43
    matric potential, 44
    maximum retentive capacity, 44
    micropores, 43
    permanent wilting point, 44
Soil wood, 54
Sorenson's index, 132
Spatial scale, 17
Species diversity, 123, 128
  alpha diversity, 128
  beta diversity, 128
  Brillouin's index, 130
  Gamma diversity, 128

Shannon-Weaver index, 129
Simpson's index, 128
Species evenness, 130
modified Hill's ratio, 130
Species richness, 123
Spodic, 49
Stage height, 92
Stand, 143
Standard error of the mean, 319
Stand table, 192
State historic preservation programs, 263
State laws, 236, 238
Commercial Forest Act, 239
Statistical distribution, 321
Stock table, 193
Stormflow, 99
Stratification, 329
Stratified random sampling, 329
Streamflow, 92
duration, 98
ephemeral, 98
intermittent, 98
perennial, 98
Stream order, 99
Stream pattern, 98
Subangular blocky, 40
Surface rights, 35

Takings, 234
Tax Reform Act, 236
Temperature, 64
average, 64
cooling degree day, 67
degree day, 65
expected freeze dates, 65
frost free period, 65
growing degree day, 66
growing season length, 65
heating degree day, 67
maximum daily temperature, 65
minimum temperature, 64
Temporal scale, 7
Thornthwaite's equation, 72, 89
Threatened species, 219, 234
Topographic scale, 165
Track counts, 211
Transpiration, 85
Tree, 160
basal area, 161
breast height, 160
crown position, 169
diameter, 160
grade, 165
height, 163
location, 171
measurement, 159
merchantable height, 163
sampling design, 190
scaling, 144
soundness, 165
species, 160
stump height, 163
total height, 163
Trim allowance, 145

Underrun, 156
Universal soil loss equation, 53
Universal Transverse Mercator, 275
User accuracy, 285

Variable radius plots, 181
Vector-based, 275
Vegetative associations, 118
Vegetative community, 115
Viewpoint index, 257
Visual impact, 255
Volume table, 165

Water balance, 88
Water Quality Act, 235, 237
Water quality measures, 106
dissolved oxygen, 106
total dissolved solids, 107
total suspended solids, 106
Watershed, 34, 85
basin relief, 87
compactness coefficient, 87
divides, 85

Watershed (*cont.*)
  drainage basin slope, 87
  drainage density, 101
  hydrologic response, 99
  maximum basin relief, 87
  relief ratio, 87
  shape, 86
Water storage, 85
Water table, 101
Weather, 15, 62
Wedge prism, 185
Weight scaling, 144
Wetlands, 103
Wilderness Act, 235
Wildlife, 18, 196
Wild and Scenic Rivers Act, 235
Wind, 75
Wind rose, 75
World heritage list, 263